Drought Prediction Under Different Agro-Climatic Zones

About the Author

Dr. M.L. Jat was born into an agriculture family on January I[st], 1958 in Bhilwara district of Rajasthan state, India. Graduating from CTAE, Udaipur, MPUAT, Udaipur (Rajasthan). He obtained M.E. (Ag.) in Irrigation water management from RAU, Bikaner. He got Ph.D in soil and water conservation engineering from MPUAT, Udaipur (Rajasthan). He joined university services as Assistant Prof. in the year 1984. Presently Dr. Jat is working as Associate Professor (SWE) in AICRP on Dryland Agriculture at DFRS, Arjia, Bhilwara, MPUAT, Udaipur (Rajasthan).

The author has a vast experience in the field of prediction of drought and management and also in ground water. while working in AICRP on dryland agriculture and agrometeorology.

He is life member of several technical societies. He has published about 50 scientific research papers and articles in several reputed international and national scientific journals and conferences, prepared various technical reports.

Drought Prediction Under Different Agro Climatic Zones

Mohan Lal Jat

2024

Daya Publishing House®

A Division of

Astral International Pvt. Ltd.

New Delhi – 110 002

First Published, 2007

Reprinted, 2024

ISBN 978-81-7035-414-7

Published by : **Daya Publishing House®**
A Division of
Astral International Pvt. Ltd.
– ISO 9001:2015 Certified Company –
4736/23, Ansari Road, Darya Ganj
New Delhi-110 002
Ph. 011-43549197, 23278134
E-mail: info@astralint.com
Website: www.astralint.com

Foreword

One of the worst natural calamities that affects India is the large scale incidence of drought during the summer monsoon season (June to September) . In fact, the most common natural cause of Indian famines in the past has been the occurrence of widespread droughts. If the drought is extensive and much prolonged, the cumulative effect is often disastrous. Though we have entered the twenty first century, yet drought continues to affect the Indian economy in countless ways. In virtually, we are still discovering the complex inter-relationship between drought and agriculture society and grappling with response and mitigation strategies that will lessen impacts to reduce vulnerability for future generations. Future efforts will require not only in depth understanding of the nature and functioning of the components of drought but also how it interact and influence the system as a whole. Right understating, prediction and mitigation of drought at a particular location in an agro-climatic zone in a precise manner is the need of hour.

I am happy to see the present efforts of Dr. M.L. Jat, Associate Professor, Dryland Farming Research Station, MPUAT, Udaipur for bringing out a good work on " Drought Prediction under Different Agro-climatic Zones". To my knowledge, this is the first comprehensive work containing update data and information on various aspects of drought prediction in different Agro-climatic Zones of Rajasthan. The subject matter coverage is both wide and intensive. Areas covered through individual chapters include Introduction, Rainfall pattern, Water balance, Definition and Classification of Drought, Frequency of Drought, Stochastic Modeling of Water Deficit, Crop Planning based on Rainfall and Water Balance Analysis and Conclusion. It is hoped that the technologies presented and consolidated in this publication will be very much useful in supplementing the efforts being undertaken by the Department of Agriculture as well as Rural Department and other developmental agencies and NGO's in the state. I have no doubt that the present efforts will be an important reference material for those who need greater exposure to sectoral problem and in addition to subject matter training for many years to come.

Pratap Singh
Director Research
MPUAT, Udaipur

Preface

Drought is a natural climatic phenomenon common to many parts of the world. It is a pre cursor of famine and undoubtedly man's worst natural enemy. The amount of attention paid to definition of drought and classification of areas as drought prone areas often leaves a misleading impression that drought is a problem conferred to drought prone regions only. It may sound ironical but often occurrence of drought is recognized as when the impact of monsoon failure, shortage of rainfall, moisture deficiency extends beyond. The will recognized drought prone areas. Therefore a thorough understanding different practical aspect of dry spell of drought in different agro climatic situation is sin-qua non.

A lot of research is being carried out to generate knowledge and evolve techniques for study of magnitude frequency and prediction of drought but practical aspects with respect to calculation of dry spell, aridity, water deficit based on local weather conditions has been provided in this book to cater the needs of students and weakness related to drought management. My long experience of research work in drought management has helped me to understand and appreciate the requirements of students, scientific community and field staff working in agriculture and allied sectors for appropriate study material.

Drought predication when properly used assist both the farmers at the field level and planners at the regional/national level with better cost benefit ratio for all climate related recommended practices. The basic limitation is non availability of long term continuous and accurate climatic data over well distributed network of stations to cover all possible climatic/ ecological zones of a given region/country.

This book is therefore designed to fulfill the need for an up to date comprehensive text on drought prediction for students and workers of agriculture and allied sectors of different universities and colleges adopted different courses on this aspect at under graduate and post graduate level.

The text comprised eight chapters which cover different aspects of drought prediction and management of drought through crop planning based on rainfall and drought analysis and water balance. An introducing chapter describes the basics of drought and different agro-climatic situations taking example from state of Rajasthan. The coverage in other chapter include–

The scope of book makes it a useful reference for students of agricultural engineering, agricultural, physics and environmental sciences as well. It may also serve as a guide book to persons in NGOs engaged in agricultural extension activities and those in government working as agricultural development and extension officers in rural areas is tool in the area of short and long term agricultural planning as well as on the crop early warning and the drought management with references to dry land agriculture in developing countries.

The advent of sophisticated computers has made it easy to handle large amount of data series even at daily interval. In the light of this the present book presents some of the techniques for crop early warning drought prediction and drought management for long term drought strategies covering both mitigation and emergency preparedness agricultural planning and at local and or regional level that can be used with data series at shorter interval (weekly or daily) to assists farmers to improve crop production as well as to provide assistance to public in general and farmers in specific by government and non government agencies. However some aspects that relate to these activities indirectly such as systematic studies of rainfall pattern, characteristics of rainy season and rainfall probabilities and predication on drought in terms of water deficit, dry spell, aridity and crop planning are discussed based on rainfall and water balance analysis.

I express my sincere gratitude to Prof R.V. Singh -Prof Pratap Singh, Director Research, MPUAT, Udaipur, Dr. V. K. Shrivastva, Dean, CTAE, Udaipur, Dr. R. C. Purohit, Prof. & Head, Deptt. SWC, CTAE, Udaipur, Dr. B.R.Ranwah, Associate Professor (PBG), RCA, Udaipur, Dr. P.M. Jain, Chief Scientist and Dr. S.K. Sharma, Dr. B.S.Kumpawat Associate Professor (Agronomy) and Dr. J.K. Balyan at DFRS Bhilwara for extending me valuable help from time to time during preparation of manuscript and also I highly acknowledged to my all well wishers for their direct or indirect corporation.

Contents

List of Tables

List of Figures

List of Appendices

List of Signs and Symbols

σ	Standard deviation
μ	Mean value of S_t
τ	Standard statistic measure
ρ	density of air, kg m^{-3}
$\Gamma(\gamma)$	Gama function of
γ^*	Psychomertic constant
ρ_1	Auto correlation coefficient at log^1
χ^2	Chi-squae values
℃	Degree Celsius
$\alpha_\kappa, \beta_\kappa$	Fourier coefficient of periodic mean series harmoni c,k
$\phi_{p,k}$	k^{th} auto regressive coefficient of model order P
$\phi_{p,p}$	Partial autoregressive coefficiet of model order p
A_0	Mean of historical series
AET	Actual evaportranspiration
A_k, B_k	Fourier coefficient of harmonic, k
ANN	Annual
AR	Auto Regressive
ASAE	American Society of Agricultural Engineers
a_t	Independent normally distributed residual variable
AWC	Storage capacity of soil moisture
BSW	Banswara
C_1	Auto covariance function at lag I, 1 = 0, 1, 2, ...
Coeff.	Coefficient
C_P	Specific heated constant pressure, MJ kg^{-1} C^{-1}
CV	Coefficient of variance
DEF	Water deficit
Div.	Division

DSI	Index of severity of drought
E (.)	Mathematical expectation
e.g.	Exampligratia, for example
Engg.	Engineering
et al.	Etalibi, and other
ET_0	Reference evapotranspiration
etc.	Flecetera and so forth
FAO	Food and Agricultural Organsiation
Fig.	Figure
FTR	Fatehpur
i.e.,	Id est, that is
I_a	Aridity index
I_m	Moisture index
I_{ma}	Index of moisture adequacy
IMD	India Meteorological Department
ISAE	Indian Society of Agricultural Engineers
ISE	Integral Square Error
JDP	Jodhpur
Jour.	Journal
JPR	Jaipur
K	Number of significant harmonic
Kmph	Killometer per hour
KRC Test	Kendall's Rank Correlation Test
KTA	Kota
LGS	Length of growing season
Ltd.	Limited
MAI	Moisture availability index
Max.	Maximum
Mm	Milimeter
m_r	Periodic mean, r = 1, 2, 3, ..., p
N	Number of data point
No.	Number
P	Order of autoregressive
P	Time span of periodicity
P(D)	Initial probability of a week being dry

P(W)	Initial probability of week being week
P(W/D)	Conditional probability of wet week preceded by a dry week
P(W/W)	Conditional probability of a wet week preceded by a wet week
P.	Page
PAI	Palfai aridity index
PDI	Plamar's drought index
PDSI	Palmer drought severity index
PET	Potential Evapotranspiration
PLI	Pali
PPT	Precipitation
PT	Number of turning points
P_t	Periodic component at time t, t = 1, 2, ..., N
R.F	Rainfall
r_1	Serial correlation coefficient at log 1
$S^2{}_2$ (p)	Residual variance
SCC	Serial Correlation Coefficient
SD	Standard deviation
SGN	Sriganganagar
S_t	Stochastic component at time t, 1=1, 2 ..., N
STOR	Actual storage of soil moisture
SUR	Surplus water
T	Students statistical value
TP Test	Turning point test
T_t	Trend component
UDP	Udaipur
UN	United Nations
Uz	Wind speed at height
Var	Variance
WMO	World Meteorological Organisation
X_t	Average weekly water deficit series
Y_t	Trendfree series
α	Scale parameter
β	Shape parameter
γ	Location parameter

1. Introduction

Drought has been a matter of serious concern to man since ancient times, and even today it is an outstanding example of man's helplessness before nature's large scale and formidable phenomena. It is a natural phenomenon that occurs fairly periodically all over the globe. Throughout the world, drought has become a central problem of concern. In the last decade Ethiopia, Somalia, Australia, India and other countries were more severely affected by drought. Drought is a climatic anomaly characterized by deficient supply of moisture. Such deficiency may result from sub-normal rainfall, erratic rainfall distribution, excessive water need or a combination of these factors.

Drought ranks as one of the foremost among earth's natural hazards representing the adverse effect due to shortage of water mainly from rainfall. Droughts are a threat to food production through agriculture and are still an unconquered ill. They are an outstanding example of man's limitations in the face of the broad scale formidable forces of nature. History reveals that all over the world, famines have almost been mainly consequence of droughts.

Definitions of drought are almost as numerous as there are publications on the subject. Lack of sufficient water to meet the normal water requirements of the locality is a common and a fairly satisfactory definition of drought. The perception of drought is different for different people. To the farmer, drought is a period during which his normal farm operations are hampered. To the hydrologist, with prolonged deficiency of rainfall, drought occurs with marked depletion of surface water and consequent drying up of reservoirs, lakes, streams and rivers, cessation of spring flows and fall in ground water levels. To the meteorologist, it is a situation, when there is a significant decrease of rainfall from the normal values over a given area. On the other hand, to the agricultural scientist, it occurs when the soil moisture and rainfall are inadequate during the growing season to support healthy crop growth to maturity and cause extreme water stress. Thus, droughts can be classified as hydrological, meteorological and agricultural.

Hydrological drought is the period of low stream flow and depleted reservoir storage or during which stream flows are inadequate to supply established uses under a given water management system. Meteorological drought is the period during which rainfall is below normal. Agricultural drought is a situation when the rainfall and soil moisture are inadequate to meet the water requirement of crops for their healthy growth and maturity.

Drought is not a direct result of the shortage of rainfall alone but is something related to the soil and water need of a place. Hence, any drought analysis, which is based on the rainfall criteria alone is inadequate to explain drought conditions properly. Thornthwaite and Mather (1955) also felt that drought does not begin when rain ceases but rather only when plant roots can no longer obtain moisture in the needed amount.

Though droughts are the severest catastrophes of nature, no foolproof methods have been developed either for their understanding or for the prediction. Acute water shortages. because of lack of rains over extended periods, are the prime cause of the severe and disastrous droughts. Such droughts which occur more frequently in arid and semi-arid regions seem to fallow a pattern giving to the hope that possibilities exist for their prediction, as droughts do not descend all of a sudden but are usually the ultimate result of a set of weather sequences that require extended periods to develop. Accordingly, it has been reported that droughts have an origin, spread as well as decay. Hence, synoptic and dynamic approaches to this problem are believed to yield objective results and render drought prediction possible to some extent.

In dry farming, rainfall is the main source of water for raising various crops and it is well known that the supply of water through rainfall cannot be as regular as it can be through irrigation. There is no proper way by which the occurrence of droughts can be known well in advance. The farmer in dry areas will continue to anticipate the occurrence of drought till he satisfies himself that reasonable harvest can be expected with the progress of the season. The wide spread occurrence of severe droughts throughout the world during the past and the associated human misery underscore the vulnerability of dry lands to climatic hazard and point out the need to better understand, predict and mitigate its impact.

Drought is the precursor of famine and undoubtedly man's worst natural enemy. Rainfed and drought-prone areas predominate the Indian agricultural scenario. Development of such areas assumes crucial significance since these lands contribute nearly 42 per cent to the nation's food basket. Besides, major portion of oil seeds and pulses comes from these areas. Owing to the recurrent droughts, such areas are generally believed to expose risk of crop failure and death of animals resulting in low and uncertain pay offs from farming. An understanding of the pattern of instability in farming would help to formulate polices for the development of such areas.

In Rajasthan, agriculture is primarily rain dependent. Every year its fate oscillates with the quantity, onset, progress and spatial distribution of rainfall. Thus, failure of monsoon in an area, where more than two-third of the area under cultivation is rainfed and where irrigation system is largely poor and insufficient, often creates critical fall in employment, food, fodder and drinking water for both men and cattle. Due to geographical situation and climatic conditions, drought is almost prevailing in some parts of Rajasthan like an uninvited guest. Sometimes when the situation is too severe and uncontrollable, people along with the cattle wealth even migrate to other adjacent states like Haryana, Gujarat, Punjab, Madhya Pradesh and Uttar Pradesh. Due to the unpredictability of monsoon, the people of the state, on an average have to face one drought year after every three years. In Marwar area of Rajasthan people generally expected one lean year in every three years and one famine year in every eight years (Gahlot, 1937).

The economy of Rajasthan is under the crippling grip of droughts of exceptional severity. A small deviation from normal rainfall promotes drought like conditions. As a result, rainfed crops are subjected often to sub-optimal moisture availability. Recurrent droughts and associated famines in Rajasthan over the past many decades have led to the realization that this will be a continuing phenomenon and that these droughts are likely to persist. The study of magnitude, frequency and prediction of drought in Rajasthan is essential in terms of dry spell, aridity and water deficits due to the following reasons:

(i) Based on the drought analysis, a strategy for the irrigation of crops can be chalked out.

(ii) To develop a set of basic guidelines for the development of long-term drought strategies, covering both mitigation and emergency preparedness.

(iii) To delineate the ambit of validity of the myopic rainfall analysis for the state as a whole in agriculture planning and management over different agro-climatic zones.

(iv) All understanding of the pattern of instability in farming would help to formulate policies for the development of the state.

(v) To supplement the basic information required for the selection of crops or their cultivars, specifying fertilizer schedules and adopting soil, rain-water and crop management practices under different agro-climatic zones of the state.

(vi) To have information on the periods during which dry spell conditions are likely to occur so that advance action can be taken to avoid severe moisture stress to the crops through: (a) choice of different crops/varieties, which can withstand moisture stress, (b) adoption of appropriate moisture conservation practices like mulching and land configuration, and (c) life saving irrigation by recycling of harvested runoff water wherever possible.

Some studies have been conducted in western Rajasthan for prediction of drought. However, systematic analysis for prediction of drought under different agro-climatic conditions of the state has not been done so far. Present study was, therefore undertaken for prediction of drought in Rajasthan.

1.1 General Description of the Study Area

1.1.1 Location of the Study Area

The area comes under the study is the whole of Rajasthan state. Rajasthan is the largest state of India with the geographical area of 34,224 thousand hectares. It represents 10.41 per cent area of the country. The state is situated in north-western part of India between 23° 30′ and 30° 12′ N latitude and 69° 30′ and 78° 17′ E longitude. The western and northern boundary of the state touches the eastern boundary of Pakistan. In the north and north-east, it is bounded by Punjab, Haryana, and Uttar Pradesh, the east and south-east by Madhya Pradesh and in the south-west by Gujarat. The shape of Rajasthan state resembles an irregular rhombus with the north-south diagonal about 784 km and the east-west diagonal about 850 km. Administratively, it is divided into 32 districts.

The annual normal rainfall in Rajasthan varies from 164 mm in Jaisalmer to 1004 mm in Jhalawar with mean annual rainfall of 573 mm. The annual average rainfall of western Rajasthan and eastern Rajasthan is 311 mm and 704 mm respectively.

1.1.2 Agro-climatic Zones of Rajasthan State

Based on rainfall distribution, soil characteristics, cropping pattern and other physical characteristics, the state of Rajasthan is divided into nine agro-climatic zones (Fig. 1.1), which are briefly described as under:

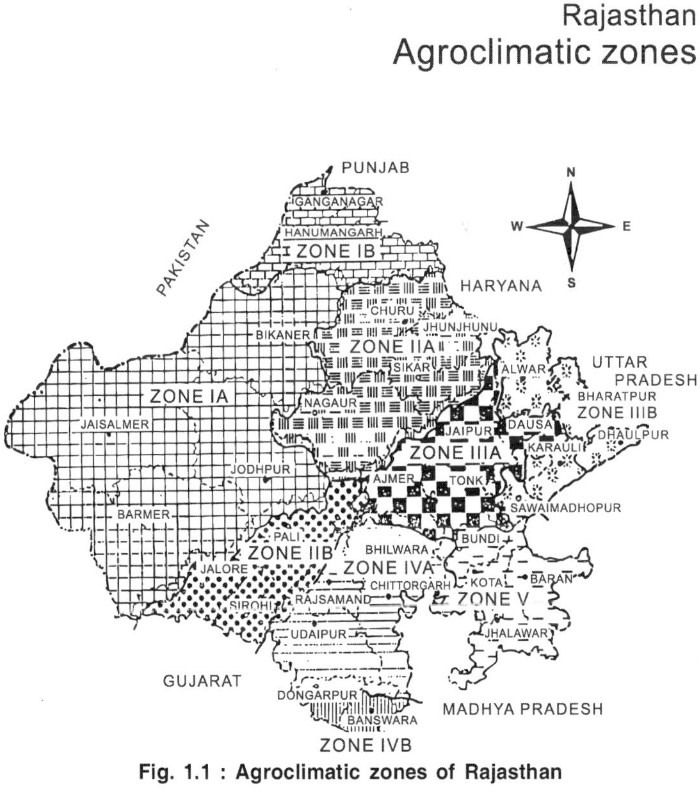

Fig. 1.1 : Agroclimatic zones of Rajasthan

1.1.2.1 Arid western plain zone (I-A)

The zone covers Jaisalmer, western Barmer, western Jodhpur, Bikaner and western Churu districts excluding the command area of Indira Gandhi Canal falling in Bikaner and Jaisalmer districts. It has geographical area of 12.44 Mha. Desert soils and sand dunes occupy the major area in the zone. The mean daily temperatures in the zone range from maximum of 40°C in June to a minimum of 8°C in January. The rainfall in the zone ranges from 100 mm in west to about 300 mm in the east. The common crops grown are pearlmillet, green gram, moth bean and clusterbean. Whenever ground water is available for irrigation, salt resistant Rabi cereals are also grown.

1.1.2.2 Irrigated north western plain zone (I-B)

The zone covers Sriganganagar district and the Indira Gandhi Canal command area falling in western parts of Bikaner and Jaisalmer districts. It has a geographical area of about 2.06 Mha. The soils are alluvial deposits of river Ghagger in the flood plains of Sriganganagar, yellowish brown in colour, loam to silty clay loam in texture and are calcareous in nature. At many places they are intermixed with sandy material. Generally, the soils have very high content of soluble salts and exchangeable sodium. The rainfall ranges from about 100 mm in the south-west to about 350 mm in the east. The mean daily maximum temperature in the zone ranges from 20.5°C in January to 42.1°C in June. Likewise, mean daily minimum temperature ranges from 4.7°C in January to 28.0°C in June. With the introduction of canal irrigation, a variety of field crops *viz.* wheat, gram, mustard, ground nut, cotton and sugarcane are now grown in the area.

1.1.2.3 Transitional plain of inland drainage zone (II - A)

The zone covers Sikar, Jhunjhunu, Nagaur and eastern part of Churn districts. This zone has an area of about 3.59 Mha. The soils are sandy loams to clay loam and cultivation is restricted due to shallow depth and stony surface. Red desert soils, saline and lithosols are found in depressions. The rainfall ranges form 300 mm in west to about 500 mm in the east. The mean daily minimum temperature in the zone ranges from 5.3°C in January to 27.5°C in June. Due to condition of erratic precipitation, large area is put under pearlmillet and Kharif pulses, particularly moth-bean and clusterbean. In addition, groundnut, occupies a sizeable area. In Rabi, irrigated wheat and barley are important crops.

1.1.2.4 Transitional plain of Luni basin zone (II-B)

The zone covers western Sirohi, Barmer, eastern Jodhpur, Pali and Jalore districts and the area lies west of Aravali ranges. It covers a geographical area of about 2.94 Mha. In Jodhpur, Jalore, Pali and Barmer districts red desert soils are found. Saline soils are found in Ranns of Barmer district. Sierozems are found in Pali and Sirohi district. About 27 per cent of the cropped area is under irrigation mainly through canals and wells. The rainfall ranges from 300 mm in the west to 500 mm in the east. The principal crops of the zone are pearlmillet, Kharif pulses, sesamum, maize, wheat, barley, mustard and gram. The area is flood prone by the river Luni with 16 years flood cycle. The area under cotton is increasing in the region with the increasing irrigation facilities.

1.1.2.5 Semi-arid eastern plain zone (III-A)

The zone includes Ajmer, Jaipur and Tonk districts. It covers a geographical area of 2.94 Mha. The soils are sierozems in the western parts of Ajmer and Jaipur districts. The alluvial soils of western and north-western parts of Jaipur and Ajmer are modified due to increase of aeolian material flows through gaps in the Aravalli ranges. Old alluviums in south-eastern parts and lithosols near the foothills in Ajmer are also found. Brown soils are found in parts of Tonk district. The rainfall ranges from about 500 mm in the west and north-west to about 600 mm in south-east. The mean daily maximum temperatures range from 22°C in January to 40.6°C in May and daily minimum temperature from 8.3°C in January to 27.3°C in June. The principal crops of this zone are pearlmillet, sorghum, clusterbean, cowpea, green gram and groundnut during Kharif and wheat, barley, mustard and gram during Rabi maize and cotton are also grown in Ajmer district.

1.1.2.6 Flood prone eastern plain zone (III - B)

The zone includes eastern Alwar, Bharatpur, Dholpur and southern Sawai Madhopur districts. It covers a geographical area of about 2.37 Mha. Soils are mainly alluvial, subject to recurring floods and waterlogging. The rainfall ranges from about 500 mm in the north-west to about 600 mm in the south-east. The principal crops of the zone are gram, mustard, pearlmillet, wheat and pigeonpea.

1.1.2.7 Sub humid southern plain zone (IV-A)

The zone includes eastern Sirohi, Udaipur, Bhilwara and Chittorgarh districts. The zone covers a geographical area of about 3.36 Mha. The terrain is irregular, the soils are lithosols at the foothills in Udaipur and Chittorgarh districts, but those away from the hills are old alluviums. The rainfall ranges from 500 mm in west and north-west to about 700 mm in the south-west. The mean daily maximum temperatures range from 24.2°C in January to 38.6°C in June. The principal crops of the zone are maize, sorghum, groundnut, cotton, blackgram, wheat, barley, gram and mustard. In the Kharif season, the major area is covered under maize and in Rabi, by wheat and barely.

1.1.2.8 Humid southern plain zone (IV-B)

The zone includes Dungarpur, Banswara, south eastern part of Udaipur and southern Chittorgarh districts. It covers a geographical area of about 1.72 Mha. Predominantly the soil of this zone are reddish in colour, medium textured, well drained and calcareous. There are shallow soils on hill slopes and deep soils in the valleys. Maize, paddy, wheat, cotton are the major crops of the zone. Small millets, Kharif pulses and gram are also grown.

1.1.2.9 Humid south eastern plain zone (V)

The zone includes Jhalawar, Kota, Bundi, eastern Chittorgarh and western Sawai Madhopur districts. It covers a geographical area of 2.70 Mha. The soils of this zone are predominantly black soils of alluvial origin. These are generally clay loam to clay. In some pockets soil salinity and ground water salinity have been encountered. The rainfall ranges from 650 mm in the north-west to about 1000 mm in the south-east. The mean daily maximum temperature ranges from 24.5°C in January to 42.5°C in May. Similarly the mean daily minimum temperature ranges from 10.6°C in January to 29.7°C in May. The principal crops of the zone are sorghum, maize, cotton, paddy, sugarcane, wheat, gram, mustard and linseed.

1.2 Collection of Meteorological and Soils Data

Eight stations *viz.* Jodhpur, Sriganganagar, Fatehpur, Pali, Jaipur, Udaipur, Banswara and Kota were selected for the present study on the basis of availability of long term data. Latitude, longitude and height above mean sea level of the stations are given in Table 1.1.

The daily data of rainfall, air temperature, relative humidity, wind speed, sunshine hours and pan evaporation were collected for a period of 15 years (1981-95) from Meteorological Observatory of College of Technology and Engineering, Udaipur for Udaipur station. For Jodhpur and Pali stations the data were collected from the Central Arid Zone Research Institute Jodhpur. For Sriganganagar, Fatehpur, Jaipur, Banswara and Kota stations the data were obtained from their respective Agricultural Research Stations and also from India Meteorological Department, Pune. The meteorological data were compiled in three categories *viz.* (i) Weekly (ii) Mean monthly and (iii) Annual.

Table 1.1 : Location of stations under different agro-climatic zones of Rajasthan

Station	Agro-climatic zone	Latitude	Longitude	Height above mean sea level (m)
Jodhpur	Arid western plain (I-A)	26º18' N	73º01' E	224
Sriganga-nagar	Irrigated north western plain (I-B)	29º55' N	73º53' E	177
Fatehpur	Transitional plain of inland drainage (II-A)	27º37' N	75º08' E	432
Pali	Transitional plain of Luni basin (II-B)	25º47' N	73º20' E	212
Jaipur	Semi-arid eastern plain (III-A)	26º51' N	75º47' E	436
Udaipur	Sub-humid southern plain (IV-A)	24º35' N	73º42' E	582
Banswara	Humid southern plain (IV-B)	23º33' N	74º27' E	218
Kota	Humid south-eastern plain (V)	25º11' N	75º50' E	257

Soils data were collected from annual reports of respective Agricultural Research Stations in the state. Physical properties of soils of different stations are given in Table 1.2.

Table 1.2 : Physical properties of soils at different stations

Station	Soil texture	Field Capacity (%)	Wilting point (%)	Maximum soil depth (cm)	Available water holding capacity at maximum soil depth (mm)
Jodhpur	Loamy sand	10.20	2.80	100	100
Sriganga-nagar	Sandy loam to loam	16.60	4.84	100	100
Fatehpur	Sandy loam to loam	11.00	2.50	120	100
Pali	Sandy clay loam	29.8	9.10	90	90
Jaipur	Sandy to sandy loam	10.80	4.47	200	130
Udaipur	Sandy clay loam	17.50	8.40	150	150
Banswara	Clay loam	25.0	10.00	200	270
Kota	Loam to clay loam	30.0	10.00	120	200

2. Rainfall Pattern

2.1 Introduction

The most important climatic element is the rainfall to determine the severity of a drought. India receives its rainfall mainly from the south-west monsoon and partly from the north-east monsoon. The average rainfall of the country is about 1200 mm. However, the distribution across the country varies from less than 100 mm in extreme arid areas of western Rajasthan to greater than 3600 mm in the north-eastern states and 1000 mm from east coast to 2500–3000 mm in the west coast. Nearly 70 per cent of the annual rainfall in India is recorded in the south-west monsoon. Deficient rainfall during the south-west monsoon leads to drought. It is, therefore, essential to study rainfall pattern of a particular location for prediction of drought.

Handa and Mishra (1968) analysed 75 years rainfall data of Kota in Rajasthan state. They worked out variability of annual rainfall, its trend, annual and monthly distribution together with the maximum intensity of different storms ranging from 5 minutes to 60 minutes duration. The analysis indicated a slight increasing trend in annual precipitation. While there was a wide variability in rainfall from year to year and from month to month, the study showed a better consistency in the distribution of rainfall during Kharif (monsoon) season, while during Rabi (winter) season the variability in rainfall distribution was much more.

Krishnan and Thanvi (1977) studied rainfall in western Rajasthan and found that the mean annual rainfall in western Rajasthan varies from 100 mm in the north-western Rajasthan of Jaisalmer district to 600 mm in south-eastern part of Pali district. The general decreasing gradient was from south-east to north-west. There was a considerable year to year variation in rainfall as revealed by the high standard deviation values. The coefficient of variation of annual rainfall varied from less than 40 per cent in Sikar and Jhunjhunu districts to more than 70 per cent in Jaisalmer and Barmer districts. From the rainfall variability point of view, western Rajasthan can be classified in the subregions. Apart from high variability of rainfall in western Rajasthan, there was a considerable skewness in the distribution of annual rainfall.

Choudhaury *et al.,* (1979) analysed 70 years (1901-1970) rainfall data of 26 stations in Maharashtra state. Two types of dry days were classified, *viz.,* a day receiving less than 2.5 mm of rainfall and a day receiving less than 6.3 mm of rainfall. Dry spells for both the categories have been computed and analysed. Similarly, probabilities of occurrence of dry spells of different lengths have been analysed. A theoretical distribution was also fitted to dry spells. The analysis revealed that logarithmic distribution nearly represents both categories of dry days.

Jeevarathnam and Jayakumar (1979) made probability analysis of rainfall by using 60 year's rainfall data (1918–1977) for Ootacamund, Tamil Nadu. Three probability distribution functions, namely, Gumbel distribution, log-Pearson Type III distribution and log normal distribution have been compared with the observed data. The log Pearson Type III distribution gave the close fit to the observed data.

Agnihotri and Bansal (1983) analysed daily data of Chandigarh from 1958 to 1979 to study wet spells monthwise by applying the logarithmic model. A computer programme was developed for finding out the parameters. On an average Chandigarh had 33 wet spells out of which 19 occurred during June to September. The average number of rainy days in a year were 64, out of which 42 occurred from June to September. Maximum length of wet spell was 14 days during the month of August. There was about 50 per cent chance for any of the days being rainy during July and August.

Kumar and Rao (1983) studied the annual rainfall at seven stations in Andhra Pradesh. Values of mean, median, coefficient of variation, standard deviation, coefficient of skewness and decile of annual rainfall were determined. Surplus and deficient rainfall patterns in the state were discussed. The amounts of surplus and deficient rainfall which can be expected once in five and ten years were also determined.

Krishnan (1984) analysed daily rainfall data (1901–75) of Karnataka and Tamil Nadu states of southern peninsular India. A comparison of rainfall normal (1901-50) was made with the rainfall amounts recorded at different stations in these states during 1951-75. The decreases in annual rainfall have occurred in the latter period for many regions in Tamil Nadu which receive rainfall predominantly during north-east monsoon, whereas increases in annual rainfall have occurred in Malard area, coastal Karnataka and north interior Karnataka which receive rainfall mostly during south-west monsoon.

Rao (1987) used a generalized power transformation to normalize weekly rainfall series by an iterative procedure. Weekly rainfall estimates at specified probability of incidence obtained from the normalized series were found to be comparable to the corresponding values obtained after fitting the incomplete gamma distribution to the weekly rainfall data.

Mishra and Satyanarayan (1987) analysed the rainfall data for 26 years, (1956-1981) at Kharagpur. The different statistical parameters such as mean, standard deviation and coefficient of variation have been worked out for the weekly, monthly and yearly rainfall data. The extreme events in the rainfall have been traced out and their effects on the statistical parameters have been discussed.

Yadav *et al.,* (1987) analysed the rainfall data of nine stations in Balason catchment of Darjeeling hills to study the distribution of rainfall within the catchement with a view to understand variability and trend of mean annual rainfall which may be useful in planning and development of water resources in the catchment. The mean annual rainfall varied from 241 cm to 403 cm with coefficient of variation ranging from 9.76 to 11.65 per cent. The rainfall data were analysed and the results of trend, oscillation, persistence and normality were presented.

Patil and Patil (1991) analysed the rainfall data for 24 years (1956-1988). The mean, standard deviation and coefficient of variation were computed for the monthly, seasonal and yearly time series. The lowest coefficient of variation of 48.56 per cent in the wettest month of July and the highest coefficient of variation of 278.95 per cent in the driest month of February were recorded. The coefficient of variation of annual rainfall was 25.95 per cent. The extreme events occurred in March 1967, May 1971, February 1978 and April 1985. The coefficient of uniformity of rainfall was 0.74 in the Kharif season and 0.78 on the annual basis.

Tripathi (1992) used Markov-chain model to predict the sequences of wet and dry days. The analysis was based on rainfall data of 22 years at Pantnagar. Weekly probability and probability of particular day in the week being wet or dry showed that wheat planted in the first week of December has a 50–70 per cent chance of getting rain at crown root initiation and flowering stages and a sure chance of rain at jointing and milk stages.

Kumar (1995) analysed daily rainfall data for a period of 28 years to evaluate rainfall variation pattern at Pantnagar. The triad rainfalls were computed by summation of three days rainfall from June to October period and Oliver's precipitation index was calculated. It was revealed that there is no difference in number of rain spells during sub-periods 1970-79 and 1980-89.

Tripathi (1996) analysed the weather data (1960-1981) of Pantnagar for the probability of occurrence of annual and weekly rainfall and length of dry spells and showed that the annual and total rainfall can be predicted fairly accurately by the log-normal distribution model. Results revealed that at least 861 mm of rainfall can be expected every year, which could rise to 1209 mm in alternate years. Highest annual rainfall, exceeding 2000 mm has a return period of II years. Probability of time to the next wet day calculated at 70 per cent level compared best with the observed values. From the analysis, the onset of monsoon at Pantnagar can be expected in the 24th week of the calendar year which may last till the 37th week in conformity with the traditional forecasts.

Chakraborty (1998) analysed rainfall data of 10 years (1983–1992) at Regional Research Station, Kakdwip (West Bengal). He used Weibull's method for analysis of rainfall data and worked out computed rainfall amounts at different return periods.

Rao (1999) studied the variability of monsoon rainfall over the Godavari river basin, for different time scales. The monsoon rainfall in the Godavari river basin was more variable (17%) than the all India monsoon rainfall (11%) during the period of study (1951-90). Similarly inter-annual variability of monsoon rainfall on smaller time scales was still higher and increased from seasonal to daily scales. The intra-seasonal variability of monsoon rainfall had a significant negative relationship with the total seasonal rainfall in the basin.

Kumar (2000) analysed daily rainfall data (1977-1985) of Rani Chauri, on using Weibull's formula, log Pearson Type III distribution and log normal distribution at different probability levels. He used Chi-square test of goodness of fit and found that log-normal distribution gave closest fit to the observed values.

2.2 Types of Models

2.2.1 Determination of Start and End of Rainy Season by Forward and Backward Accumulation

Start of rainy season was computed from weekly rainfall data using Moris and Zandstra, (1979) method. In this method weekly rainfall can be summed by forward accumulation (22 + 23 + ... + 52 weeks) until a certain amount of rainfall is accumulated. If this process is repeated for a long-term period, then the probability of given amounts of rain can be determined for each week. An amount of 75 mm accumulation of rainfall, at the onset time for the given season was selected as it may be sufficient for land preparation and sowing of dry seeded crops.

The end of the rainy season was determined by backward accumulation of rainfall (52+51+...... + 41 weeks) data, 10 mm of accumulation of rainfall was choosen for end of the rainy season, which may be sufficient for ploughing of fields after harvesting of crops. Then the years were assigned with rank number, *i.e.,* 1981 to 1995 as 1 to 15. The probability (P) of each rank was calculated by the following Weibull's formula:

$$P = \frac{m}{N+1} \times 100 \qquad ...(2.1)$$

where, m is the rank number and N is number of years. For forward accumulation the rank order and the probability level were arranged in ascending order and the corresponding week numbers were arranged in the same manner. Similarly for backward accumulation the rank order and the probability level were arranged in descending order and the corresponding week numbers were arranged in the same way.

2.2.2 Incomplete Gamma Distribution for Assured Rainfall Analysis

The probability density function [f(x)] of the two parameter gamma distribution function was determined from the following equation:

$$f(x) = \frac{x^{\gamma-1} e^{-x/\beta}}{\beta^{\gamma} \Pi(\gamma)}, \text{ for } x>0, \gamma, \beta>0 \quad ...(2.2)$$

$$= 0, \text{ for } x \leq 0$$

where,

β = scale parameter,

= shape parameter,

$\Pi(\gamma)$ = gamma function of γ

when, $\gamma \rightarrow \infty$, gamma distribution approaches to normal distribution. However, if γ is large, distribution is Guassian and can be represented by the following expression:

$$\gamma = \left(1 + \frac{1}{4A}\right)\left[1 + \left\{1 + \left(\frac{4A}{3}\right)\right\}^{\frac{1}{2}}\right] \quad ...(2.2a)$$

$$A = L_n\bar{x} - \Sigma(\text{in } \bar{x}/N); \quad ...(2.2b)$$

$$\beta = \bar{x}/\gamma \quad ...(2.2c)$$

where

$\bar{x}$ = arithmetic mean of x_1, x_2, x_3 ... x_n, (the non zero values of variable x),

L_n = natural logarithms,

n = rank number

N = number of values in the data series greater than zero.

Rainfall data of different stations were used for fitting the Gamma distribution.

2.2.3 Estimation of Rainfall Probabilities by Markov Chain Process

The concept of estimating probabilities with respect to a given amount of rainfall is extremely useful for agricultural planning. In a crop growing season, many times decisions have to be taken based on the probability of receiving certain amount of rainfall during a given week [P(W)], which are called "initial probabilities". Then the probability of rain during next week, if rain occurred this week [P(W/W)]; and the probability of rain next week being wet, if this week is dry [P(W/D)] are very important and are called "conditional probabilities". Analogously, initial and conditional probabilities for a dry week can be defined. These initial and conditional probabilities would help in determining the relative chance of occurrence of a given amount of rainfall. This becomes the basis for the analysis of rainfall using Markov chain process. The parameters estimated for the analysis were as follows:

a. initial probabilities of a week being wet, [P(W)],
b. conditional probability of a wet week preceded by a wet week, [P(W/W)], and
c. conditional probability of a wet week preceded by a dry week, [P(W/D)].

The following formulae were employed for calculation by considering following two phenomena:

a. dry condition, D_j at the ith period, and
b. wet condition, W_j at the jth period.

The probabilities can be defined as follows:

$$P(W_j) = \frac{N(W_j)}{N} \qquad \text{...(2.3)}$$

where,

$N(W_j)$ = number of occurrences of W in jth period,
$N(D_j)$ = number of occurrences of D in jth period, and

$$N = N(W_j) + N(D_j)$$

Similarly,

$$P(D_i) = \frac{N(D_i)}{N} \qquad \text{...(2.4)}$$

Also knowing the probability of ith dry period, given jth wet period *i.e.*, $P(D_j/W_j)$, the change in the probability of W_j gives the additional information that D_j has actually occurred and can be written as,

$$P(W_jD_i) = \frac{(P(W_j)\ P(D_i/W_j)}{P(D_i)} \qquad \text{...(2.5)}$$

Further knowing the joint probability distribution of $P(W_j, D_j)$

$$P(W_j/D_i) = \frac{(P(W_j, D_j)}{P(D_j)} \qquad \text{...(2.6)}$$

The general formula for a dry spell of *n* weeks (n consecutive weeks) is given by:

$$P(D, D, D, ..., n) = P(D)\ \ P(D/D)^{n-1} \qquad \text{...(2.6a)}$$

Similarly, for wet spell of *n* weeks is given as:

$$P(W, W, W, \ldots, n) = P(W) \quad P(W/W)^{n-1} \qquad \ldots(2.6b)$$

The probability of [P(2D)] 2 consecutive dry weeks is calculated as follows:

$$P(2D) = P(D) \quad P(D/D)^2 \qquad \ldots(2.6c)$$

where,

$P(2D)$ = probability of two consecutive dry weeks,

$P(D)$ = probability of the week being dry (I week),

$P(D/D)_2$ = probability of 2nd consecutive dry week given the preceding week being dry.

Similarly, the P(2W) have been estimated in the same manner as P(2D) and the notations used as having same meaning except the week being wet.

Using the above relationships the initial and conditional probabilities were calculated for all the stations during each week (standard weeks 1–52) for considering less than 20 mm rainfall as a dry week (Pandarinath, 1991). A computer programme in FORTRAN *77* was used for the analysis. The computer programme is given in Appendices C1-C2.

2.3 Discussions

2.3.1 Analysis of Rainfall Pattern Under Different Agro-climatic Zones of Rajasthan

Rainfall being the only source of water under dry land agriculture, a more complete and quantitative understanding of the time and spatial variation of this natural resource, provides a lot more insight for proper planning with respect to crop production. The daily and weekly rainfall data of eight stations under different agro-climatic zones of Rajasthan for 15 years period (1981–95) were analysed, for annual, seasonal and weekly patterns and the results are presented as under:

2.3.2 Annual Rainfall

Annual rainfall data of different stations were analysed statistically and their different statistical parameters like mean, standard deviation, median, coefficient of variation and coefficient of skewness were calculated which are shown in Table 2.1. It is evident from Table 2.1 that Banswara is the highest rainfall receiving zone in the state having an average annual rainfall of 913.9 mm with standard deviation of 313.4 mm and coefficient of variation of 34.3 per cent. Sriganganagar receives the lowest amount of rainfall, its mean annual rainfall being 325.8 mm with standard deviation of 96.3 mm and coefficient of variation of 29.6 per cent. Minimum and maximum annual values of rainfall of different stations are also given in Table 2.1. Table 2.1 shows that Jodhpur has minimum rainfall annual value of 110.7 mm and Jaipur has got the maximum annual rainfall (1622.1 mm). It may be pointed out that Jaipur received maximum rainfall due to cloud brust in the 1982 which caused heavy damage. It is an exceptional case. However, Banswara remains the maximum rainfall receiving zone in the state. Moreover, it is seen that the minimum range of minimum and maximum rainfall was observed for Kota zone about 1.81 times minimum rainfall with lowest coefficient of variation of 16.5 per cent. Jaipur has got highest coefficient of variation of 51.1 per cent as it has maximum range of minimum and maximum annual rainfall.

Table 2.1 Annual rainfall parameters under different agro-climatic zones of Rajasthan

Satation	Minimum annual rainfall (mm)	Maxi-mum annual rainfall (mm)	Mean annual rainfall (mm)	Stan-dard deviation (mm)	Coeffi-cient of vari-ation (%)	Median (mm)	Coeffi-cient of skew-ness
Jodhpur	110.7	844.0	351.8	178.9	50.9	313.6	0.64
Sriganganagar	200.0	595.2	325.8	96.3	29.6	317.5	0.26
Fatehpur	242.0	569.3	400.9	88.3	22.0	393.7	0.24
Pali	152.7	923.3	389.9	197.0	50.5	348.2	0.64
Jaipur	235.8	1622.1	687.0	350.9	51.1	611.9	0.64
Udaipur	333.7	895.4	584.8	154.9	26.5	598.4	–0.26
Banswara	396.7	1473.1	913.9	313.4	34.3	861.3	0.50
Kota	469.1	848.8	685.1	113.1	16.5	722.1	–0.98

It would also be seen from Table 2.1 that for most of the agro-climatic zones in Rajasthan, median values are much lower than the mean values thereby showing presence of large skewness in the distribution of annual rainfall. The values of coefficients of sknewness are also presented in Table 2.1. The highest skewness (with coefficient exceeding +0.5) is noticed in respect of Jodhpur, Pali and Jaipur. In Sriganganagar, Fatehpur and Banswara zones the skewness coefficient values vary from + 0.24 to 0.5. The least positive skewness coefficient is noticed in respect of Fatehpur zone which considerably has less variability in the annual rainfall. The analysis of annual rainfall data shows that Udaipur and Kota zones are negatively skewed. Table 2.1 also shows that the standard deviation of annual rainfall ranges from 88.3 to 350.9 mm and coefficient of variation ranges from 16.5 to 51.1 per cent for different agro-climatic zones of the state. These significant variations of rainfall indicate that the annual rainfall pattern in the state is very much erratic from zone to zone.

2.3.3 Seasonal Rainfall

Different parameters of seasonal rainfall received from June to September under different agro-climatic zones of Rajasthan are shown in Table 2.2. It is clear from Table 2.2 that Banswara receives the highest seasonal rainfall in the state having an average of 833.2 mm with coefficient of variation of 39:14 per cent and contributes to 91.5 per cent annual rainfall. Sriganganagar receives the lowest amount of seasonal rainfall, its average being 248.9 mm and coefficient of variation of 34.99 per cent and its contribution to annual rainfall is 76.26 per cent. From Table 2.2 it is evident that Jodhpur has got minimum seasonal rainfall (64.4 mm) and Banswara has got maximum seasonal rainfall (1436.2 mm). Moreover, it is seen that the range of minimum seasonal rainfall of all the zones is very high which results in higher dispersion and hence deviation in the rainfall data.

The analysis of rainfall data presented in Table 2.2 shows that the coefficient of variation ranges from 19.12 per cent (Kota) to 58.58 per cent (Jodhpur). These significant variations of rainfall indicate that the seasonal rainfall pattern in the state is very much erratic form zone to zone. It also reveals that as the difference between minimum and maximum rainfall increases the coefficient of variation increases. The rains from south-west monsoon season were found to be more reliable as coefficient of variation is less than 40 per cent for Kota, Fatehpur, Sriganganagar, Udaipur and Banswara whereas the south-west monsoon seasonal rains were not found to be much dependable for Jodhpur, Pali and Jaipur as coefficient of variation is more than 40 per cent.

Table 2.2 : Seasonal rainfall (June-September) parameters under different agroclimatic zones of Rajasthan

Station	Minimum seasonal rainfall (mm)	Maximum seasonal rainfall (mm)	Mean seasonal rainfall (mm)	Coefficient of variation (%)	Contribution of seasonal rainfal to annual rainfall (%)
Jodhpur	64.4	795.1	301.7	58.58	85.76
Sriganganagar	75.2	404.2	248.9	34.99	76.26
Fatehpur	241.1	473.3	324.1	22.10	82.47
Pali	120.0	848.9	299.5	50.39	76.81
Jaipur	203.6	1424.6	609.4	49.59	88.70
Udaipur	290.1	895.4	529.9	32.31	90.61
Banswara	374.9	1436.2	833.2	39.14	91.05
Kota	381.0	795.7	605.8	19.12	88.40

2.3.4 Weekly Rainfall

The data of mean weekly rainfall and their coefficient of variation for each standard meteorological week for all the stations are presented in Appendix A.1. Since coefficient of variation gives an idea of variation of rainfall amount about its mean, the adequacy of the measure depends upon the use to which it is put. For modem agriculture, information on climatic features is required for a shorter period, as the growing period for most of the crops is about 100 days. Even the use of monthly rainfall data suffers from many shortcomings. Therefore, in this study weekly rainfall pattern is considered. Weekly maximum, minimum and mean rainfall, weeks receiving rainfall more than 10 mm, 20 mm and 40 mm and also the range of coefficient of variation are presented in Table 2.3.

It is evident from Table 2.3 that the maximum mean weekly rainfall of 139.1 mm was recorded at Banswara (31st week) among all the stations. The maximum mean weekly rainfall ranged from 29.1 mm (Sriganganagar) to 139.1 mm (Banswara) during the study period. The minimum mean weekly rainfall of 0.0 mm was recorded at all the eight stations. It is interesting to note that in western Rajasthan all four stations, *viz.,* Jodhpur, Sriganganagar, Fatehpur and Pali, recorded the maximum mean weekly rainfall in 30th standard meteorological week, whereas in eastern Rajasthan the maximum mean weekly rainfall recorded in 34th, 32nd, 31st and 30th weeks at Jaipur, Udaipur, Banswara and Kota respectively. However, during 48th week no rainfall was recorded at all the stations except at Sriganganagar.

Table 2.3 : Weekly rainfall and coefficient of variation for different stations

Station	Maximum weekly rainfall (mm)	Minimum weekly rainfall (mm)	Weekly rainfall >10 mm	Weekly rainfall >20 mm	Weekly rainfall >40 mm	Range of coefficient of variation (%)
Jodhpur	48.4 (30)	0.0 (45, 47-49, 51)	27, 29-36	27, 29-33 and 36	27 and 30	0.0-416.7
Sriganganagar	29.1 (30)	0.0 (43, 44, 46)	24, 26-33 and 36	27-31 and 33	–	0.0-380.0
Fatehpur	46.2 (30)	0.0 (43, 48, 50)	25, 27-35	27-32 and 34	29-30	0.0-481.8
Pali	55.5 (30)	0.0 (43, 48, 50)	24, 26-36 and 39	27-34	27 and 30	0.0-400.0
Jaipur	102.2 (34)	0.0 (42, 48, 49)	25-36	26, 28-36	29-30, 32, 34-35	0.0-500.0
Udaipur	57.8 (32)	0.0 (13, 45, 48, 49)	24-35 and 38, 39	26-35	28-34	0.0-396.7
Banswara	139.1 (31)	0.0 (6-8, 14-15, 20-21, 48-49 and 51)	24-39, 41	24-36, 39, 41	27-34, 36	0.0-400.0
Kota	82.1 (30)	0.0 (47-48)	25-36, 38, 41	26-36	29-36	0.0-450.0

Figure in parenthesis indicates standard meteorological weeks.

Table 2.3 indicates that the rainfall more than 10 mm was recorded only in 9 weeks at Sriganganagar (24, 26-23 and 36), whereas at Banswara the amount was recorded in 17 weeks (24-39 and 41). The rainfall more than 20 mm was recorded in 6 weeks at Sriganganagar (27-31 and 33), whereas at Banswara the amount was recorded in 16 weeks (24-36, 39 and 41). Similarly, the rainfall more than 40 mm was not recorded in any week at Sriganganagar, whereas at Banswara the amount was recorded in 14 weeks (27-34 and 36). Table 2.3 also reveals that the least rainfall with shortest duration was recorded in north-western part of Rajasthan (Sriganganagar) and highest rainfall with longest duration in south-eastern part of Rajasthan (Banswara). The highest range for coefficient of variation was obtained at Jaipur (0.0-500.0). This might be due to high variation of weekly rainfall. The lowest range for coefficient of variation was obtained at Sriganganagar.

2.3.5 Characteristics of Rainy Season

Start, end and length of the rainy season were worked out by forward and backward accumulation of weekly rainfall (Morris and Zandstra, 1979) for different agroclimatic zones of Rajasthan and are presented in Table 2.4.

2.3.5.1 Start of the rainy season

In the beginning of the rainy season, there should be adequate rainfall for land preparation and sowing of crops. In Rajasthan, the rainfall is received due to south-west monsoon (June-September) and north-east monsoon (October-December). Before the start of the south-west monsoon season, some rainfall may be expected due to pre-monsoon thunder showers. The start of the rainy season is considered as the date by which the rainfall accumulates to 75 mm including the pre-monsoon showers at the beginning of the season for growing of dry seeded crops. Considerable variation in the start of the rainy season occurs from year to year. Therefore; the mean week of start of the rainy season, the earliest week during which the start of the rainy season has commenced in the past and the week during which rainy season started late during some of the years in the past are given zone wise for the state in Table 2.4. From Table 2.4, the following inferences can be made with regard to start of the rainy season:

Table 2.4 : Characteristics of rainy season in different agro-climatic zones of Rajasthan

Station	Start of rainy season			End of rainy season			Length of rainy season (weeks)		
	Early	Late	Mean	Early	Late	Mean	Maximum	Minimum	Mean
	Standard meteorological weeks								
Jodhpur	26.0	33.0	29.6	33.0	44.0	38.1	17.0	3.0	8.4
Sriganganagar	25.0	33.0	28.5	35.0	45.0	38.9	16.0	4.0	10.4
Fatehpur	25.0	31.0	28.0	34.0	47.0	39.2	20.0	4.0	11.2
Pali	26.0	31.0	28.3	33.0	50.0	39.0	22.0	3.0	10.7
Jaipur	25.0	32.0	28.1	32.0	52.0	39.2	23.0	3.0	11.1
Udaipur	24.0	30.0	27.1	36.0	50.0	40.7	24.0	8.0	13.6
Banswara	24.0	33.0	27.1	34.0	52.0	41.7	24.0	6.0	14.7
Kota	25.0	30.0	24.8	33.0	52.0	41.2	23.0	5.0	13.4

(i) In south-east zones early start of the rainy season occurs during 24th week (June 11-17). During the years with delay in start of rainy season, the season may start as late as 33rd week (Aug. 13-19) in Banswara zone and 30th week (July 23-29) in Udaipur and Kota zones. Under normal conditions, the rainy season starts by 27.1th week (July 2-8) in Udaipur and Banswara zones and 27.8th week (July 2-8), in Kota zone.

(ii) In western Rajasthan, start of the rainy season may be as early as 25th week (June 18-24) for Jodhpur and Pali zones and 26th week (June 25-1) for Sriganganagar and Fatehpur zones. When the start of rainy season is delayed it may be as late as 33rd week (August, 13-19) in Jodhpur and Sriganganagar zones and by 31st week (Aug 30-5) in Fatehpur and Pali zones. Under normal conditions, the season may commence by 28th week (July 9-15) in Fatehpur zone, 28.3th week (July 9-15) in Pali zone, 28.5th week (July 9-15) in Sriganganagar zone and 29.6th week (July 16-22) in Jodhpur zone.

(iii) In Jaipur zone start of rainy season may be as early as 25th week (June 18-24) and under delayed conditions, start of the season may be possible by 32nd week (Aug. 6-12). Under normal conditions, the rainy season starts by 28th or 29th week (July 9-15 or July 16-22).

The percentage probabilities for start of the rainy season during different standard meteorological weeks in various zones of Rajasthan are presented in Table 2.5. Start of the rainy season may be expected by 28th week in Udaipur, 29th week in Banswara, 30th week in Sriganganagar and 31st week in Jodhpur zones with a probability of 75 per cent *i.e.,* the rainy season can be expected to start on the average in 3 out of 4 years. Similarly, the start of rainy season can be expected by 30th week in Fatehpur, Pali and Jaipur with a probability of 87.50 per cent. In Kota zone the start of rainy can be expected by 29th week with a probability of 81.25 per cent.

2.3.5.2 End of the rainy season

End of the rainy season is determined by backward accumulation of rainfall from 52nd week accounting to an amount of 10 mm rainfall that can be expected during the remaining period

Table 2.5 : Probabilities of the start of rainy season during standard meteorological weeks under different agro-climatic zones of Rajasthan

Station	Probability (%) for start of rainy season by the end of standard meteorological week									
	24	25	26	27	28	29	30	31	32	33
Jodhpur	–	–	12.5	18.75	25.0	50.0	68.75	75.0	87.50	93.75
Sriganganagar	–	6.25	25.0	37.5	43.75	62.50	75.0	87.50	–	93.75
Fatehpur	–	6.25	–	31.25	68.25	81.25	87.50	93.75	–	–
Pali	–	–	12.5	37.5	50.0	62.50	87.50	93.75	–	–
Jaipur	–	12.5	25.0	–	43.75	81.25	87.50	–	93.75	–
Udaipur	6.25	12.5	37.5	50.0	75.0	87.50	93.50	–	–	–
Banswara	6.25	37.5	56.25	–	68.75	75.00	81.25	87.25	93.75	–
Kota	–	6.25	18.75	37.50	62.50	81.25	93.75	–	–	–

Table 2.6 : Probabilities of the end of rainy season during standard meteorological weeks under different agro-climatic zones of Rajasthan

Station	Probability (%) for the end of rainy season during standard meteorological weeks													
	32	33	34	35	36	37	38	39	40	41	42	43	44	45
Jodhpur	–	93.75	87.50	81.25	68.75	56.25	43.75	37.50	25.0	18.75	–	12.5	–	–
Sriganganagar	–	–	–	93.75	87.50	75.0	62.50	43.75	–	–	37.50	–	–	25.0
Fatehpur	–	–	93.75	–	81.25	75.0	62.50	56.25	31.50	31.25	18.75	–	12.5	–
Pali	–	93.75	–	87.50	81.25	62.50	–	56.25	31.25	18.75	–	–	–	12.5
Jaipur	93.75	87.50	–	81.25	68.75	50.0	–	37.50	–	31.25	–	–	–	–
Udaipur	–	–	–	–	93.75	–	87.50	81.25	50.00	31.25	25.0	–	18.75	–
Banswara	–	–	93.75	87.50	–	–	81.25	–	62.50	56.25	50.0	37.50	31.25	25.0
Kota	–	93.75	–	–	87.50	68.75	62.50	56.25	650.0	43.75	–	–	31.25	25.0

of calendar year. Table 2.4 shows the end of rainy season for different stations. The percentage probabilities for the end of rainy season for different agro-climatic zones of Rajasthan are given in Table 2.6.

Table 2.6 reveals that during the years with early withdrawal of south-west monsoon the rainy season may end by 42nd week in Jaipur with 93.75 per cent probability. Under normal conditions, the rainy season, ends by 38.1th week in Jodhpur with 43.75 per cent probability. During some of the years, the north-east monsoon rainfall may extend the rainy season up to 44th week in Jodhpur and 52nd weeks in Sriganganagar, Jaipur, Banswara and Kota.

Table 2.6 also indicates that during the years with early withdrawal of south-west monsoon, the latest end of rainy season was observed by the end of 36th week at Udaipur with 93.75 per cent probability. It can be inferred from Table 2.6 that during the years with early withdrawal of south-west monsoon rainy season may end by 34th week in Fatehpur and Banswara with 93.75 per cent probability.

2.3.5.3 Length of the rainy season

The .length of the rainy season is the period between start and end of the rainy season. The length of the rainy season gets prolonged either due to better pre-monsoon rainfall and/or extended north-east monsoon rainfall. Similarly, length of rainy season may be curtailed either due to failure of pre-monsoon showers, early withdrawal of south-west monsoon and/or failure of north-east monsoon. The length of the rainy season for different stations is presented in Table 2.4.

Table 2.4 reveals that the minimum length of rainy season of 3 weeks was observed at Jodhpur, Pali and Jaipur. While the maximum length of rainy season of 24 weeks was recorded at Udaipur and Banswara. The mean minimum length of rainy season was of 8.4 weeks recorded at Jodhpur. The mean maximum length of rainy season was of 14.7 weeks recorded at Banswara. The mean length of rainy season was 8.4–12.4 weeks in western Rajasthan and 11.1-14.7 weeks in eastern Rajasthan.

2.3.6 Dry and Wet Spell Analysis by Markov Chain Process

Initial and conditional probabilities and probable periods of occurrence of two consecutive dry weeks were analysed by Markov Chain process for all the stations under study.

2.3.6.1 Initial and conditional probability

The initial and conditional probabilities of rainfall for amounts 20 mm (Pandarinath, 1991) during 52 weeks were estimated by Equations 3.3 to 3.6. The details of the relationships used for computation of initial and conditional probabilities are given in Section 3.5. The results are given in Table 2.7.

It is clear from Table 2.7 that the probability of occurrence of dry week [P(D)] is high at Jodhpur (86-100%), Sriganganagar (80-100%), Fatehpur (80-100%), Pali (66.67-100%), Jaipur (73.33-100%) and Kota (80-100%) from 22nd to 26th week then it falls in subsequent weeks of the year. The probability of occurrence of dry weeks again increases from 26th to 28th week of the year at Jodhpur and Sriganganagar. While, at Fatehpur and Pali the probability of occurrence of dry week decreases from 26th to 28th week. Similarly, the probability of occurrence of dry week decreases from 26th to 28th week at, Jaipur, Udaipur and Kota, while for Banswara it increases.

The conditional probability of dry week preceded by a dry week [P(D/D)] is also high up to 26th weeks at Jodhpur (84.62-100%), Sriganganagar (80.00-100%). Fatehpur (76.92-100%), Pali (66.67-100%), Jaipur (76.92-100%), Udaipur (54.55-92.31%), Banswara (50.00-93.33%) and Kota (71.43-100.00%). The probability of occurrence of dry week preceded by a dry week increases from 26th to 28th week at Jodhpur but at all other stations it decreases. It is evident from Table 2.7 that the rainfall distribution is highly erratic and low, at Jodhpur, Sriganganagar, and Pali and no week was observed as wet week at 70 per cent probability of occurrence.

Similarly one wet week preceeded by a wet week was at Sriganganagar and Pali. It is also inferred from Table 2.7 that the conditional probability of occurrence of wet week preceded by a wet week [P(W/W)] is observed in one to four weeks at all the stations, which indicated that rainfall distribution during the study period is highly temporal.

2.3.6.2 Probable period of occurrence of two consecutive dry weeks

Dry spells due to inadequate rainfall may occur through out the rainy season. The weekly evaporative demand of different agro-climatic zones of Rajasthan varies from 40 mm during the beginning of the rainy season to 30 mm during active rainy season.

Table 2.7: Probability (%) of dry and wet week under different agro-climatic zones of Rajasthan

Standard meteoro-logical week	P(D)	P(D/D)	P(2D)	P(W)	P(W/W)	P(2W)	Standard meteoro-logical week	P(D)	P(D/D)	P(2D)	P(W)	P(W/W)	P(2W)
			Jodhpur								Sriganganagar		
22	86.67	85.71	86.67	13.33	0.00	0.00	22	100.00	100.00	80.00	0.00	0.00	0.00
23	100.00	100.00	86.67	0.00	0.00	0.00	23	80.00	80.00	66.67	20.00	0.00	13.33
24	86.67	86.66	73.33	13.33	0.00	0.00	24	73.33	83.33	66.67	26.67	66.67	6.67
25	86.67	84.62	80.00	13.33	0.00	6.67	25	86.67	90.91	66.67	13.33	25.00	0.00
26	86.67	92.31	60.00	13.33	50.00	0.00	26	80.00	76.92	53.33	20.00	0.00	0.00
27	73.33	69.23	73.33	26.67	0.00	6.67	27	73.33	66.67	53.33	26.67	0.00	13.33
28	93.33	100.00	40.00	6.67	25.00	6.67	28	66.67	72.73	26.67	33.33	50.00	0.00
29	40.00	42.86	20.00	60.00	100.00	33.33	29	60.00	40.00	46.67	40.00	0.00	26.67
30	46.66	50.00	26.67	53.33	55.56	13.33	30	60.00	77.78	33.33	40.00	66.67	0.00
31	66.67	57.14	46.67	33.33	25.00	20.00	31	73.33	55.56	53.33	26.67	0.00	6.67
32	60.00	70.00	33.33	40.00	60.00	13.33	32	73.33	72.73	60.00	26.67	25.00	6.67
33	60.00	55.56	26.67	40.00	33.33	6.67	33	80.00	81.82	66.67	20.00	25.00	0.00
34	60.00	44.44	46.67	40.00	16.67	13.33	34	86.87	83.33	66.67	13.32	0.00	0.00
35	73.33	77.78	66.67	26.67	33.33	13.33	35	80.00	76.92	73.33	20.00	0.00	6.67
36	80.00	90.91	66.67	20.00	50.00	6.67	36	86.67	91.67	80.00	13.33	33.33	6.67
37	80.00	83.33	73.33	20.00	33.33	0.00	37	86.67	92.31	66.67	13.33	50.00	0.00
38	93.33	91.67	80.00	6.67	0.00	0.00	38	80.00	76.92	73.33	20.00	0.00	0.00
39	86.67	85.71	86.67	13.33	0.00	0.00	39	93.33	91.67	93.33	6.67	0.00	0.00

Table 2.7– *contd...*

Standard meteoro-logical week	P(D)	P(D/D)	P(2D)	P(W)	P(W/W)	P(2W)	Standard meteoro-logical week	P(D)	P(D/D)	P(2D)	P(W)	P(W/W)	P(2W)
			Fatehpur							Pali			
22	100.00	100.00	80.00	0.00	0.00	0.00	22	86.67	86.67	86.67	13.33	0.00	0.00
23	80.00	80.00	66.67	20.00	0.00	0.00	23	100.00	100.00	66.67	0.00	0.00	0.00
24	86.67	83.33	66.67	13.33	0.00	0.00	24	66.67	66.67	53.33	33.33	0.00	6.67
25	80.00	76.92	80.00	20.00	0.00	0.00	25	80.00	80.00	60.00	20.00	20.00	6.67
26	100.00	100.00	60.00	0.00	0.00	0.00	26	73.33	75.00	60.00	26.67	33.33	13.33
27	60.00	60.00	26.67	40.00	0.00	0.00	27	73.33	81.82	46.67	26.67	50.00	6.67
28	46.67	44.44	40.00	53.33	50.00	40.00	28	66.67	63.64	20.00	33.33	25.00	20.00
29	53.33	85.71	20.00	46.67	75.00	40.00	29	33.33	30.00	20.00	66.67	60.00	40.00
30	26.67	37.50	26.67	73.33	85.71	26.67	30	46.67	60.00	20.00	53.33	60.00	13.33
31	73.33	100.00	33.33	26.67	36.36	6.67	31	60.00	42.86	46.67	40.00	25.00	20.00
32	53.33	45.45	40.00	46.67	25.00	13.33	32	66.67	77.78	40.00	33.33	50.00	6.67
33	73.33	75.00	46.67	26.67	28.57	20.00	33	66.67	60.00	46.67	33.33	20.00	26.67
34	53.33	63.64	46.67	46.67	75.00	13.33	34	53.33	70.00	46.67	46.67	80.00	26.67
35	80.00	87.50	66.67	20.00	28.57	0.00	35	66.67	87.50	53.33	33.33	57.14	20.00
36	86.67	83.33	80.00	13.33	0.00	6.67	36	66.67	80.00	53.33	33.33	60.00	0.00
37	86.67	92.31	80.00	13.33	50.00	0.00	37	86.67	80.00	86.67	13.33	0.00	0.00
38	93.33	92.31	80.00	6.67	0.00	6.67	38	100.00	100.00	73.33	0.00	0.00	0.00
39	80.00	85.71	80.00	20.00	100.00	0.00	39	73.33	73.33	73.33	26.87	0.00	0.00

Table 2.7– *contd...*

Standard meteoro-logical week	P(D)	P(D/D)	P(2D)	P(W)	P(W/W)	P(2W)	Standard meteoro-logical week	P(D)	P(D/D)	P(2D)	P(W)	P(W/W)	P(2W)
	Jaipur							Udaipur					
22	86.67	85.71	86.67	13.33	0.00	0.00	22	86.67	86.67	80.00	13.33	0.00	0.00
23	100.00	100.00	86.67	0.00	0.00	0.00	23	93.33	92.31	73.33	6.67	0.00	0.00
24	86.67	86.67	73.33	13.33	0.00	0.00	24	80.00	78.57	53.33	20.00	0.00	0.00
25	86.67	84.62	66.67	13.33	0.00	6.67	25	73.33	66.67	40.00	26.67	0.00	6.67
26	73.33	76.92	66.67	26.67	50.00	13.33	26	60.00	54.55	33.33	40.00	25.00	13.33
27	80.00	90.91	20.00	20.00	50.00	13.33	27	60.00	55.56	20.00	40.00	33.00	13.33
28	26.67	25.00	0.00	73.33	66.67	66.67	28	46.67	33.33	6.67	53.33	33.33	33.33
29	6.67	0.00	0.00	93.33	90.91	53.33	29	26.67	14.19	13.33	73.33	62.50	46.67
30	40.00	0.00	20.00	60.00	57.14	53.33	30	33.33	25.00	6.67	66.67	63.64	40.00
31	26.67	50.00	6.67	73.33	88.89	53.33	31	40.00	40.00	6.67	60.00	60.00	53.33
32	26.67	25.00	0.00	73.33	72.73	40.00	32	13.33	16.67	26.67	86.67	88.89	60.00
33	33.33	0.00	20.00	66.67	54.55	53.33	33	33.33	50.00	53.33	66.67	69.23	33.32
34	33.33	60.00	20.00	66.67	80.00	53.33	34	40.00	20.00	60.00	60.00	50.00	33.33
35	33.33	60.00	26.67	66.67	80.00	33.33	35	53.33	66.67	53.33	46.67	55.56	26.67
36	60.00	80.00	60.00	40.00	50.00	20.00	36	73.33	100.00	53.33	26.67	57.14	6.67
37	80.00	100.00	66.67	20.00	50.00	0.00	37	80.00	81.82	60.00	20.00	25.00	6.67
38	86.67	83.33	80.00	13.33	0.00	0.00	38	66.67	66.67	93.33	33.33	33.33	20.00
39	93.33	92.31	93.33	6.67	0.00	0.00	39	66.67	80.00	60.00	33.33	60.00	0.00

Table 2.7– *contd...*

Standard meteoro-logical week	P(D)	P(D/D)	P(2D)	P(W)	P(W/W)	P(2W)	Standard meteoro-logical week	P(D)	P(D/D)	P(2D)	P(W)	P(W/W)	P(2W)
			Banswara							Kota			
22	93.33	93.33	86.67	6.67	0.00	0.00	22	100.00	100.00	93.33	0.00	0.00	0.00
23	93.33	92.86	53.33	6.67	0.00	6.67	23	93.33	93.33	86.67	6.67	0.00	0.00
24	53.33	57.14	26.67	46.67	100.00	20.00	24	93.33	92.86	66.67	6.67	0.00	0.00
25	53.33	50.00	33.33	46.67	42.86	33.33	25	73.33	71.43	73.33	26.67	0.00	20.00
26	46.67	62.50	46.67	53.33	71.43	40.00	26	80.00	100.00	40.00	20.00	75.00	6.67
27	60.00	100.00	26.67	40.00	75.00	26.67	27	53.33	50.00	6.67	46.67	33.33	20.00
28	40.00	44.44	13.33	60.00	66.67	60.00	28	33.33	12.50	6.67	66.67	42.86	53.33
29	13.33	33.33	6.67	86.67	100.00	80.00	29	20.00	20.00	6.67	80.00	80.00	60.00
30	13.33	50.00	13.33	86.67	92.31	80.00	30	26.67	33.33	13.33	73.33	75.00	66.67
31	20.00	100.00	6.67	80.00	92.31	60.00	31	20.00	50.00	6.67	80.00	92.91	73.33
32	26.67	33.33	26.67	73.33	75.00	66.67	32	13.33	33.33	0.00	86.67	91.67	66.67
33	33.33	100.00	6.67	66.67	90.91	33.33	33	20.00	0.00	0.00	80.00	76.92	40.00
34	40.00	20.00	33.33	60.00	50.00	40.00	34	40.00	0.00	26.67	60.00	50.00	46.67
35	53.33	83.33	40.00	46.67	66.67	26.67	35	40.00	66.67	20.00	60.00	77.78	33.33
36	60.00	75.00	33.33	40.00	57.14	13.33	36	46.67	50.00	46.67	53.33	77.78	20.00
37	60.00	55.56	40.00	40.00	33.33	6.67	37	80.00	100.00	66.67	20.00	37.50	6.67
38	73.33	66.67	60.00	26.67	16.67	13.33	38	80.00	83.33	66.67	20.00	33.33	6.67
39	73.33	81.82	66.67	26.67	50.00	6.67	39	80.00	83.33	73.33	20.00	33.33	6.67

A week receiving rainfall of about 20 mm will be able to meet 0.5 to 0.75 times the evaporative demand. During the early stages of growth, the crop water requirements will be about half the evaporative demand and subsequently increase during the reproductive stage of crop growth. Therefore, a week with rainfall less than 20 mm was considered as a dry week (Panderinath, 1991). However, during dry week, the crop may meet its water requirements through the moisture available in the soil. If the rainfall is less than 20 mm per week for two or more consecutive weeks, the crops are likely to be subject to moisture stress in the absence of adequate stored soil moisture. The standard meteorological weeks during which dry spells of two consecutive weeks are likely to commence with probability less than 10 per cent in different agro-climatic zones of Rajasthan are given in Table 2.8.

Table 2.8 : Probability of occurrence of two consecutive dry weeks under different agro-climatic zones of Rajasthan

Station	Dry weeks with rainfall less than 20 mm per week occurring consecutively with different probability levels			
	< 10%	10-25%	25-50%	> 50%
Jodhpur	–	29	28-30, 34	23-27, 35 onwards
Sriganganagar	–	–	28-29	23-27, 31 onwards
Fatehpur	–	29	27-30, 34	23-27, 31 onwards
Pali	–	28-30	27-31, 34	23-26, 35 onwards
Jaipur	28-29, 31-32	27, 30,33-35	23-26, 35	23-26, 36 onwards
Udaipur	28-29, 31-32	27, 30	25-26, 34	23-24,35 onwards
Banswara	29,31,33	28, 30	24-27, 32, 34-37	23,38 onwards
Kota	27-29, 31-33	30, 35	26, 36	23-25, 37 onwards

It would be seen from Table 2.8 that in Jaipur, Udaipur, Banswara, and Kota probability of occurrence of rainfall less than 20 mm per week for two consecutive weeks is less than 10 per cent during 28-29, 28-29, 29 and 27-29 weeks respectively, once the rainy season commences. Mid season droughts are likely to occur at Jaipur, Udaipur, Banswara and Kota during the period 31-32, 31-32, 31 or 33 and 31-33 respectively with a probability of less than 10 per cent and also during the period 33-35, 30, 30 and 30-35 with a probability of 10-25 per cent due to break in monsoon conditions.

2.3.7 Probability of Occurrence of Weekly Rainfall

Rainfall amounts at different probability levels (10, 25, 50, 75 and 90 per cent) were computed by using Gamma distribution (Equation 2.2), for each week (1-52 standard meteorological weeks). Rainfall at different probability levels at different stations are depicted in Figs. 2.1 to 2.8 and also given in Appendices B1 to B8.

Figure 2.1 reveals that, there was probability of 25-50 per cent for receiving mean weekly rainfall during 24 to 39 weeks. The probability lies between 10-25 per cent for onset of monsoon weeks (27-30) and withdrawal of monsoon weeks (36 and 37) at Jodhpur. The results at different probabilities showed that a continuous period from 29-36 week had a reasonable degree (50%) of dependability for receiving rainfall at Jodhpur.

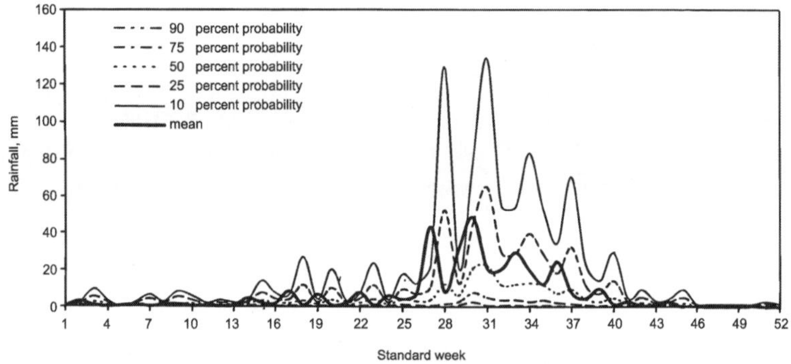

Fig. 2.1 : Rainfall distribution at different probability levels by Gamma distribution at Jodhpur

Figure 2.2 indicates that the probability of receiving mean weekly rainfall form 24-39 weeks lies at 25-50 per cent probability levels at Sriganganagar. Figure 4.2 also shows that a continuous period from 30 to 31 week had a reasonable degree (50%) of dependability for receiving rainfall at Sriganganagar.

Figure 2.3 shows that there was probability of 25-50 per cent for receiving mean weekly rainfall during 19-39 weeks at Fatehpur. The probabilities lies between 10-25 per cent for onset of monsoon week (27 and 28) and withdrawal of monsoon, weeks, (36 and 37). It can also be seen that week 28 to 30 and 32 had a reasonable degree (50%) of dependability for receiving rainfall at Fatehpur.

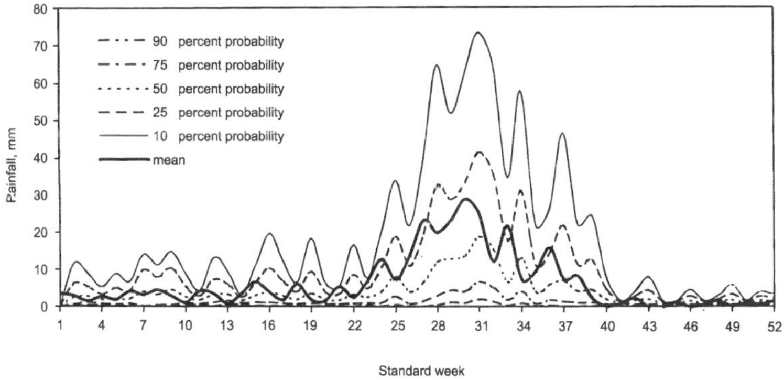

Fig. 2.2 : Rainfall distribution at different probability levels by Gamma distribution at Sriganganagar

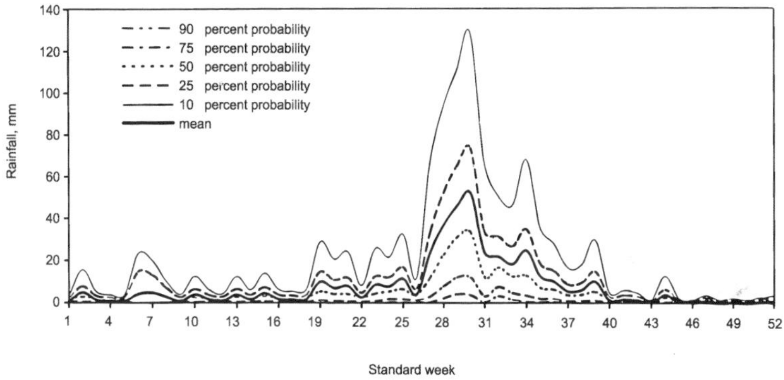

Fig. 2.3 : Rainfall distribution at different probability levels by Gamma distribution at Fatehpur

Figure 2.4 shows that, there was probability of 25-50 per cent for receiving mean weekly rainfall during 22 to 39 weeks at Pali. The probabilities lies between 10-25 per cent for onset of monsoon weeks (27 and 28) and withdrawal of monsoon weeks (36 and 37). It is also inferred that a continuous period from 29-32 week had a reasonable degree (50%) of dependability for receiving rainfall at Pali.

Figure 2.5 revels that, there was probability of 25-50 per cent for receiving mean weekly rainfall during 21-39 weeks at Jaipur. The probability lies between 10-25 per cent for onset of monsoon weeks (25 and 26) and withdrawal of monsoon weeks (36 and 37).

Figure 4.5 also shows that a continuous period from 28-35 weeks had a reasonable degree (50%) of dependability for receiving rainfall at Jaipur.

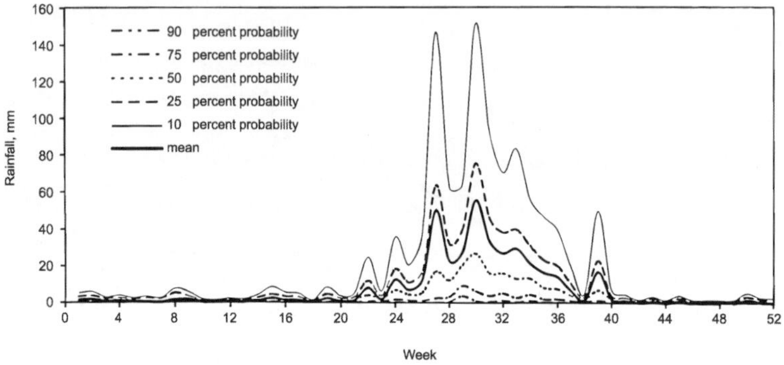

Fig. 2.4 : Rainfall distribution at different probability levels by Gamma distribution at Pali

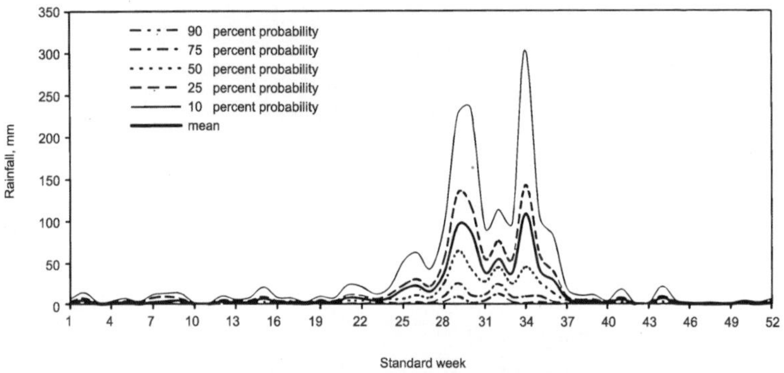

Fig. 2.5 : Rainfall distribution at different probability levels by Gamma distribution at Jaipur

Figure 2.6 indicates that, there was probability of 25-50 per cent for receiving mean weekly rainfall during 22-41 weeks at Udaipur. The probability lies between 10-25 per cent for onset of monsoon weeks (25 and 26) and withdrawal of monsoon weeks (39 and 40). The results at different probabilities indicated that a continuous period from 27-35 weeks had a reasonable degree (50%) of dependability for receiving rainfall at Udaipur.

Figure 2.7 reveals that, there was probability of 25-50 per cent for receiving mean weekly rainfall during 24-42 weeks at

Banswara. The probability of 10-25 per cent was observed for onset of monsoon weeks (25 and 26) and withdrawal weeks (41 and 42). Figure 2.7 also reveals that a continuous period from 28-36 week had a reasonable degree (50%) of dependability for receiving rainfall at Banswara.

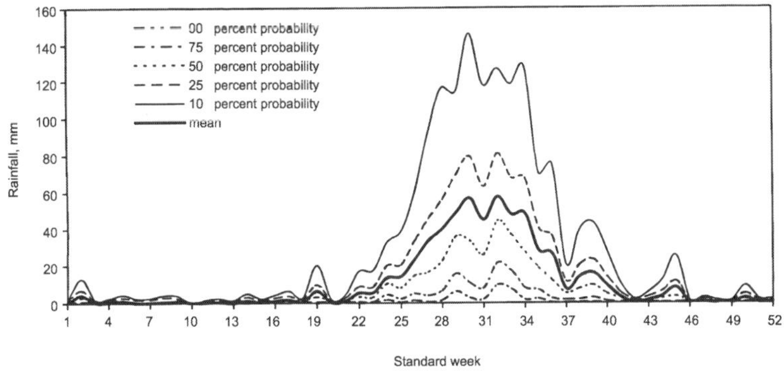

Fig. 2.6 : Rainfall distribution at different probability levels by Gamma distribution at Udaipur

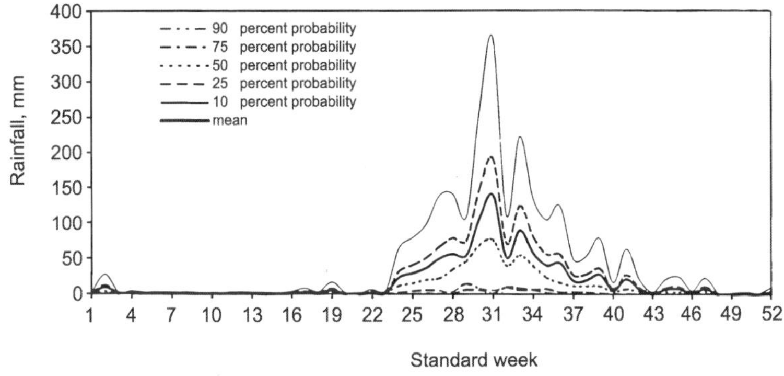

Fig. 2.7 : Rainfall distribution at different probability levels by Gamma distribution at Banswara

Figure 2.8 indicates that, there was probability of 25-25 per cent for receiving mean weekly rainfall during 21-42 weeks at Kota. The probability of 10-25 per cent was observed for onset of monsoon weeks (25 and 26) and withdrawal of monsoon weeks (41 and 42). Figure 2.8 also reveals that a continuous period from 28-36 week had a reasonable degree (50%) of dependability for receiving rainfall at Kota.

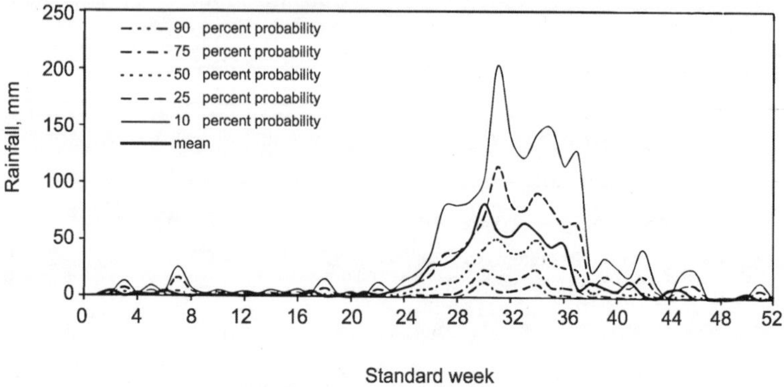

Fig. 2.8 : Rainfall distribution at different probability levels by Gamma distribution at Kota

It is also interesting to note that, the probability of 25-50 per cent for receiving mean annual rainfall was observed at all the stations under study except at Kota.

3. Water Balance

3.1 Reference Evapotranspiration

Evapotranspiration, which includes evaporation from land and water surfaces and transpiration by vegetation, continues to be of foremost importance in prediction of drought. Evapotranspiration from vegetated surface is the result of several processes like radiation exchange, vapour transport and biological growth operating within a system involving the atmosphere, plant and soil. The literature on evapotranspiration is extensive. Some excellent reviews and books are available on the subject (Rosenberg *et al.,* 1968; Brustsaert, 1982; Rosenberg *et al.,* 1983; Sharma, 1985, Jensen *et al.,* 1990; Allen *et al.,* 1998; Singh, 2000; Bhakar, 2000). A brief review related to the present study is presented as under:

Evapotranspiration is a basic component of the hydrological cycle. Evapotranspiration is influenced by micrometeorological, plant and soil factors. The micrometeorological influence may be considered as the atmospheric demand or more precisely the potential evapotranspiration. A large number of empirical or semi empirical formulas have been proposed for estimation of potential evapotranspiration using readily available climatic data. Penman (1948) was the first to use energy balance and aerodynamic considerations to derive combination equation for estimation of potential evapotranspiration from a short green crop, completely shading the ground and never short of water. The principles underlying the combination method have been reviewed and discussed a number of times and different modifications have been proposed for estimation of potential evapotranspiration.

Because of ambiguities involved in the interpretation of potential evapotranspiration the term reference evapotranspiration (ET_0) is frequently being used. Doorenbos and Pruitt (1977) defined reference evapotranspiration as the rate of evapotranspiration from an extensive surface of 8 to 15 cm, green grass cover of uniform height, actively growing, completely shading the ground and not short of water.

Reference evapotranspiration (ET_0) is essentially equivalent to potential evapotranspiration with the additional specifications that it represents evapotranspiration from a convenient well watered crop with full canopy cover. Many investigators have used potential evapotranspiration from alfalfa (lurcene) to represent reference evapotranspiration. In many other cases, researchers have used potential evapotranspiration from grass as a reference evapotranspiration.

The main purpose of developing reference evapotranspiration equations is to approximate, a reference evapotranspiration by which a crop coefficient is multiplied in order to obtain an estimate of crop evapotranspiration. Since publication on Food and Agriculture Organisation (FAO) Irrigation and Drainage Paper 24 by Doorenbos and Pruitt (1977), advances in research and more accurate assessment of crop water use have revealed the need to update the FAO methodologies for calculating ET_0.

In May, 1990, the FAO organized a consultation of experts and researchers in collaboration with the International Commission on Irrigation and Drainage and with the World Meteorological Organisation, to review the FAO methodologies on crop water requirements and to advise on the revision and update of procedures.

The panel of experts recommended the adoption of the Penmen-Monteith combination method as a new standard for reference evapotranspiration and advised procedures for calculating the various parameters. The FAO Penmen-Monteith method was developed by defining the reference crop as a hypothetical crop with an assumed height of 0.12 m with a surface resistance of 70 sm^{-1} and an albedo of 0.23 closely resembling the evaporation from an extensive surface of green grass of uniform height, actively growing and adequately watered (Allen *et al.*, 1998). The method overcomes the shortcomings of the FAO-24 Penman-method (Doorenbos and Pruitt, 1977) and provides values that are more consistent with actual crop water use data worldwide.

Recently the FAO published Irrigation and Drainage Paper, 56 on Crop Evapotranspiration, Guidelines for Computing Crop Water Requirement's by Allen *et al.,* (1998). The Penmen-Monteith method for estimation of ET_0 may be referred as FAO-56 Penman-Monteith method.

Various methods of estimating potential evapotranspiration have been proposed. Numerous formulas have been developed that relate potential evapotranspiration and climatological data. These formulas are used to transpose observed potential evapotranspiration data from one area to other areas where only climatological records are available. Some of the important work on potential evapotranspiration is as under:

Thornthwaite (1948) determined the water need of vegetation over large areas in order to carry out broad climatic classification. He estimated potential evapotranspiration on a monthly basis from readily available climatologically data such as mean air temperature and day length. The potential evapotranspiration values obtained were adjusted to the number of days in a month and the maximum duration of sunshine depending on season and latitude of the place.

Blaney and Criddle (1962) developed and proposed a simplified empirical formula to compute evapotranspiration (consumptive use) using a correlation, which utilized mean monthly air temperature, amount of daylight and a crop factor showing seasonal variation. They noted that the efficient design and operation of water supply project depends on awareness of the quantity of water that is lost through the process of evapotranspiration.

Jensen and Haise (1963) estimated potential evapotranspiration from solar radiation using energy balance approach, based on long-term data and indicated that reasonable and reliable estimates of evapotranspiration can be made using the solar radiation as the main parameter. They presented suitable curves for different field crops as the ratio of measured evapotranspiration to solar radiation, which reflected the combined effects of short wave reflectance (albedo), and relative efects of efective thermal radiation, sensible heat flux to air and soil, plus other minor energy balance components at various stages of crop growth. They concluded that the approach might be used for estimating mean weekly, monthly or seasonal evapotranspiration of crops.

Hargreaves (1968) conducted several important experiments and developed methods to compute evapotranspiration from pan evaporation data. The method and the procedure presented are applicable to both arid and humid climates and to locations where data are quite complete as well as to those locations for which only temperature and rainfall data are available.

Allen *et al.,* (1983) studied evapotranspiration estimates with reference to weather station siting and documented the effect of weather station aridity on evapotranspiration estimates. For the purpose of the study, three sites in irrigated and two sites in nonirrigated areas were instrumented with weather station. It was found that air temperature was higher and vapour pressures were lower at the arid sites. Results indicated the importance of weather site evaluation and adjustment of siting effects on weather before evapotranspiration estimates are made. They concluded that the environment of a weather station site is important in estimating evapotranspiration by irrigated crops and evapotranspiration may be overestimated by air temperature and vapour pressure data from a weather station with an arid local environment.

Hatfield and Allen (1997) studied different forms of the reference evapotranspiration equations that included Priestly-Taylor and Penman-Monteith. To estimate actual evapotranspiration, the Priestly-Taylor with unadjusted coefficient for available soil water and the Penman-Monteith with a variable soil surface resistance were compared to water use for grain sorghum, cotton and grass forage at three locations. Both models gave acceptable results, however, the Penman-Monteith model with daily meteorological data input provided more consistent results over the growing season.

Grismer (2000) studied the variability of orchard water use from season to season. Multiple water use evaluations were conducted including direct crop evapotranspiration measurements, soil moisture monitoring and 15 year water budget assessment. The crop evapotranspiration measurements indicated that the mature trees utilize approximately 60 per cent of reference evapotranspiration, with the avocado evapotranspiration being slightly greater than that for lemons presumably due to the greater canopy coverage.

Samani (2000) introduced a procedure to estimate solar radiation and subsequently reference crop evapotranspiration using the minimum climatologically data. A modification was made to the original equation that uses minimum and maximum temperature to estimate solar radiation and reference crop evapotranspiration. The proposed modification allowed for the correction of the errors associated with indirect climatologically parameters affecting the local temperature range.

3.2 Reference Evapotranspiration

Drought is evaluated in a more meaningful manner by water balance approach taking into account the rainfall pattern, evapotranspiration and moisture characteristics of soil. The prime factor controlling the water balance of the plant-soil-environment is the water supply available to the plant. The water balance equation can be represented in the following way:

$$P + I = Q + U + ET + \Delta W \qquad ...(2.1)$$

where, P is the precipitation, I is the irrigation water added, Q is runoff, U is the deep drainage passing beyond the root zone, ET is the actual evapotranspiration and, ΔW is the change in soil water storage.

Krishnan and Thanvi (1969) worked out monthly water balance from meteorological data (1941–60) for arid zone of Rajasthan. They found variation in the water balance pattern from year to year. Normal rainfall values varied from 100 to 450 mm. High water surplus which would produce runoff and recharging of aquifers beyond the soil mantle generally occurcd once in 20 years. Even in the wettest year, the moisture deficiency during January to onset of monsoon was high.

Ratnam *et al.,* (1973) studied climatic shifts at Dharwar using moisture index values obtained from the water balance technique. The climate migrated to 10 times into the drier side during the 23 years (1950-1973). The water balance for driest and wettest years were presented and discussed. Ratnam and Joshi (1977) also studied the weekly water balances, for the Kharif season, of the normal and severe drought years at Dharwar. In the drought year, water deficiency occurred during 12 out of the 17 weeks of rainy season, whereas, the corresponding figures for a normal year were 5 out of 17 weeks. The role of a 3 week spell of above normal rains, after a prolonged period of water deficiency in saving the Kharif crop was limited.

Subramaniam and Rao (1980a) worked out the broad scale pattern of water balance with particular reference to its variation with longitude for Rajasthan. The elements of water balance for climatically normal, most wet and most dry years were determined during the period 1901 to 1977. The variation of aridity index with longitude has been discussed. Thc variation of doughtiness factor and soil moisture storage have been studied in relation to longitude. Soil moisture storage increased towards the east of the state even in the most wet years.

Subramaniam and Uma Devi (1983) made an attempt to estimate water availability periods in Orissa. For this, the water balance parameters such as rainfall, potential evapotranspiration and actual evapotranspiration, were estimated according to Thronthwaite (1955) scheme for 11 stations in Orissa during the period 1920 to 1978. As it is a known fact that length and duration of the water availability help to suggest crop management with suitable agronomic practices, the commencement and ending of humid period were also analysed.

Ramana Rao *et al.,* (1984) studied the moisture variability in the Indian arid zone for the period from 1901 to 1980. The coefficient of variation of moisture index decreases with decrease in moisture index. The stations with low moisture index have stable climate while the stations with high moisture index have large fluctuations in climatic pattern. The conditional probabilities were computed using first order Markov chain model to describe the year to year variation in moisture regime as related to the mean moisture index.

Subramaniam and Narasimha Raju (1986) worked out the water availability periods for nine stations in north coastal Andhra Pradesh by computing water balance. The region experienced more number of moist days in high land area than in coastal belt. The onset and cessation of water availability periods were also presented. It was found that humid period begins from 10th May (in plain land) and ends by the last week of November (along the coast).

Rao *et al.,* (1990) carried out water balance studies for Kerala and Tamil Nadu states in rubber growing areas. The results indicated that total rainfall, moisture deficiency and water surplus increased by two to three times form south to north along the

Western Ghats. The water deficit at any time during the year significantly reduced the cumulative yield. Thus, moisture availability during winter, pre and post-monsoon periods significantly enhanced the yield. Under normal conditions, about 93 per cent of the yield variations in different agroclimatic conditions could be explained by the annual index of moisture adequacy.

Victor *et al.,* (1991) carried out weekly water balance for the years 1911–85 considering different available water holding capacities ranging from 25–200 mm to examine the extent to which the moisture regime and aridity are dependent on water holding capacity of the soil. The analysis revealed that the aridity is accentuated by the shallowness of the soil coupled with low rainfall. The number of years with drought were found to vary from 3 to 6 per decades. The water harvesting may be useful either for late sown Kharif crops or for establishment of tree plantations in this region. The annual average water balance of the entire state during climatically extreme years have also been illustrated and discussed.

3.3 Procedure for Computing the Water Balance

The soil water balance was computed following the procedure introduced by Thronthwaite and Mather (1955, 1957). (Fig. 3.1).

Thornthwaite and Mather (1955) suggested the use of potential evapotranspiration (PET) values for computation of soil water balance. Because of ambiguities in the interpretation of potential evapotranspiration, the term reference evapotranspiration (ET_0) is used through out the world. Therefore, the original equation of Thornthwaite and Mather (1955) was modified by using ET_0 in place of PET. The actual storage of soil moisture can be represented by the following modified equation:

$$\text{STOR} = \text{AWC}\ e^{\frac{\text{ACC}(P-ET_0)}{\text{AWC}}} \qquad ...(3.7)$$

where,

STOR = actual storage of soil moisture, mm

AWC = moisture storage capacity of soil, mm

P = precipitation, mm

ET_0 = reference evapotranspiration, mm

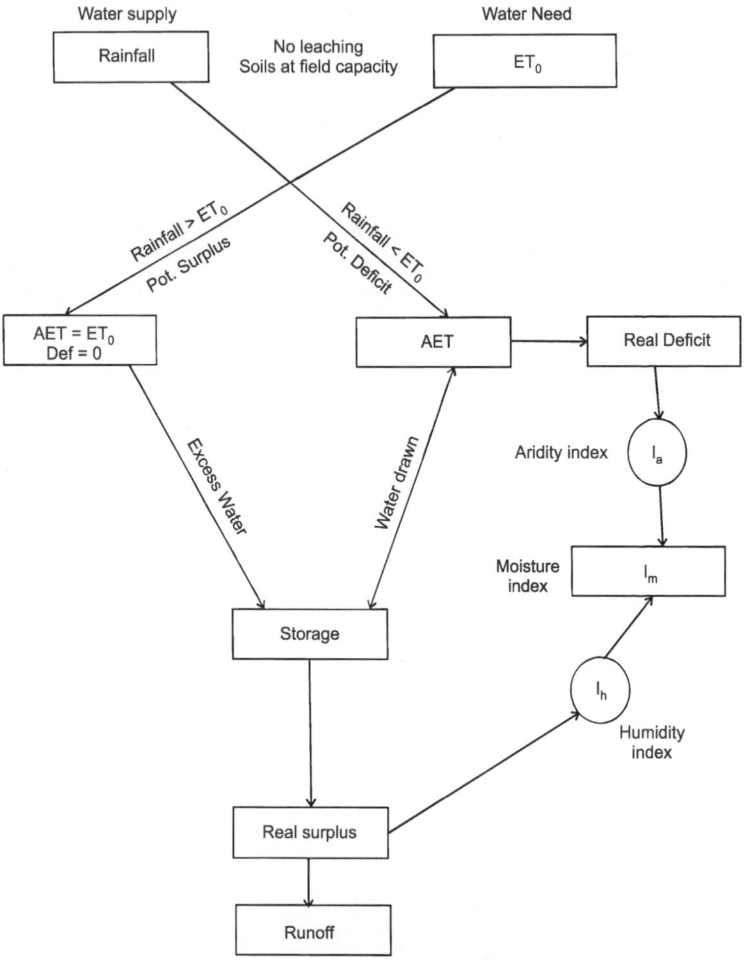

Fig. 3.1 : A generalized flow diagram of the climatic water balance

Weekly soil water balance values of eight stations were computed by employing following steps:

Step 1

Reference evapotranspiration (ET_0) was computed as suggested by Allen *et al.,* (1998) by Penman-Monteith FAO-56 method. Allen *et al.,* (1998) gave the following equation for estimation of reference evapotranspiration:

$$ET_0 = \frac{0.408\Delta\,(R_n\text{-}G)+\gamma \dfrac{900}{T+273}\,U_2(e_a\text{-}e_d)}{\Delta+\gamma\,(1+0.34U_2)} \qquad ...(3.8b)$$

where,

ET_0 = reference evapotranspiration (mm day^{-1}),
R_n = net radiation (MJ m^{-2} d^{-1}) = $R_{ns} - R_{nl}$,
R_{ns} = net short wave radiation (MJ m^{-2} d^{-1}),
R_{nl} = net long wave radiation (MJ m^{-2} d^{-1}),
T = mean daily air temperature (°C),
γ = psychrometric constant (kPa°C^{-1}),
Δ = slope of the saturation vapour pressure function (kPa°C^{-1}),
e_a = saturation vapour pressure at temperature T (kPa),
e_d = saturation vapour pressure at dew point (kPa),
U_2 = average daily wind speed at 2 m height (ms^{-1}),

$$U_2 = U_z \frac{\ln \dfrac{(Z_2 - d)}{(Z_{om})}}{\ln \dfrac{(Z_m - d)}{(Z_{om})}} \qquad ...(3.8a)$$

U_z = mean wind speed measurement at height (ms^{-1}),
Z_m = height of wind speed measurement (m),
Z_2 = standard height of wind speed measurement (m) = 2m
d = zero plane displacement of wind profile (m) = 0.08 m
Z_{om} = roughness parameter for momentum (m) = 0.015 m.

A computer programme written in Fortran 77 was used for computation of reference evapotranspiration (Bhakar and Singh 2001). The computer programme is given in Appendix–D.

Step II

Estimation of weekly water excess as (+) and deficit as (–) P over ET_0 was made.

Step III

Computation of accumulated values of P – ET° for each week was made to account for gross potential deficit and gross potential surplus for all the stations.

Step IV

Computation of actual storage of soil moisture for each week was done for all the stations by using Equation (3.7).

Step V

Computation of change of actual storage form week to week (ΔSTOR) was done for all stations. When the storage remains at capacity level, the STOR = 0. When STOR reaches values of less than the capacity, then STOR was calculated as the substraction of STOR for the present week from STOR of the previous week. The STOR can be negative and implies extraction of water from the soil to be used for evapotranspiration. The STOR can be positive which implies infiltration of water in the soil and its addition to the soil moisture storage.

Step VI

Weekly actual evapotranspiration (AET) values for all the stations were computed from the following equations:

(i) when STOR is negative

$$AET = STOR + P \quad ...(3.9a)$$

(ii) when STOR is positive

$$AET = ET^{o} \quad ...(3.9b)$$

Step VII

Weekly water deficit (DEF) values for all the stations were determined from the following expression:

$$DEF = ET_0 - AET \quad ...(3.10)$$

Step VIII

The values of weekly surplus water (SUR) for all the stations were computed by the following equations:

$$SUR = (P - ET_0) - STOR \quad ...(3.11a)$$

or

$$SUR = P - AET \quad ...(3.11b)$$

The SUR is the amount of water percolating to the water table and becoming available for run off for the underground system. It occurs only when (P–ET_0) is positive and when storage has reached the capacity level.

Step IX

On the basis of above parameters, climatological indices such as humidity index (I_h), aridity index (I_a), moisture index (I_m) and index of moisture adequacy (I_{ma}) were computed for all the stations under the study by using following expressions:

$$I_h = \frac{SUR}{ET_0} \times 100 \qquad ...(3.12)$$

$$I_a = \frac{DFF}{ET_0} \times 100 \qquad ...(3.13)$$

$$I_{ma} = \frac{AET}{ET_0} \times 100 \qquad ...(3.14)$$

and

$$I_m = I_h - I_a \qquad ...(3.15)$$

3.4 Discussion

3.4.1 Reference Evapotranspiration Under Different Agro-Climatic Zones of Rajasthan

Reference evapotranspiration (ET_0) refers to total water losses through evaporation from soil surface and transpiration from vegetations under unlimited supply of water. Recently Allen *et al.,* (1998) modified the concept of reference evapotranspiration and suggested Penmen-Monteith equation for estimation of ET_0. Annual and weekly reference evapotranspiration values were computed by using Equation (3.8) under different agro-climatic zones of Rajasthan.

3.4.1.1 Annual reference evapotranspiration

The annual reference evapotranspiration values for different stations are presented in Table 3.1. Table 3.1 indicates that the maximum reference evapotranspiration values vary from 1176.8 mm in Banswara in the south-eastern Rajasthan to 1792.0 mm in Jodhpur in the western Rajasthan. Similarly, the mean annual reference evapotranspiration values vary from 1363.8 mm with coefficient of variation of 6.80 per cent in Banswara to

1628.9 mm with coefficient of variation of 7.83 per cent in Jodhpur. Comparison of rainfall with reference evapotranspiration indicates that the evaporative need is several times higher than the rainfall received in the state. Table 3.1 also shows that the least coefficient of variation is observed at Udaipur (2.97 percent), which indicates that the reference evapotranspiration is more stable at Udaipur than at the other stations. It is evident from Table 3.1 that at Sriganganagar reference evapotranspiration values are lowest, which may be due to fog, high humidity and cloudy weather during the winter season.

Table 3.1 : Annual reference evapotranspiration (ET_0) under different agro-climatic zones of Rajasthan

Station	Maximum ET_0 (mm)	Minimum ET_0 (mm)	Mean ET_0 (mm)	Standard deviation (mm)	Coefficient of variation (%)
Jodhpur	1792.0	1423.8	1628.9	127.5	7.83
Sriganganagar	1560.4	1312.7	1405.3	69.9	5.26
Fatehpur	1692.4	1456.0	1529.1	66.2	4.30
Pali	1748.8	1496.9	1614.8	69.2	4.30
Jaipur	1710.8	1444.8	1549.7	82.9	5.35
Udaipur	1474.3	1308.4	1379.4	41.0	2.97
Banswara	1517.9	1176.8	1363.8	92.7	6.80
Kota	1678.1	1431.6	1538.1	60.3	3.92

3.4.1.2 Weekly reference evapotranspiration

The weekly reference evapotranspiration computed by Penman-Monteith equation for different stations are shown in Fig. 3.2 presented in Appendices E-l to E-8. It can be seen from Fig. 3.2 that weekly reference evapotranspiration values vary from 8.8 mm in 2nd week at Sriganganagar to 54.6 mm in 22nd week at Jaipur. Figure 3.2 shows that the low values of weekly reference evapotranspiration observed in winter season at Sriganganagar may be due to the fog and high humidity. It is interesting to note that the highest annual reference evapotranspiration values were obtained at Jodhpur but the highest weekly reference evapotranspiration values were observed at Jaipur. It is evident from Fig. 3.2 that low values of weekly reference evapo-

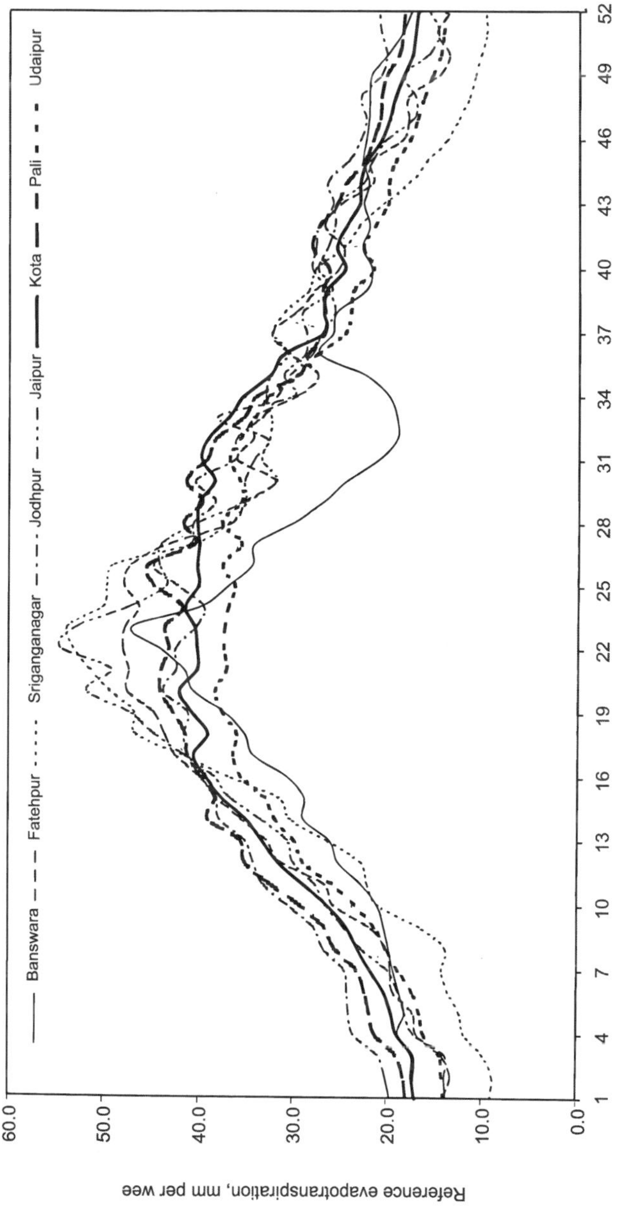

Fig. 3.2 : Variation of weekly reference evapotranspiration at different stations (1981-1995)

transpiration were obtained during the rainy season at Banswara as compared to other stations. This may be due to high humidity conditions during the rainy season. High values of weekly reference evapotranspiration were observed even in winter season at Jodhpur as compared to other stations. This may be due to dry weather conditions at Jaipur. Fig. 3.2 also shows that the high values of weekly reference evapotranspiration were obtained between 20th-26th week at all the stations. The mean weekly reference evapotranspiration values vary from 25.72 mm at Banswara to 31.26 mm at Jodhpur.

3.4.2 Moisture Status Under Different Agro-Climatic Zones of Rajasthan

The concept of water balance has come to be understood universally as a basis for understanding the moisture status of a place. The weekly water balance elements for different stations under study have been computed by Thornthwaite and Mather (1955) method.

3.4.2.1 Annual moisture status

The elements of annual water balance have been computed for the period 1981-1995 and are given for normal, wettest and driest years in Table 3.2. Table 3.2 indicates that the reference evapotranspiration during the normal years at different stations varies from 1337.8 mm (Banswara) to 1625.6 mm (Jodhpur) while during the wettest and driest years it ranges from 1176.8 mm (Banswara, 1994) to 1572.1 mm (Jodhpur, 1994) and 1408.6 mm (Udaipur, 1987) to 1792.0 mm (Jodhpur, 1987) respectively. The value of actual evapotranspiration during the normal years ranges from 326.4 mm at Sriganganagar to 643.5 mm at Kota, while in wettest and driest years it ranges from 354.4 mm (Fatehpur, 1981) to 695.7 mm (Kota, 1982) and 111.1 mm (Jodhpur, 1987) to 449.5 mm (Kota, 1987) respectively.

It is interesting, to note that though the annual rainfall in the normal and the wettest years was more at Banswara as compared to Kota, but higher values of actual evapotranspiration have been observed at Kota. This is due to low reference evapotranspiration at Banswara during the rainy season.

Table 3.2 : Annual water balance under different agro-climatic zones of Rajasthan

Item	Rainfall (mm)	Reference evapotrans-piration (mm)	Actual evapotrans-piration (mm)	Water deficit (mm)	Water surplus (mm)
Jodhpur					
Normal year	352.3	1625.6	352.2	1273.4	0.0
Wettest year(1990)	844.0	1572.1	463.1	1109.0	378.4
Driest year (1987)	110.7	1792.0	111.1	1680.9	0.0
Sriganganagar					
Normal year	326.4	1431.4	326.4	1105.0	0.0
Wettest year (1983)	595.2	1359.9	542.2	817.7	50.4
Driest year (1987)	200.0	1417.2	200.5	1216.7	0.0
Fatehpur					
Normal year	393.0	1534.5	392.5	1142.0	0.0
Wettest year (1981)	569.3	1456.3	354.4	1101.9	203.0
Driest year (1989)	242.0	1598.6	250.1	1348.5	0.0
Pali					
Normal year	389.8	1619.7	389.6	1230.1	0.0
Wettest year (1990)	923.3	1550.3	552.7	997.6	368.0
Driest year (1987)	152.7	1749.1	152.9	1596.2	0.0
Jaipur					
Normal year	680.1	1565.9	576.4	989.9	92.9
Wettest year (1982)	1622.1	1470.0	734.0	735.4	880.0
Driest year (1987)	235.8	1710.8	236.1	1474.7	0.0
Udaipur					
Normal year	581.5	1379.0	562.4	816.6	0.0
Wettest year (1989)	895.7	1414.4	596.9	817.5	283.5
Driest year (1987)	377.9	1408.6	361.9	1046.7	0.0
Banswara					
Normal year	915.1	1337.8	629.2	708.6	170.9
Wettest year (1994)	1473.1	1176.8	542.7	634.1	924.9
Driest year (1987)	396.7	1517.9	394.4	1123.5	0.0
Kota					
Normal year	684.4	1538.1	643.5	900.2	0.0
Wettest year (1982)	826.8	1473.8	695.7	778.1	84.3
Driest year (1987)	469.1	1669.2	449.5	1219.7	0.0

The value of annual water deficit during the normal years ranges from 708.6 mm at Banswara to 1273.4 mm at Jodhpur, while in the wettest and driest years it ranges from 634.1 mm (Banswara, 1994) to 1109 mm (Jodhpur, 1994) and 1046.7 mm (Udaipur, 1987) to 1680.9 mm (Jodhpur, 1987) respectively.

Table 3.2 also shows that 92.9 mm and 170.9 mm of water appears as surplus at Jaipur and Banswara stations respectively, while at other stations no surplus water has been appeared during the normal years. No surplus water was registered at any station during the driest year 1987. The surplus water in different stations during the wettest years ranges from 50.4 mm (Sriganganagar, 1983) to 924.9 mm (Banswara, 1994).

3.4.2.2 Meriditional variation of moisture status in Rajasthan

Rajasthan is the largest state in the country and is located between the longitudes 70°E and 78°E in north-western parts of India. The variation of water balance with longitude is presented in Figs. 3.3 and 3.4 for the normal year and 1987 year respectively.

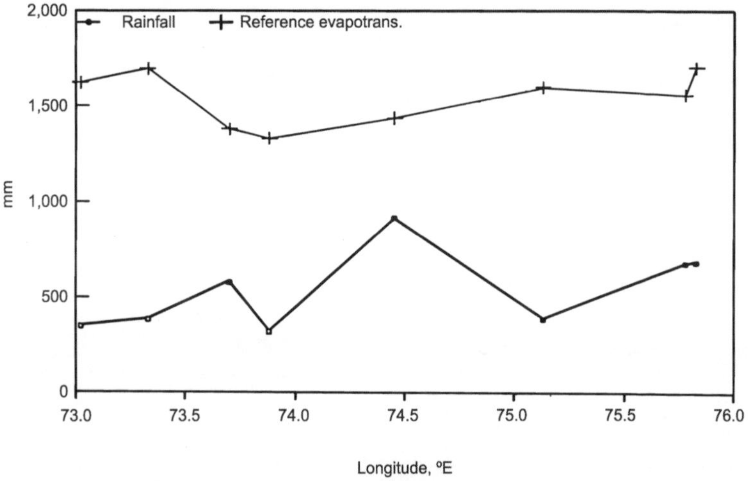

Fig. 3.3 : Normal meriditional variation of moisture status in Rajasthan state

Figures 3.3 and 3.4 show that the water need (ET_0) is almost constant at all the longitudes. It can be observed that the rainfall increases from west to east of the state and highest rainfall is observed at 74.5°E longitude. The amount of water need and the entire rainfall is generally used up as actual reference evapotranspiration (Subramanyiom and Rao, 1980). Figure 3.10 also shows that the water deficiency decreases from 73°E to 74.5°E longitude and increases from 74.5°E to 75.5°E longitude. The least amount of water deficiency was observed at 74.5°E.

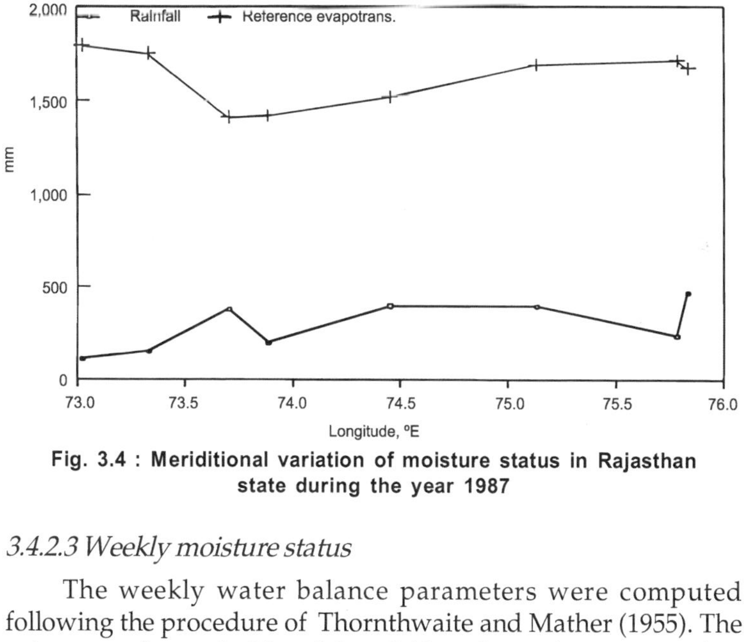

Fig. 3.4 : Meriditional variation of moisture status in Rajasthan state during the year 1987

3.4.2.3 Weekly moisture status

The weekly water balance parameters were computed following the procedure of Thornthwaite and Mather (1955). The values are shown in Figs. 3.3 to 3.10 and are also presented in Appendix E1 to E8 for different stations.

Figure 3.5 shows the water balance at Jodhpur. It is observed that the highest weekly rainfall is 48.4 mm in the 30th week with zero deficit and zero surplus water. Again Fig. 3.5 reveals that rainfall exceeds ET_0 for two weeks (27th and 30th). Out of the total deficit of 1273.4 mm, the 306.9 mm (24.10%) deficit appears from 23rd to 29th week, while rainfall was 298.1 mm (84.6%). On an annual basis the station has a water need of 1625.6 mm, whereas the rainfall was 352.3 mm.

At Sriganganagar no surplus water appears in any week during the rainy season as shown in Fig. 3.5. It can also be seen from Fig. 3.6 that rainfall did not exceed in any week during the normal year. The highest weekly rainfall is 29.1 mm in the 30th week with 2.7 mm deficit. The 384.8 mm deficit appears from 23rd-

39th weeks, while total deficit during the normal year was of 1105.0 mm. On an annual basis the station has a water need of 1431.4 mm, whereas the rainfall was 326.4 mm. The water deficiency is well distributed form 23rd to 26th week and from 34th to 39th week.

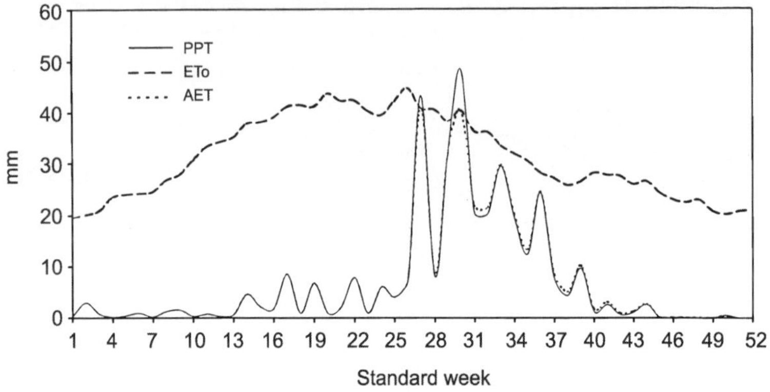

Fig. 3.5 : Normal weekly water balance at Jodhpur

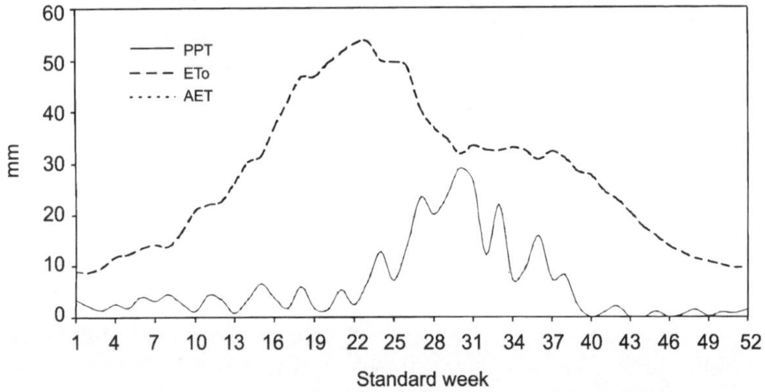

Fig. 3.6 : Normal weekly water balance at Sriganganagar

Figure 3.7 shows weekly water balance at Fatehpur. It is observed from Figure 3.7 that the highest weekly rainfall is 46.2 mm in the 30th week with zero deficit. No surplus water appears in any week of the normal year. It can be seen that the rainfall exceeds reference evapotranspiration for three weeks (28th to 30th). Out of total deficit of 1142.0 mm, 309.0 mm deficit (27.06%) appeared from 23rd to 39th week, while the rainfall was 314.4 mm (80%) during that period. On an annual basis the station has a water need of 1534.5 mm, whereas the rainfall was 393.0 mm.

At Pali, it can be seen that the highest weekly rainfall is 55.5 mm in the 30th week with zero deficit. No surplus water appears in any week during the normal year (Figure 3.8). Figure 3.8 also indicates that the rainfall exceeds reference evapotranspiration for two weeks (27th and 30th). The deficit is well distributed from 24th to 26th week and 34th to 37th occurring week. Out of the total deficit of 1230.1 mm, 274.9 mm deficit (22.32%) appeared from 23rd to 39th week, while the rainfall was 355.0 mm (91.07%). On an annual basis the station has a water need of 1619.7 mm, whereas the rainfall was 389.8 mm.

Figure 3.9 shows normal weekly water balance for Jaipur. It is observed from Figure 3.9 that highest weekly rainfall of 102.2 mm exceeds in the 34th week with 72.1 mm surplus water. The soil gets started to recharge from 29th to 31st weeks to its water holding capacity. The surplus water was recorded as 92.9 mm from 32nd to 36th weeks. Again it is revealed that rainfall exceeds ET_0 for 8 weeks (29th to 36th). The water deficiency is uniformally distributed from 25th to 27th weeks and from to 38th to 39th weeks during the rainy season. Out of the total deficit of 989.5 mm, the 182.8 mm (18.4%) deficit appeared from 23rd to 39th week, while rainfall was 606.9 mm (89.24%) On an annual basis the station has a water need of 1565.9mm, whereas rainfall was 680.1 mm. It can also be seen that rainfall distribution at Jaipur is bimodal, as it has more than one peak.

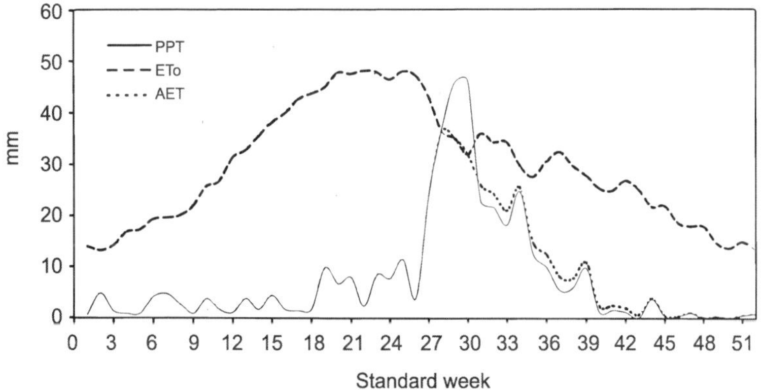

Fig. 3.7 : Normal weekly water balance at Fatehpur

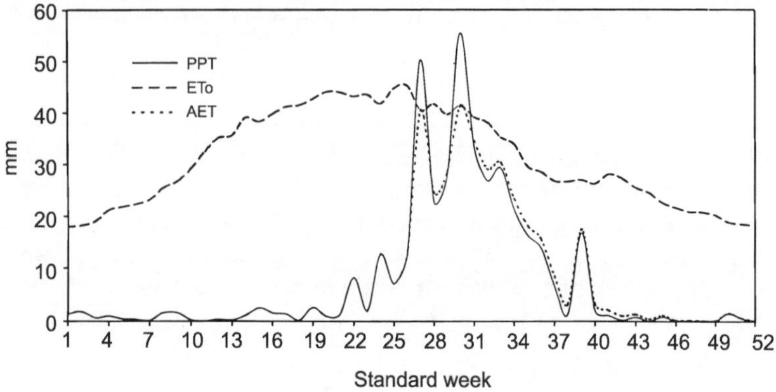

Fig. 3.8 : Normal weekly water balance at Pali

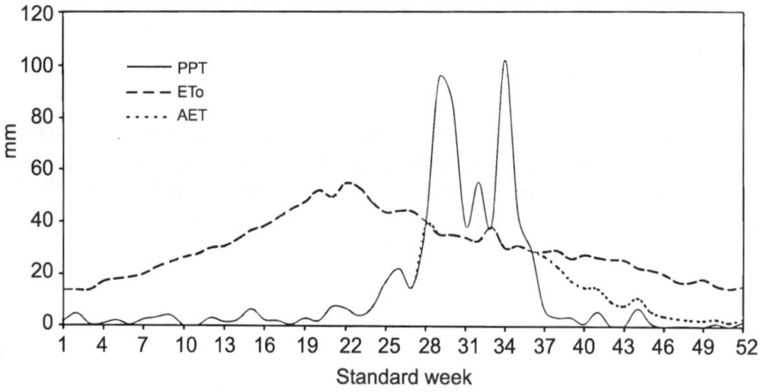

Fig. 3.9 : Normal weekly water balance at Jaipur

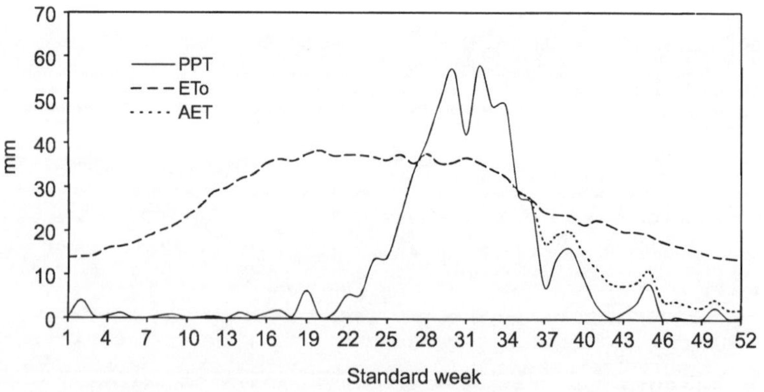

Fig. 3.10 : Normal weekly water balance at Udaipur

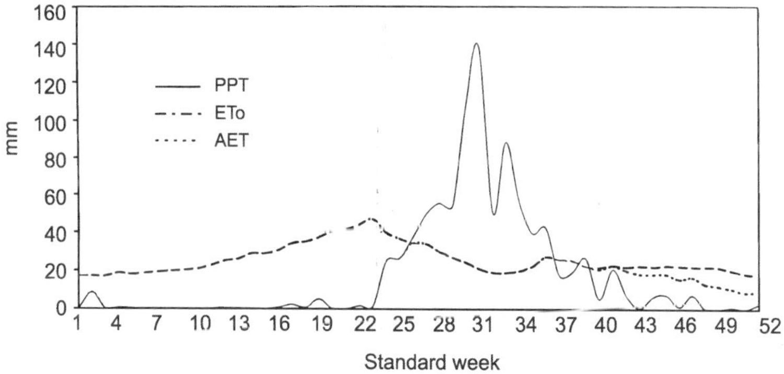

Fig. 3.11 : Normal weekly water balance at Banswara

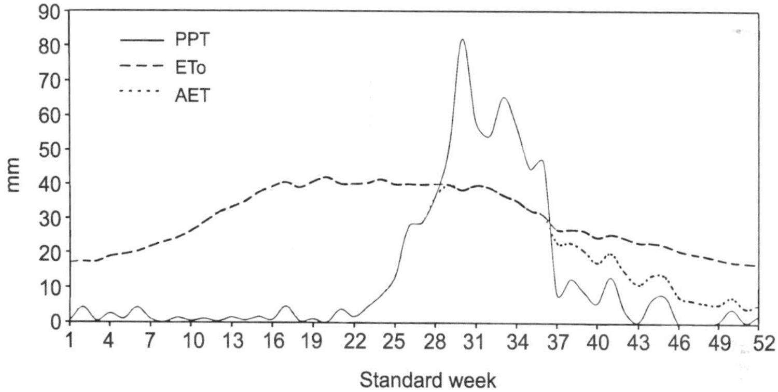

Fig. 3.12 : Normal weekly water balance at Kota

At Udaipur, the highest mean weekly rainfall is 57.8mm with zero deficit and zero surplus water observed in 32nd week as shown in Fig. 3.10. It can be seen from Fig. 3.10 that rainfall exceeds ET_0 for 7 weeks (28th to 34th). No surplus water was observed in any week during the normal year. The water deficiency was uniformly distributed from 37th to 39th week during the rainy season. Out of the total deficit of 816.6 mm, the 109.5 mm deficit. (13.41%) occurred in the rainy season, while the rainfall was 526.0 mm (90.46%). It can also be seen that rainfall distribution at Udaipur is bimodal, as it has more than one peak. On an annual basis the station has a water need of 1379.0 mm, whereas rainfall was 581.5 mm.

Figure 3.11 shows normal weekly water balance for Banswara. It was observed that the highest rainfall of 139.1 mm with 0.5 mm surplus water in 31st week, whereas minimum surplus water of 69.6 mm was registered in 33rd week. The soil recharged from 26th week. Total surplus water of 170.0 mm was recorded from 31st-36th week during rainy season. It can also be seen from Fig. 3.11 that, rainfall exceeds ET_0 for 11 weeks (26th to 36th and 39th) and water deficiency was well distributed from 24th to 25th weeks and from 37th to 38th weeks. Out of the total deficit of 708.6 mm recorded during the normal year, 71.3 mm (10.06%) was appeared during the rainy season, while rainfall was recorded as, 836.5 mm (91.41%). It can also be seen that at Banswara, the weekly rainfall distribution is bimodal. On an annual basis the station has a water need of 1337 mm against 915.1 mm rainfall.

It can be seen from Fig. 3.12 that at Kota, the highest rainfall of 82.1 mm with zero deficit recorded in 30th week. No surplus water was recorded in any week during the normal year. Again it is observed that rainfall exceeds reference evapotranspiration for 8 weeks. (29th to 36th) The water deficiency was well distributed from 27th to 28th weeks and from 37th to 39th weeks. Out of the total deficit of 900.2 mm, 176 mm (19.56%) occurred from 23th to 39th week, while rainfall was 607.0 mm (88.69%). It can also be seen that the weekly rainfall distribution at Kota is bimodal as it has more than one peak. On an annual basis the station has a water need of 1538.1 mm against 684.4 mm rainfall.

3.4.2.4 Mean aridity index and droughtiness factor

The mean aridity index and droughtiness factor were worked out as $(I_{max} - I_{min})/I_{mean}$. Table 3.3 shows the mean aridity index and droughtiness factor for different stations.

Table 3.3 indicates that the highest mean aridity index of 79.96 per cent with coefficient of variation of 8.13 per cent has been observed at Jodhpur, while lowest aridity index has been observed at Banswara (57.78%) with a coefficient of variation of 12.18 per cent. Table 3.3 also shows that the mean aridity index decreases from west to east and also from north to south of the state.

The mean minimum aridity index value was observed in south-east (Banswara) part of the state due to moist sub-humid climatic condition. It can be seen that the droughtiness factor increases with the increase in the coefficient of variation of aridity index and also increases with increase in rainfall variability.

Figures 3.13 to 3.14 show the meriditional variation of droughtiness factor and mean aridity index of the state. A comparison of Fig. 3.12 and Fig. 3.13 shows that the low values of droughtiness factor did not coincide with low values of aridity index and high values of droughtiness factor coincide with low values of aridity index probably due to the high variability of rainfall with respect to longitudes. The aridity index was found to be nearly constant from 73.0-73.25°E longitudes.

Table 3.3 : Mean aridity index and droughtiness factor under different agro-climatic zones of Rajasthan

Station	Mean aridity index (%)	Standard deviation (%)	Coefficient of variation (%)	Droughtiness factor
Jodhpur	79.96	6.50	8.13	0.34
Sriganganagar	77.80	6.20	7.97	0.33
Fatehpur	75.86	4.60	6.06	0.20
Pali	79.70	7.40	9.28	0.34
Jaipur	69.85	10.40	14.89	0.54
Udaipur	63.49	6.21	9.78	0.30
Banswara	57.78	7.04	12.18	0.43
Kota	60.60	5.70	9.41	0.33

3.4.2.5 Climatic shifts under different agro-climatic zones of Rajasthan

The moisture index, which is the basis for climatic classification has been studied for year to year change in climate. These indicate the fluctuations of moisture index in different years. Climatic conditions at different stations are given in Table 3.4.

From Table 3.4 it can be seen that aridity was stable at Jodhpur, Sriganganagar, Fatehpur and Pali as 86.66% of the years at Jodhpur, Fatehpur and Pali were arid while 93.33% of the years were arid at Sriganganagar. Whereas in the case of Jodhpur and Pali the swing from arid to semi-arid conditions was only in two years, which was 13.33% of the period for which analysis is made. In case of Jaipur the swing from semi arid to arid conditions was in 6 years, which was 40% of the period. In Sriganganagar the swing from arid to semi-arid condition was 6.6% and in Fatehpur the swing from arid to semi-arid conditions was 13.33% of the period for which analysis is made. The swing from semi-arid to dry sub-humid and from semi-arid to moist sub-humid condition was one year respectively at Jaipur, which was 6.66% of the period under study. In case of Udaipur the swing from semi-arid to arid conditions was 26.67% of the period and none of the years under study experienced dry humid or moist sub-humid conditions.

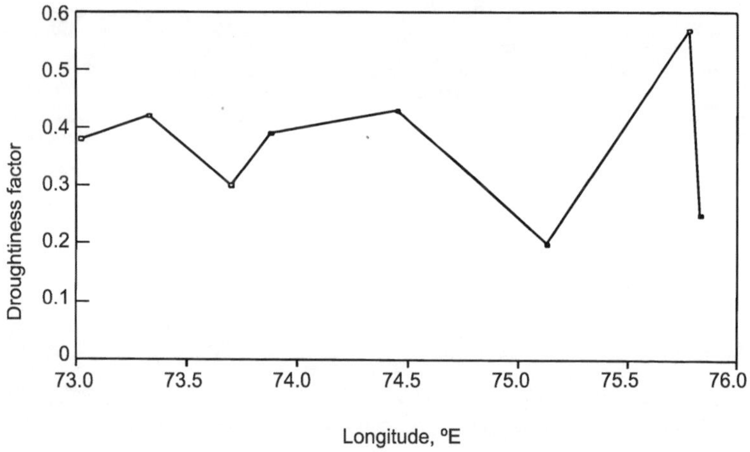

Fig. 3.13 : Meriditional variation of droughtiness factor in Rajasthan state

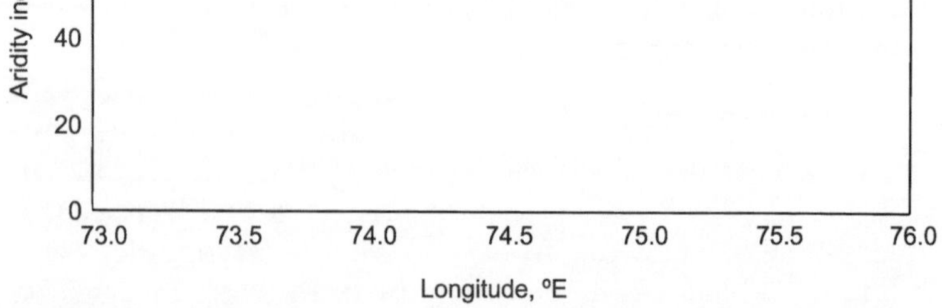

Fig. 3.14 : Meriditional variation of aridity index in Rajasthan state

Table 3.4 : Climatic conditions at different stations during the period 1981-1995

Station	Arid	Semi-arid	Dry sub-humid	Moist sub-humid	Total
	Number of years of shift				
Jodhpur	13	2	–	–	15
Sriganganagar	14	1	–	–	15
Fatehpur	13	2	–	–	15
Pali	13	2	–	–	15
Jaipur	6	7	1	1	15
Udaipur	4	11	–	–	15
Banswara	1	8	4	2	15
Kota	3	12	–	–	15

It is interesting to note that while Udaipur, Banswara and Kota are normally stated to be of having sub-humid type climate 11, 8, 12 years out of 15 years under study experienced semi-arid climatic condition at Udaipur, Banswara and Kota respectively.

In case of Banswara the swing from semi-arid to dry sub-humid condition was in 4 years which was 26.6 per cent of the period and from semi-arid to moist sub-humid condition was in 2 years which was 13.33 per cent of the period while in Kota none of the years experienced dry sub-humid or moist sub-humid climatic conditions during the period under study.

Thus, Jodhpur, Sriganganagar, Fatehpur and Pali were found to have predominantly arid climate during the period 1981-1995 and. Jaipur remained mostly to have arid to semi-arid climatic conditions. Similarly, Udaipur and Kota remained mostly to have semi-arid climate. Banswara remained to have semi-arid to dry sub-humid climatic conditions.

3.4.2.6 Surplus water under different agro-climatic zones of Rajasthan

The percentage of frequencies of occurrence of surplus water (which includes both runoff and deep percolation) of 20, 40, 60, 80 and 100 mm during different standard weeks of the main rainy season are computed using the Equations 3.3 and 3.4 for different agro-climatic zones of Rajasthan and are presented in Tables 3.5 to 3.12.

Table 3.5 : Probability of surplus water during rainy season at Jodhpur

Week	Date	Surplus water 20 mm	40 mm	60 mm	80 mm	100 mm
		Probability %				
26	June 25-1	0.0	0.0	0.0	0.0	0.0
27	July 2-8	6.7	6.7	6.7	6.7	6.7
28	9-15	0.0	0.0	0.0	0.0	0.0
29	16-22	0.0	0.0	0.0	0.0	0.0
30	23-29	0.0	0.0	0.0	0.0	0.0
31	30-5	0.0	0.0	0.0	0.0	0.0
32	Aug. 6-12	0.0	0.0	0.0	0.0	0.0
33	13-19	6.7	0.0	0.0	0.0	0.0
34	20-26	0.0	0.0	0.0	0.0	0.0
35	27-2	0.0	0.0	0.0	0.0	0.0
36	Sept. 3-9	6.7	6.7	0.0	0.0	0.0
37	10-16	0.0	0.0	0.0	0.0	0.0
38	17-23	0.0	0.0	0.0	0.0	0.0
39	24-30	0.0	0.0	0.0	0.0	0.0
40	Oct. 1-7	0.0	0.0	0.0	0.0	0.0

Table 3.5 shows that the surplus water at Jodhpur is obtained in 27, 33 and 36 weeks. The surplus water is likely to exceed 20 mm in 27th and 36th weeks with probability of 6.7 per cent. The surplus water is likely to exceed 40 mm in 27th and 36th weeks with probability of 6.7 per cent only. Therefore, the water may be useful is this zone for late shown Kharif crops or for establishment of tree plantation. Table 3.6 shows that no surplus water is available at Sriganganagar in any week during the rainy season with an amount of 20 mm or more.

Table 3.7 reveals that the surplus water at Fatehpur exceeds 20 mm during 29th to 31st weeks during the main cropping season. The 100 mm surplus water is obtained in 29th week with 6.67 per cent probability, while 80 mm surplus water is obtained with 6.67 per cent probability in 29th and 31st weeks. The 30th week has the highest probability of 13.33 per cent of obtaining 20 mm to 60 mm surplus water. Table 3.8 shows that at Pali surplus water is likely to exceed 20 mm in 27th, 30th, 32nd and 35th week during the main cropping season. The surplus water is likely to exceed 20 mm in 27th and 32nd weeks. During the 27th week the surplus water may exceed 80 mm with 6.67 per cent probability.

Table 3.6 : Probability of surplus water during rainy season at Sriganganagar

Week	Date		Surplus water (mm)				
			20 mm	40 mm	60 mm	80 mm	100 mm
			Probability %				
26	June	25-1	0.00	0.00	0.00	0.00	0.00
27	July	2-8	0.00	0.00	0.00	0.00	0.00
28		9-15	0.00	0.00	0.00	0.00	0.00
29		16-22	0.00	0.00	0.00	0.00	0.00
30		23-29	0.00	0.00	0.00	0.00	0.00
31		30-5	0.00	0.00	0.00	0.00	0.00
32	Aug.	6-12	0.00	0.00	0.00	0.00	0.00
33		13-19	0.00	0.00	0.00	0.00	0.00
34		20-26	0.00	0.00	0.00	0.00	0.00
35	Sept.	27-2	0.00	0.00	0.00	0.00	0.00
36		3-9	0.00	0.00	0.00	0.00	0.00
37		10-16	0.00	0.00	0.00	0.00	0.00
38		17-23	0.00	0.00	0.00	0.00	0.00
39		24-30	0.00	0.00	0.00	0.00	0.00
40	Oct.	1-7	0.00	0.00	0.00	0.00	0.00

Table 3.7 : Probability of surplus water during rainy season at Fatehpur

Week	Date		Surplus water				
			20 mm	40 mm	60 mm	80 mm	100 mm
			Probability %				
26	June	25-1	0.00	0.00	0.00	0.00	0.00
27	July	2-8	0.00	0.00	0.00	0.00	0.00
28		9-15	0.00	0.00	0.00	0.00	0.00
29		16-22	6.67	6.67	6.67	6.67	6.67
30		23-29	13.33	13.33	13.33	0.00	0.00
31		30-5	6.67	6.67	6.67	6.67	0.00
32	Aug.	6-12	0.00	0.00	0.00	0.00	0.00
33		13-19	0.00	0.00	0.00	0.00	0.00
34		20-26	0.00	0.00	0.00	0.00	0.00
35	Sept.	27-2	0.00	0.00	0.00	0.00	0.00
36		3-9	0.00	0.00	0.00	0.00	0.00
37		10-16	0.00	0.00	0.00	0.00	0.00
38		17-23	0.00	0.00	0.00	0.00	0.00
39		24-30	0.00	0.00	0.00	0.00	0.00
40	Oct.	1-7	0.00	0.00	0.00	0.00	0.00
41		8-14	0.00	0.00	0.00	0.00	0.00

Table 3.9 reveals that at Jaipur the surplus water is likely to exceed 20 mm from 29th to 36th weeks with 6.67 to 26.67 per cent probabilities. During the weeks from 29th to 31st and 33rd to 36th surplus water may exceed 80 mm with 6.67 per cent probability.

Table 3.10 reveals that at Udaipur, the surplus water is likely to exceed 20 mm from 30th to 31st week with probability of more than 13.33 per cent. During the 34th week, the surplus water may exceed 80 mm with 6.67 per cent probability and surplus water may exceed 60 mm during 33rd and 34th week with 6.67 per cent probability.

Table 3.8 : Probability of surplus water during rainy season at Pali

Week	Date	Surplus water				
		20 mm	40 mm	60 mm	80 mm	100 mm
		Probability %				
26	June 25-1	0.00	0.00	0.00	0.00	0.00
27	July 2-8	13.33	6.67	6.67	6.67	6.67
28	9-15	0.00	0.00	0.00	0.00	0.00
29	16-22	0.00	0.00	0.00	0.00	0.00
30	23-29	6.67	0.00	0.00	0.00	0.00
31	30-5	0.00	0.00	0.00	0.00	0.00
32	Aug. 6-12	6.67	6.67	6.67	6.67	0.00
33	13-19	0.00	0.00	0.00	0.00	0.00
34	20-26	0.00	0.00	0.00	0.00	0.00
35	Sept. 27-2	6.67	0.00	0.00	0.00	0.00
36	3-9	0.00	0.00	0.00	0.00	0.00
37	10-16	0.00	0.00	0.00	0.00	0.00
38	17-23	0.00	0.00	0.00	0.00	0.00
39	24-30	0.00	0.00	0.00	0.00	0.00
40	Oct. 1-7	0.00	0.00	0.00	0.00	0.00

Table 3.11 shows that at Banswara the surplus water may likely to exceed 20 mm from 28th to 36th week, 39th week and 41st week with 13.33 per cent probability or more than 13.33 per cent probability. The surplus water may likely to exceed 80 mm from 28th to 35th weeks and during 39th week with 6.67 per cent probability or more than 6.67 per cent probability. Surplus water may likely to exceed 100 mm in 31st and 33rd week with 20 per cent probability.

Table 3.9 : Probability of surplus water during rainy season at Jaipur

Week	Date	Surplus water 20 mm	40 mm	60 mm	80 mm	100 mm
		Probability %				
26	June 25-1	0.00	0.00	0.00	0.00	0.00
27	July 2-8	0.00	0.00	0.00	0.00	0.00
28	9-15	0.00	0.00	0.00	0.00	0.00
29	16-22	6.67	6.67	6.67	6.67	0.00
30	23-29	13.33	13.33	13.33	13.33	6.67
31	30-5	13.3	13.33	6.67	6.67	0.00
32	Aug. 6-12	26.67	26.67	13.13	6.67	6.67
33	13-19	13.33	6.67	6.67	6.67	0.00
34	20-26	26.67	26.67	20.00	6.67	6.67
35	Sept. 27-2	20.00	13.33	13.33	13.33	6.67
36	3-9	6.67	6.67	6.67	6.67	6.67
37	10-16	0.00	0.00	0.00	0.00	0.00
38	17-23	0.00	0.00	0.00	0.00	0.00
39	24-30	0.00	0.00	0.00	0.00	0.00
40	Oct. 1-7	0.00	0.00	0.00	0.00	0.00

Table 3.10 : Probability of surplus water during rainy season at Udaipur

Week	Date	Surplus water 20 mm	40 mm	60 mm	80 mm	100 mm
		Probability %				
26	June 25-1	0.00	0.00	0.00	0.00	0.00
27	July 2-8	6.67	0.00	0.00	0.00	0.00
28	9-15	0.00	0.00	0.00	0.00	0.00
29	16-22	13.33	0.00	0.00	0.00	0.00
30	23-29	13.33	13.33	6.67	0.00	0.00
31	30-5	13.33	13.33	0.00	0.00	0.00
32	Aug. 6-12	20.00	13.33	0.00	0.00	0.00
33	13-19	13.33	6.67	6.67	6.67	0.00
34	20-26	6.67	6.67	6.67	6.67	6.67
35	Sept. 27-2	13.33	6.67	0.00	0.00	0.00
36	3-9	20.00	20.00	13.33	0.00	0.00
37	10-16	0.00	0.00	0.00	0.00	0.00
38	17-23	6.67	0.00	0.00	0.00	0.00
39	24-30	6.67	0.00	0.00	0.00	0.00
40	Oct. 1-7	0.00	0.00	0.00	0.00	0.00
41	8-14	0.00	0.00	0.00	0.00	0.00

Table 3.11 : Probability of surplus water during rainy season at Banswara

Week	Date		Surplus water 20 mm	40 mm	60 mm	80 mm	100 mm
			Probability %				
26	June	25-1	0.00	0.00	0.00	0.00	0.00
27	July	2-8	6.67	0.00	0.00	0.00	0.00
28		9-15	20.00	13.33	13.33	13.33	6.67
29		16-22	20.00	6.67	6.67	6.67	6.67
30		23-29	20.00	6.67	6.67	6.67	6.67
31		30-5	26.67	26.67	20.00	20.00	20.00
32	Aug.	6-12	33.33	13.33	6.67	6.67	6.67
33		13-19	53.33	46.67	33.33	20.00	20.00
34		20-26	20.00	20.00	20.00	6.67	6.67
35	Sept.	27-2	20.00	13.33	13.33	6.67	6.67
36		3-9	26.67	26.67	26.67	13.33	13.33
37		10-16	6.67	6.67	0.00	0.00	0.00
38		17-23	6.67	6.67	6.67	0.00	0.00
39		24-30	14.33	13.33	13.33	6.67	6.67
40	Oct.	1-7	0.00	0.00	0.00	0.00	0.00
41		8-14	13.33	6.67	6.67	0.00	0.00

Table 3.12 : Probability of surplus water during rainy season at Kota

Week	Date		Surplus water 20 mm	40 mm	60 mm	80 mm	100 mm
			Probability %				
26	June	25-1	0.00	0.00	0.00	0.00	0.00
27	July	2-8	0.00	0.00	0.00	0.00	0.00
28		9-15	0.00	0.00	0.00	0.00	0.00
29		16-22	0.00	0.00	0.00	0.00	0.00
30		23-29	26.67	20.00	13.33	13.33	13.33
31		30-5	33.33	13.00	13.00	0.00	0.00
32	Aug.	6-12	20.00	20.00	6.67	6.67	0.00
33		13-19	40.00	33.33	33.33	13.33	6.67
34		20-26	33.33	33.33	33.33	26.67	13.33
35	Sept.	27-2	33.33	20.00	20.00	6.67	6.67
36		3-9	33.33	26.67	20.00	13.33	13.33
37		10-16	0.00	0.00	0.00	0.00	0.00
38		17-23	6.67	0.00	0.00	0.00	0.00
39		24-30	0.00	0.00	0.00	0.00	0.00
40	Oct.	1-7	0.00	0.00	0.00	0.00	0.00
41		8-14	6.67	6.67	0.00	0.00	0.00

Table 3.12 reveals that at Kota the surplus water is likely to exceed 20 mm from 30th to 36th week with probability of 26.67 per cent or more than 26.67 per cent probability. During the 30th, 32nd, 33rd and 35th week the surplus water may exceed 80 mm with 13.33 to 26.67 per cent probability levels. The surplus water may likely to exceed 100 mm in 32nd, 33rd and 35th weeks with 13.33 per cent probability.

The surplus water is available at all the stations except at Sriganganagar during the rainy season. Therefore, surplus water may be harvested and used for supplemented irrigation to rainfed crops during the dry spell period.

4. Definition and Classification of Drought

4.1 Introduction

The question of proper definition of drought has been addressed time and again and it has often been stated that no universal definition of drought exists. There is a no gain by reproducing a long list of conflicting definitions that merely illustrates the diverse interest of those who investigate drought. The word 'drought' in its usual sense implies dryness. The word drought is related to the term 'aridity'. It is the dryness that in its extremely intense form and long duration contributes to the aridity. The term aridity express a deficit of water arising from poor precipitation or resulting from excess of water loss as compared to the water received.

Early workers defined drought as a prolonged period without rainfall. Air Ministry Meteorological Office (1936) defined absolute drought as a period of at least 5 consecutive days none of which had rainfall of 0.25 mm or more.

Ramdas and Malik (1948) defined drought as a week with actual rainfall equal to half the normal or less.

According to Ramdas (1960) drought was taken as a situation when the seasonal rainfall is deficient by more than twice the mean deviation. In the USSR, drought was defined as a period of 10 days with a total rainfall not exceeding 5 mm. Gibbs and Maher (1967) made a study of drought in Australia by using annual rainfall deciles as drought indicators. The U.S. Weather Bureau has defined drought as a period of dry weather of sufficient length and severity to cause at least partial crop failure.

India Meteorological Department, defined drought year, as the year with annual rainfall of 75 per cent or less of normal and 50 per cent of less as a severe drought year.

Drought is, however, evaluated in a more meaningful manner by water balance computation taking into account the rainfall pattern, evapotranspiration, moisture characteristics of a soil, rooting depth and stage of crop growth.

Failure of the monsoon is the most important cause for famines in India and Rajasthan while occasionally failure of crops due to attack of insects and pests is also responsible for drought condition.

Sen (1970) analysed the data on droughts and famines in India from 1800 to 1966 and concluded that severe and widespread drought usually led to a sharp fall in food production in the country as a whole and gave rise to famine conditions. According to him, famine is due to wide spread drought condition. Lack of sufficient water to meet the normal water requirements of a locality may be considered as a common and a fairly satisfactory definition of drought.

Drought is broadly classified in three categories *viz.* (i) meteorological drought, (ii) hydrological drought, and (iii) agricultural drought. A brief review on different types of drought is as under:

4.1.1 Meteorological Drought

Meteorological drought is a situation when there is a significant (25%) decrease of rainfall from the normal value over a given area.

Banerji and Chhabra (1963) studied departure from normal of seasonal rainfall (June to September) of Telengana Division of Andhra Pradesh to assess the frequency and intensity of droughts. Frequency distribution of dry spell was considered for analysis, since it was felt that the drought conditions are not so much dependent on the total seasonal rainfall as on the length of the dry spells. Probability of dry spells and their average expected number per season were calculated and it has been submitted that a 25 per cent increase in 10 day dry spell may be deemed to constitute a drought condition. Drought conditions of various magnitudes as described by Banerji and Chabbra (1963) is presented in Table 4.1.

Table 4.1 : Drought conditions based on rainfall deficit

Drought condition	Rainfall deficit
Slight drought	11-25%
Moderate drought	26-50%
Severe drought	> 50%

Herbst *et al.* (1966) described a method based on monthly rainfall data to determine the duration and intensity of droughts and their onset and termination. A drought index was also calculated. The method has been programmed for computer for facilitating the analysis of data from a large number of rainfall stations.

Landberg (1974) summarized the global research work on droughts and pointed out that no well defined cycles of rainfall can be identified to enable forecasting of droughts.

George *et al.* (1975) found aerial extent of drought in India by computing linear correlation between south-west monsoon rainfall of a particular station (master station), when it gets below median rainfall and corresponding monsoon rainfall of each of 120 other stations forming a well distributed net work over the country. This was done for 12 master stations selected over different parts of the country. The study showed that droughts in the western parts of the country have comparatively greater aerial extent.

Das (1983) evaluated the frequency of famines and rainfall deficiency and described for the last quarter of the nineteenth century. Historical records suggested that there were 14 famines between the 11th and 17th centuries and another 12 famines in 90-years period from 1869 to 1958, with a sharp increase in numbers between 1869 and 1908. But, famines did not always coincide with year of drought.

Awade *et al.* (1984) reported that in drought years there was a large divergence of flume of sensible heat in sub-tropics, while there was a large divergence of momentum in good monsoon years. In the upper troposphere during good monsoon years waves were greater source of energy during rainy months.

Bhalme *et al.* (1984) repolted that the significant correlation between the areal extent of drought and pressure index is mostly due to the oscillations in the range of 3 to 6 years. Furthermore, 3 year oscillation in the areal extent of drought arise primarily due to the influence of the southern oscillation. The southern oscillation appears to be one possible causal climatic phenomenon for introduction of a common period from 3 to 6 years for the recurrence of large scale drought over India.

Chowdhury and Sarwade (1984) utilized the monthly values of total global and diffuse radiation from January to May for drought prediction in India. From radiation parameters it was found that the departures of global radiation during January and February perhaps provides some prior indication on the activity of monsoon during succeeding June to September months.

Mooley *et al.* (1984) studied droughts, which affected the peninsular India during the period 1861-1980 with well distributed rainfall data of 31 raingauge stations. The criterion used for identification of drought was the specific thresh hold value of rainfall. There were 10-large scale drought years during the period 1876, 1877, 1891, 1899, 1905. 1913, 1916, 1920, 1952 and 1972. Statistical tests showed that the interval between two successive drought years was randomly distributed in time continuum. The worst drought occurred in the years 1899 and 1918 when more than 50% of the stations showed drought conditions. The circulation features like position of monsoon through the depression and storms were examined with a view to understand the anomalies in rainfall during the large scale drought years.

Rao and Phulari (1984) conducted extreme probability analysis of drought periods on daily rainfall from 1944 to 1976 for the frequencies 7, 10, 15, 20 and 25 days to adjust crop phases according to their water need. The study indicated that the first fortnight of the September was not suitable for sowing, flowering, pod formation, panicle initiation, ear head emergence and grain filling of Kharif crops, respectively. It was found that if crop phases are adjusted as per their water need with the available depth of rainfall, three years of very successful cropping, five years with success and the crops may undergo severe moisture stress in two years period. Further, the study revealed that if proper sequence is chosen, the crops would have good harvest in a year particularly where the soil depth exceeds a metre.

Thapliyal (1984) conducted a study for prediction of Indian drought with lower stratospheric winds and forecasted the severe drought conditions that occur due to large-scale monsoon failure over India. He also found that the winter, in which the mean longitudinal length of the pair of trough and ridge over the northern hemisphere is more than 140 degree, followed by severe drought during the subsequent summer monsoon.

Karl *et al.* (1987) reported that the annual precipitation cycle and the probability of receiving substantial excess precipitation above normal are two important characteristics of drought termination and amelioration ill the Palmer drought model.

Ray *et al.* (1987) analysed the rainfall data of 70 years of Gopalpur district of Orissa and frequency analysis has been made to predict monthly, seasonal and annual rainfall at different percent chances during drought months.

Chowdhary *et al.* (1989) made an attempt to quantify drought for the country as a whole and identified drought years by developing drought index based on rainfall for 113 years (1875–1987). For this purpose total area of India receiving monsoon rainfall less than 75 per cent of the normal was obtained for each of the 113 years. From this, mean drought affected area and its standard deviation were worked out. The difference between area affected by drought in any year and the 113 year's mean drought were divided by the standard deviation for determination of drought index. The series of drought years, were also subjected to return period analysis.

Dabral (1996) analysed rainfall data from the year 1957 to 1995 to study the weekly, monthly, seasonal and yearly drought of Kanke, Bihar. The observed frequency of drought was maximum in 23rd week, November and December months and post monsoon season. During the 39 years period, 3 drought years were experienced. The severe droughts occurred in the years 1962 and 1979 when total annual rainfall was 68.97 per cent and 69.22 per cent, respectively of the average annual rainfall.

Ghosh and Tripathy (1996) studied the drought in the arid and semi-arid region of Gulbarga district of Karnataka state. The statistical property of precipitation (1970–1986) has been analysed to determine the drought at 30, 60, 90 and 120 days period. A prediction has been constructed to determine the amount of water available before drought period. The measure of drought was evaluated through the analysis of the normalized data.

Kaledhonkar *et al.* (1996) applied statistical method to compute frequency distribution of longest duration and largest deficit of droughts using annual rainfall series of four stations in south-west Orissa. The validity of the method was established by comparing the distribution of longest negative run lengths of drought with the theoretical distribution obtained from independent normal process. Log normal distribution was successfully fitted to describe the distribution of longest duration and largest deficit of drought. Regression models have been proposed to estimate representative sample size and parameters of log normal distribution for long-term management of drought in the region.

Tiwari *et al.* (1998) worked out frequency of droughts of Datia district in Madhya Pradesh. Analysis of data showed that 20 per cent of the years were found to be drought years. Similarly, 20 per cent of the years were found to be abnormal years. There was a possibility of one drought year in a span of 5 years.

Agnew and Chappel (1999) studied the Sahel region of west Africa and found that the averaging period had a significant impact on precipitation occurrence that can be considered to assess meteorological drought.

4.1.2 Hydrological Drought

Hydrological drought occurs with marked depletion of surface water and consequent drying up of reservoirs, lakes, streams and rivers, cessation of spring flow in ground water levels.

Yevjevich (1967) showed drought as a set of variables affecting precipitation, run off and water storage in its many forms. He also classified drought of three types *viz.* a point drought, the small area drought and the large continental drought. He observed that the duration, areal cxtent, severity, intensity, probability of recurrence, initiation and termination of drought can better be studied with time-runs.

Sen (1978) developed a new methodology, which is referred as "autorun analysis". It is used to investigate the sequential properties of hydrologic time series. The autorun analysis yielded the same results as those of the autocorrelation analysis in the case of normally distributed random variables only. An application of the autorun analysis to some mean annual flow series has been performed.

Panu and Unny (1980) found that the feature synthesis model can be utilized to synthesize stream flow realization from the natural rivers. The practical usefulness of various techniques employed in consonance with the application of the model was explained with illustrative examples dealing with stream flow records. The salient peculiarities with the construction and operation of the model for stream flow synthesis were also presented.

Sen (1980a) presented a general method for finding the exact probability distribution function of the longest drought duration in a finite sample of any process on the basis of runs theory and enumeration technique. The methodology developed was applied to monthly river flow sequences at various parts of the Turkey.

Samani *et al.* (1995) applied time series techniques to Ghara-Aghaj flow records, in order to generate forecast values of the monthly river flows. The study of data and its correlogram showed the effect of reasonability and provided no evidence of trend.

4.1.3 Agricultural Drought

Agricultural drought occurs when the soil moisture and rainfall are inadequate during the growing season to support the healthy crop growth to maturity and cause extreme crop stress or crop wilt.

Barger and Thorn (1949) presented a method for measuring rainfall requirements of an average corn crop in selected counties in Iowa. This method furnished an estimate of the minimum amount of rainfall needed during a given number of consecutive weeks in a particular county to evaluate the drought hazard. For each county the probability of getting less than this amount (the drought base) during the period in question was estimated.

Subrahmanyam and Subrahmaniam (1964) identified that droughts in India are due to failure of monsoon. They studied drought conditions of three representative stations *viz.,* Bijapur (arid), Trichy (semi-arid) and Cuddalore (sub-humid). Yearly aridity index for each station was evaluated for a period of 65 years (1891-1955) and their time trends along with 7 years overlapping means were graphed and presented. Drought years were picked up from these diagrams and their water balance conditions were discussed in relation to the climatic situations normal to the stations.

Subrahmanyam and Sastri (1969a) taken the "aridity index" as the basic parameters for the study of drought climatology of the dry sub-humid zones of south India. Departures of aridity index from the median were calculated and using the scheme proposed by Subrahmanayam and Subrahmaniam (1964) droughts were also categorized.

Subrahmanyam and Sastri (1969b) calculated aridity index for a period of 75 years from 1891 to 1965 and categorized drought as moderate, large, severe or disastrous. Dryness as the departure of the yearly aridity index from the medium values was found to be less than $1/2\sigma$, between $1/2\sigma$ and σ, and between σ and 2σ (where σ is the standard deviation). Employing water deficit data the severity of individual drought was also determined.

Subrahmanyam and Sastri (1969c) critically examined frequency and intensity of droughts of various categories in the dry sub humid zones of south India. Four representative stations from the different sectors were selected, after studying the water balance of a number of stations in the entire region. "Aridity index" (the ratio of annual water deficiency to annual water need, expressed as a percentage) was taken as the basic parameter for the study. Departures of aridity index from the median were calculated. They also studied occurrence of droughts both from the point of view of frequency as well as its intensity.

Subrahmanyam and Sastri (1971) reported that droughts have an origin, spread as well as decay. Thus in order to study the spread aspect of drought in the central region of south India, aridity anomalies on an annual basis were obtained and regional patterns were prepared with a view to obtain a closer picture of the movement of arid patterns. Cumulative techniques were used for study of duration and severity and categorization of droughts.

Rakhecha (1974) studied the droughts based on water budget method and provided a more realistic approach than on rainfall alone. The frequency and pattern of variation of the drought situations for a group of stations in Andhra Pradesh during the last 70 years were studied by computing the aridity indices from water balance considerations. From this study it has been seen that a minimum of 3-4 effective drought years may be expected over each station in a decade and that liability to drought is maximum in Nellore. An examination of the year to year changes in climate showed that the disastrous drought years generally become arid in nature.

George and Ramasastri (1975) computed weekly water

balance of a selected stations in the country. Agricultural droughts of the 1972 Kharif season were studied. Weekly aridity index value above the climatic aridity index has been used to identify drought incidence. The study showed that prolonged drought incidence and prolonged drought spells were obtained only for short periods prevailed in the season over large areas of Madhya Pradesh, Maharashtra, Marathwada. Gujarat, Rajasthan, interior Mysore and Andhra Pradesh.

Hounam *et al.* (1975) summarized various definitions far delineating and quantifying the agricultural droughts based on soil, crop and climatic parameters.

Klugm (1978) studied the patterns and periods of variable moisture conditions, and droughts in areas east and west of the upper midwest of USA. The 1930 and 1950 years were found years of severe drought but the patterns of drought were not identical. The 1930's drought spread from the west, while the 1950's drought spread from the south and south-west.

Bhalme and Mooley (1979) found that the Palmer's index failed to explain the well-known 1918 drought in the Indian sub-continent. They also observed that the weighing factor for July in respect of Chandra district was even negative and the defective weighing factor changed the sign of the moisture departure.

Rama Krishana and Sastri (1980) carried out studies on the incidence of drought during the rainy season and indicated a pattern of drought incidence and spread based on aridity index. The studies revealed that the droughts are more discernible and occurs early in regions receiving comparatively higher rainfall than regions of lower rainfall. They also indicated that drought conditions originate first in the north east region of western Rajasthan around Churu and spread in south westerly direction into Barmer region and then dissipate into the Pali region indicating a systematic pattern in the incidence over different parts or Rajasthan. The study on the pattern of drought spread and decay indicated possibilities for the existence of a pattern in drought and to enable its prediction in advance.

Sastri and Rama Krishna (1980) found that existing drought classifications did not present a true picture of the disastrous conditions that prevailed over the extreme arid regions of western Rajasthan. They proposed a modified scheme of drought classification as departure of aridity index from the normal

< 1/20 σ drought is moderate, between 1/2 σ to σ drought is large and between σ to 1½ σ drought is severe and if > 1½ σ, the drought is disastrous (where σ is standard deviation). This scheme was found to be satisfactory in identifying the periods of disastrous drought that actually prevailed over western Rajasthan during the period 1901-1970.

Chowdhury and Hussain (1983) investigated the occurrence of monthly and annual aridity and drought conditions of Bangladesh. Rainfall and temperature data of 31 years (1949-79) have been collected and processed. "Aridity" and "drought" index have been used. Aridity condition generally exists over almost all parts of Bangladesh during November-March while the same prevails a bit longer period upto April for the north-western and western parts. Drought conditions were found irregular and spatially limited in Bangladesh.

Ram Mohan *et al.* (1983) evaluated that ecologically a mild drought of prolonged duration is more detrimental to vegetation than a short duration intense drought. Further, the water balance procedure can be used effectively to delineate periods and intensities of a drought from a eco-climatic point of view. Cumulative deviation teclmiques have been used to delineate and categorize the intensity and duration of drought spells.

Subramaniam and Rao (1983) studied agricultural drought in relation to fingermillet grown under rainfed conditions. They studied the requirements of moisture adequacy and percentage of available soil moisture during cropping period.

Thomas (1983) calculated the Palmer drought severity index (PDSI) from state average of temperature and precipitation and from single station analysis. It has been demonstrated that droughts persist longer in the interior of the United States than in areas furtherer east or west. The sensitivity of the PDSI was tested in relation to changes of derived parameters in the PDSI calculation in order to determine their effect on the spatial characteristics of drought duration.

Alley (1984) also examined the Palmer drought severity index (PDSI). Under certain conditions, the PDSI was found to be very sensitive to the criteria for ending on "established" drought. The distribution of the PDSI conditioned on the values for the previous months may often found to be bimodal.

Bishnoi (1984) studied the crop moisture index. The intensity, duration and rhythm in space and time of crop moisture index have been found to be very helpful in explaining the role of agronomic practices, crop yield variation, and agricultural drought. Palmer crop moisture index failed to explain the behaviour of crop conditions.

Biswas and Nayar (1984) quantified the drought based on moisture availability index (MAI), which is the ratio of probabilistic rainfall of a week to potential evapotranspiration. The criteria for crop growth period was used as MAI > 0.3. The mid season water stress period, *i.e.,* when MAI ≤ 0.3 has been found and this period has been used to map out drought prone area both in time and space. Severity of drought has been classified by the number of water stress period as well as their durations. A drought prone area was found extending from Aurangabad in Maharashtra to Bangalore in Karnataka.

Chowdhaury and Abhyankar (1984b) determined the conditional probabilities of drought/non-drought years of seasonal rainfall departures for the period 1875–1974 by Markov chain analysis. This has been done by selecting one meteorological sub-division from each of the climatic type, *viz.,* pre-humid, humid, sub-humid (dry and moist), semi-arid and arid regions. Expected length of climatic cycles, that is, drought spell followed by a non-drought spell and vice-versa, have also been computed and discussed.

Khambete and Biswas (1984) analysed rainfall distributions. Two state Markov chain model has been used to evaluate sequences of dry and wet weeks during south-west monsoon period over dry farming tract of Maharashtra. An index based on parameters of this model has been developed to bring out the degree of drought proneness.

Narain *et al.* (1984) made an attempt to characterize the drought in three districts falling in arid and semi-arid regions of western India in relation to sorghum crop. For this purpose the estimates of the minimum amount of rainfall which was just sufficient for crop growth during a given number of consecutive weeks were worked out. The occurrence of rainfall below these drought base values gave rise to drought conditions.

Rama Krishna *et al.* (1984) studied the climatic spectrum of the Indian arid region extends from extremely arid to semi-arid conditions by application of the theory of conditional probabilities based on first order Markov chain model during the 70 years (1901-1970). It was observed that the occurrence of extreme arid conditions in the succeeding year can be predicted with 92 per cent confidence while climatic types can be predicted with 83 per cent confidence. Studies in the incidence and spread of droughts over western Rajasthan, indicated that drought conditions originate first in the north-eastern region during the month of July, spread in a south-western direction during the month of August and dissipate with an easterly movement in the month of September. This pattern was observed to repeat during the years of severe droughts. Long range prediction of drought was not yet possible with 100 per cent confidence. Therefore, a combination approach using different techniques might provide the necessary confidence in predictions of droughts.

Sastry and Chakravarty (1984) distinguished "atmospheric drought" from "soil or agricultural drought". This situation occurred at different stages of crop growth for a week or two and not reflected in the soil water balance, but was revealed in the crop growth rates and yields. The situation of advection can be traced on a micro scale, during the "break periods" and on a local scale during the post monsoon periods. An empirical technique based on sequential dry days for assessment of atmospheric drought was also used.

Sastri (1984) critically examined the runoff water balance components on a weekly basis with a view to understand the impact both individually and collectively on general crop growth. A comparison of the water balance parameters with those of normal, wettest and driest years in the decade was determined from water budgeting.

Sarma and Ravindranath (1984) investigated proneness of the different climatic types of the south Indian region to droughts by selecting one station from each zone of the climatic spectrum and categorizing the drought years. Decenial frequency of drought years was then calculated at each station and graphically plotted to get insight into drought occurrence in the various climatic zones of the south Indian region.

Singh *et al.* (1984) studied the climatic characteristics of occurrence of dry spell over 7 meteorological subdivisions laying in semi arid parts of India by using 23 years of spatially averaged sub divisional rainfall data. Five different thresholds have been used to decide the occurrence of dry and wet day over the subdivisions. The Markov chains of order 1 and 2 have been fitted to frequency of occurrence of dry spell over the stations and on testing, the chain of order 2 was found to fit better. Composite charts of probability of, occurrence of rain around depressions located in various geographical areas have been prepared by using daily rainfall data of 220 stations for 19 years. They also, developed multiple linear regression equations by stepwise screening method to predict the daily rainfall amounts over those subdivisions by using large scale synoptic parameters.

Subramaniam *et al.* (1984) analysed the floods and drought years of 14 stations of Rajasthan during the period 1901–1977. The results showed that the year 1917 was the flood year at all the stations and the year 1946 recorded the highest surplus rainfall (1970 mm) at Jhalawar. Nearly 60 per cent of the stations experienced the worst drought period in the decade 1961-70 and the number of severe droughts were more in the state as a whole, while the number of disastrous droughts were less.

Victor and Sastry (1984) worked out weekly soil moisture budget for Kharif cropping season (July to October) for Delhi. Soil moisture index, the ratio of available water to a water holding capacity in the soil root zone was directly related to the ratio of actual and potential evapotranspiration.

Rao and Subramaniam (1986) identified commencement and cessation of different intensities of droughts during 1934-77 at 23 locations of Maharashtra state by using the modified Palmer's approach. The modified drought indices performed well for identification of various classes of droughts in the region. The analysis showed that the spatial distribution of drought prone areas during Kharif (June-September) is similar to the Rabi season (October-February) in the region. They found that during Kharif season alone severe/extreme drought conditions prevailed for 28 out of 44 years in the state. But there are few stations like Kolhapure and Sironcha where the drought situation never touched the extreme conditions.

Adhikari *et al.* (1988) calculated water deficit values using the meteorological data on rainfall and evaporation for the low rainfall tract of Bellary. They found that on an average, no single month has recorded surplus water during a year. The water deficits ranged from 161 rnm (May) to 43 mm (September). Comparatively the water deficit was less during Rabi season than the Kharif season. It's value was less than 100 mm in each of the 4 months, September to December, while it varied from 100 to 161 mm in rest of the 8 months. It was presumed that successful crop production in this area was not possible without supplemental irrigation from some other sources.

Sastri (1991) studied drought climatology of Raipur for a period from 1906 to 1980. The droughts were classified based on aridity index and on index of moisture adequacy as mild, moderate and severe for rice crop.

Singh *et al.* (1991) analysed rainfall data of 12 stations for the period 1901 to 1989 of Kutch district of Gujarat. It revealed that the rainfall has decreasing trend from south-west region to north east region. Probability of weekly rainfall distribution worked but by ranking method and revealed that interior part of the district was more prone to drought conditions as compared to coastal plains. Most of the stations had recorded highest observed rainfall values nearer to the estimated rainfall of 100 years return period except at Bhuj, Anjar and Rahapur.

Abul *et al.* (1996) studied the nature, extent and frequency of drought in the monsoon season in a rainfed low land rice system of north-west Bangladesh using the method of water balance from long term rainfall data. Probability analysis showed that in the study area drought were expected at the beginning of the crop season prior to transplantation and from the beginning of the ripening period. The seasonal relative water surplus (RWS) from rainfall in an average year was 0.79 which means that 79 per cent of the crop water requirement was available from rainfall. The RWS was expected to go down to 0.51 twice in ten years because of late transplantation due to early drought. In order to stabilize the rice yield measures of drought alleviation have been suggested.

Lohani *et al.* (1996) used a non homogeneous Markov-chain approach for analyzing drought characteristics using the Palmer drought severity index (PDSI). The probability mass functions of occurrence of different drought severity classes, their durations and times of return to a particular drought class were obtained and utilized to generate the needed statistics and forecasts. Two methods of forecasting drought severity classes for one, two and three months lead times were suggested. The methodology was applied to ninety-six years of PDSI data corresponding to two climatic divisions in Virginia, USA. A comparison between the analytical results and the empirical estimates supported the utility of the method. The method can be used in a planning mode for developing buffer storage in drought prone regions and in a operational rationing of water among competing needs as drought progresses.

Sostaric and Madar (1997) used the Palfai aridity index (PAI) to determine drought periods in Croatia. The results confirmed more dry and warm periods in east Croatia after 1981. The PAI method was found to be more precise than other methods since more factors were used in its calculation. The PAI criterion can indicate the dry years in which irrigation with economical water and energy use are required.

Barakat and Handoufe (1998) studied agricultural drought years occurring in Marocco from 1940 to 1995 and highlighted the running means of national cereal production for 5-year periods by using a regression confidence interval. There were 11 droughts over this period, with 6 occurring between 1980 and 1995. To assess the intensity of these droughts, the deficit production index was determined relative to the normal value. Drought periods were studied in eight regions where cereals were mainly cropped. During these drought periods, the rainfall deficit always exceeded 20 per cent relative to normal years. Drought was certain when this mean deficit reached 37 per cent. Annual rainfall distribution did not seem to be an important drought determining factor. The rainfall deficit index for the December-February period was used to obtain efficient short-term drought forecast.

Brignall *et al.* (1999) found that the risk of agricultural drought in Europe is likely to change with climate change. From the limited, broad scale analysis undertaken, showed that the drought hazard may alleviate in northern Europe and worsen in southern Europe. For specific crops, especially winter wheat, the combined effects of a carbon-dioxide enhanced climate and temperature change, with little change in precipitation would lead to increased mean yields.

Paul *et al.* (1999) studied climatological water balance for Bhubaneswar taking 35 years (1962–1996) of climatic data. They found that Bhubaneswar has a water need of 1619 mm, with a rainfall of 1479.5 mm, actual evapotranspiration of 1195.5 mm, water surplus of 284 mm and a water deficit of 423.5 mm. Taking the various components of water balance, aridity index and moisture index they computed the incidence of drought. Study revealed that during the period under study, Bhubaneswar has experienced only 2 disastrous, 3 severe, 8 large and 4 moderate drought years.

Battau (2000) tabulated and mapped rainfall and occurrence of severe summer drought during the period (1951–98) in Emilia Romagna, Italy. Particularly serious droughts occurred in twelve years (defined as < 600 mm of rainfall during April-September), while six years recorded < 250 mm precipitation during the period.

4.1.4 Spread of Drought

Droughts, which are more frequent in arid and semi-arid regions seem to follow a pattern giving the hope that possibilities exist for their prediction. Droughts have an origin, spread as well as decay. Study of spread of droughts helps to yield drought prediction possible. Some specific reviews related to the spread of drought is presented as under.

Subrahmanyam and Sastri (1971) studied the spread aspect of drought in the central Deccan region in south India for the period 1897-1901. The cumulative deviation technique and making use of the monthly water deficiency values obtained from the water budgeting procedure determined the spread of drought. Attempts were also made to obtain severity indices for drought spells of different durations.

Malini (1981) analysed aridity anomalies and patterns on an annual basis for the years (1966-69) and found that 1968 had experienced severe drought conditions particularly in the northern coastal Andhra Pradesh. Application of the cumulative deviation technique from which severity indices were analysed, helped to understand the duration, severity and spread of droughts. Data on crop yields in the same region during the same period substantiated the results.

Sastri and Malakar (1981) reported that though the seasonal rainfall exceeded 700 mm in many places of northwest India during the year 1979, drought conditions prevailed over the entire region causing water stress conditions to all rainfed agricultural crops. Water balance technique has been used to examine the incidence, spread and dissipation of drought.

Ram Mohan *et al.* (1984) employed the index of moisture adequacy to determine the spread of drought. Droughts of selected stations in the semi arid zone of India have been analysed. The exact duration of drought spells were analysed critically by plotting the ratio of departures of monthly index of moisture adequacy from the climatic normal, to the standard deviation of index of moisture adequacy for that month.

4.2 Procedure of Evaluation of Drought, Assessment and Spread of Drought

4.2.1 Evaluation of Drought

Drought was evaluated in terms of drought intensity for all the stations under study by the following scheme proposed by Subrahmanyam and Sastri (1969b):

Departure of aridity index from the normal	Drought intensity
< ½	Moderate
½ σ to σ	Large
σ to 2σ	Severe
> 2 σ	Disastrous

where, σ is the standard deviation of aridity index.

4.2.2 Assessment of Duration and Severity of Drought

The cumulative deviation technique of Foley (1957) was used to assess duration and severity of drought for all the stations. In this technique instead of rainfall figures water deficiency deviations have been used in preparing the cumulative deviation curves similar to residual mass graph. Departures of weekly water deficiency from the cumulative normals were taken and expressed as ratios of the normal water need in units of thousands. These units were cumulated from an arbitrary origin and plotted against successive weeks. The core period of drought is then indicated by the steepest portion of rise in the mass curve, and the gradient, therefore, gives the drought severity index (DSI) during the period. Based on the DSI the droughts were categorized for all the stations by using following scheme:

Drought Severity Index	Category
0–10	Moderate
10–20	Large
20–40	Severe
> 40	Disastrous

4.2.3 Determination of Climatic Shifts

Climatic classification of Thornthwaite and Mather (1955) was used for determination of climatic shifts for all the stations during the period under study. Based on the moisture index the type of climate for all the stations was determined by the following scheme.

Type of climate	Moisture index
A - Per humid	100 and above
B_4 - Humid	800-100
B_3 - Humid	60-80
B_2 - Humid	40-60
B_1 - Humid	20-40
C_2 - Moist sub-humid	0-20
C_1 - Dry sub-humid	–33.3-0
D - Semi-arid	–66.7 to –33.3
E - Arid	–100 to –66.7

4.2.4 Determination of Spread of Drought

For depicting the spread of drought the departure of actual weekly water deficiencies from normal values were calculated for all the stations and were expressed as a per cent of water need (ET_0) in unit of one thousand (these values are referred as "D").

4.3 Discussion

4.3.1 Evaluation, Assessment and Spread of Drought Under Different Agro-climatic Condition of Rajasthan

A drought is a climatic hazard, which has drastically adverse effect on vegetation and agriculture. Drought occurs due to deficit of water, which happens when precipitation fails to meet the water need or the reference evapotranspiration.

4.3.1.1 Evaluation of drought

For the analytical study of droughts with special reference to their intensity and frequency, the aridity index (I_a) has been used. Years of drought may easily be identified and their intensities can also be assessed. Drought years and their severities for different stations are given in Table 4.2 and shown in Fig. 4.1 to 4.8.

It can be seen from Table 4.2 that according to scheme of drought classification, year 1987 was identified as a drought year in all the stations of the state. It also inferred that the frequency of occurrence of droughts of different intensities is 7 in all the stations under study except at Udaipur, where its frequency of occurrence is 6 out of 15 years. The percentage departure of I_a from the median is higher at Banswara and Kota, which caused one disastrous drought year (1987).

It is also interesting to note that the year 1987 was either severe or disastrous drought year at all the stations except at Fatehpur where, a moderate drought was experienced during 1987.

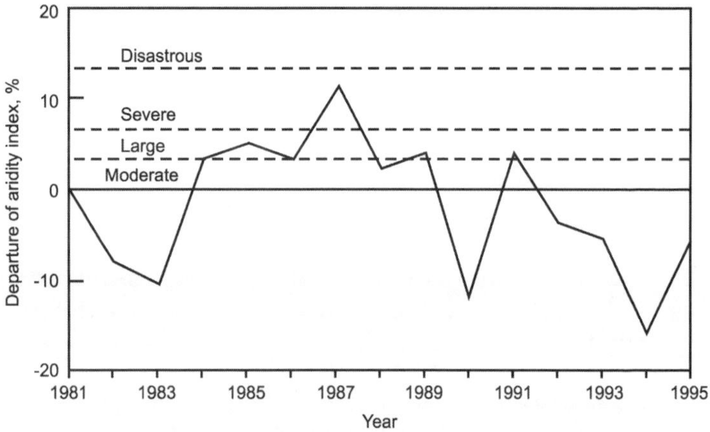

Fig. 4.1 : March of aridity index for Jodhpur (1981-95)

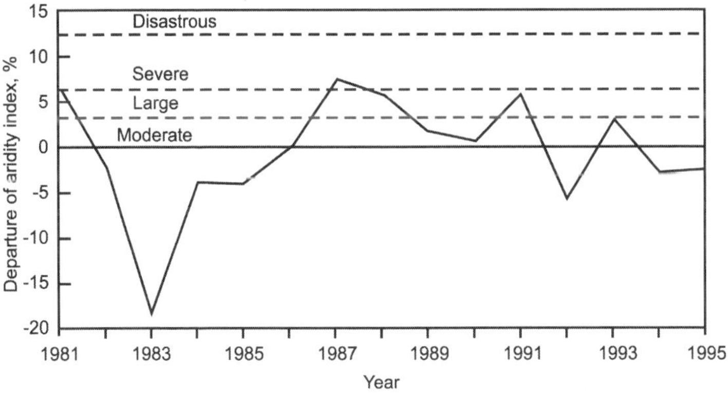

Fig. 4.2 : March of aridity index for Sriganganagar (1981-95)

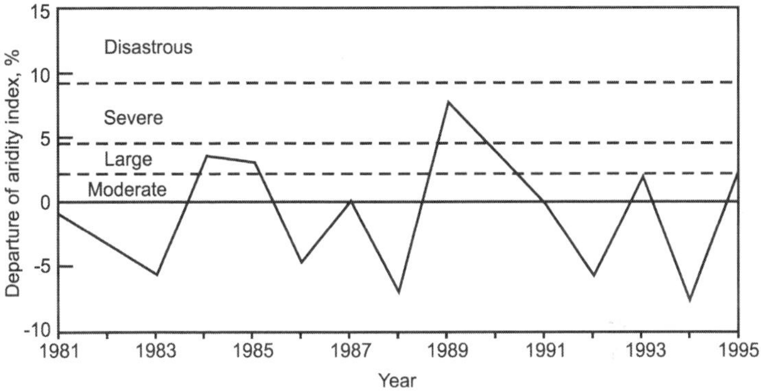

Fig. 4.3 : March of aridity index for Fatehpur (1981-95)

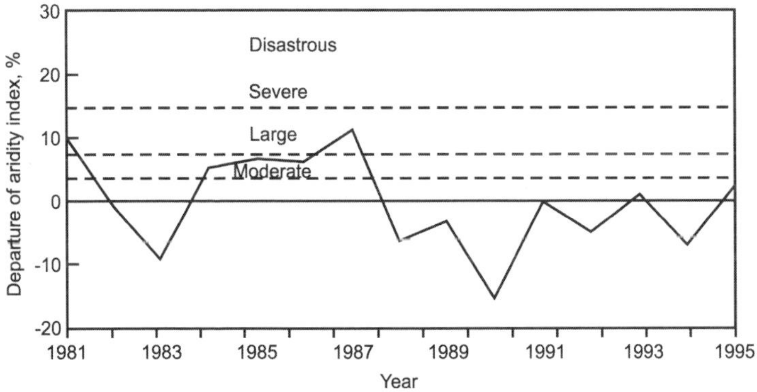

Fig. 4.4 : March of aridity index for Pali (1981-95)

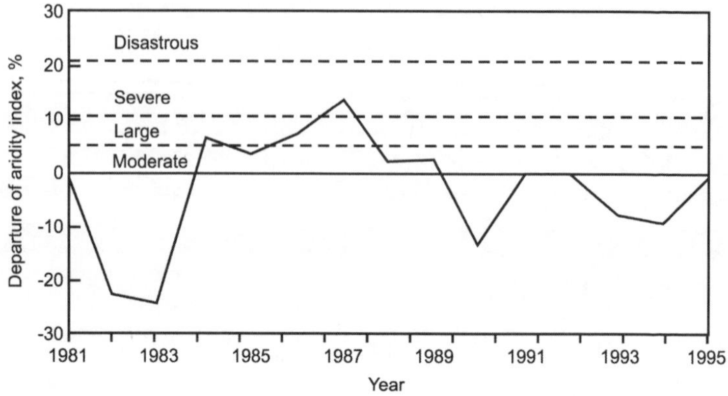

Fig. 4.5 : March of aridity index for Jaipur (1981-95)

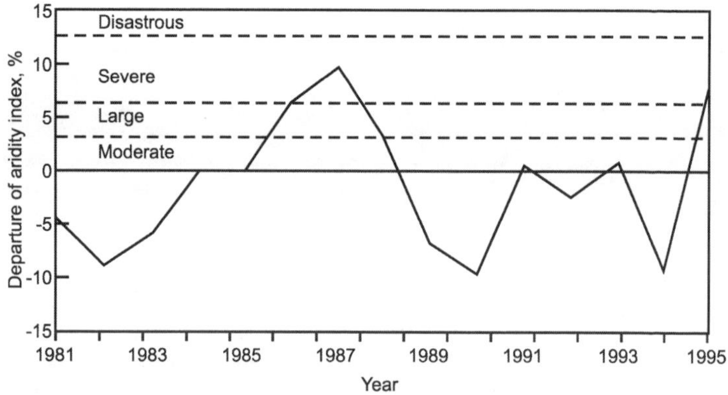

Fig. 4.6 : March of aridity index for Udaipur (1981-95)

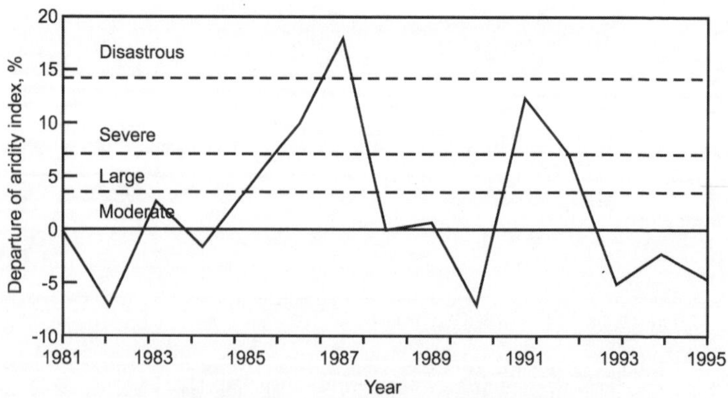

Fig. 4.7 : March of aridity index for Banswara (1981-95)

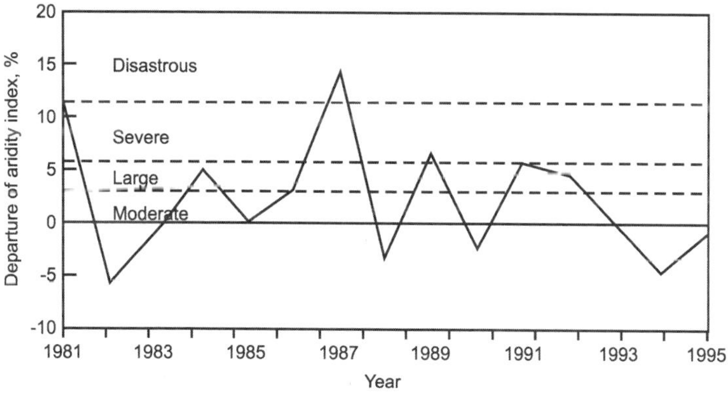

Fig. 4.8 : March of aridity index for Kota (1981-95)

Table 4.2 : Drought years and their severities during 1981-1995

Station	Severity of drought				Total
	Moderate	Large	Severe	Disas-trous	drought year
Jodhpur	1988	1984, 1985, 1986, 1989, 1991	1987	–	7
Sriganga-nagar	1989, 1990, 1993	1984, 1991	1981,1987	–	7
Fatehpur	1987, 1993	1984, 1985, 1990, 1995	1989	–	7
Pali	1993, 1995	1984, 1985, 1986	1981,1987	–	7
Jaipur	1981, 1985, 1988, 1989	1984, 1986	1987	–	7
Udaipur	1988, 1991, 1993	–	1986,1987, 1995	–	6
Banswara	1983, 1989	1985, 1992	1986, 1991	1987	7
Kota	–	1984, 1986, 1992	1981, 1989, 1991	1987	7

No moderate drought year was experienced at Kota, whereas Jaipur experienced highest moderate drought years (4). Sriganganagar, Udaipur and Banswara experienced 3 moderate drought years and Fatehpur and Pali experienced 2 moderate drought years.

The frequency of occurrence of large drought years varies from zero at Udaipur to 5 at Jodhpur. Fatehpur experienced 4 large drought years. Pali and Kota experienced 3 large drought years and Sriganganagar, Jaipur and Banswara experienced 2 large drought years.

The frequency of occurrence of severe drought years varies from 1 at Jodhpur, Fatehpur and Jaipur to 3 at Udaipur and Kota, Sriganganagar, Pali and Banswara experienced 2 severe drought years during the period under study.

Thus, the total number of drought years of all the categories are nearly equal for all the stations during the period (1981-1995) under study.

4.3.1.2 Assessment of duration of severity of drought in Rajasthan

Droughts usually result from the cumulative effect of a set of weather sequences developing over an extended period of time. Thus, drought may be either prolonged or short-lived depending upon the nature and duration of water deficiency responsible for it. The severity of drought depends upon the degree and duration of water deficiency. Therefore, the drought spells in severe or disastrous drought year (1987) were selected for detailed study of their durations and severities on weekly basis. The cumulative deviation technique of Foley (1957) was used for computing the duration and drought severity index (DSI) as mentioned in Section 3.5 and drought spells were categorized according to Malini (1981). For the analysis arbitrary origin was taken as 23rd week and 30th week was considered as the end of rainy season. Weekly, cumulative deviation curves during 1987 for different stations are shown in Figures 4.9 to 4.16. Duration and severity indices of drought spells during 1987 are given in Table 4.3.

Table 4.3 indicates that the minimum duration of large drought spell of 4 weeks occurred at Fatehpur during the year 1987, whereas maximum duration of large drought spell of 15 weeks occurred at Sriganganagar. It is seen from Table 4.3 that Jodhpur and Sriganganagar experienced one drought spell of long duration of large category. It is also confirmed that at Fatehpur, the year 1987 experienced a moderate drought as it experienced shortest duration (4 weeks), of large drought spell.

At Pali 2 weeks severe drought spell (26th–27th week) of 21.99, severity index occurred, whereas no severe drought spell occurred at any other stations under study. Two drought spells have been experienced at Fatehpur and Udaipur of large and moderate categories. Jaipur experienced three drought spells of large category.

Table 4.3 shows that three drought spells of large category (1) and moderate (2) experienced at Banswara and Kota. It is also clear from Table 4.22 that no disastrous drought spell has occurred at any stations under study.

The highest severity index of large drought spell occurred at Pali (19.76) followed by Banswara (1951). Similarly, the highest severity index of moderate drought spell occurred at Udaipur (9.47) followed by Kota (8.32).

Thus all the stations had at least one drought spell of 2 weeks or more during the year 1987. Ecologically, moderate or large drought of prolonged duration is more detrimental to vegetation than a short duration intense drought. During the study period, there were prolonged spells of moderate and large droughts at all the stations except at Fatehpur, which were capable of crippling the economy of the zones due to their adverse impact on agriculture production.

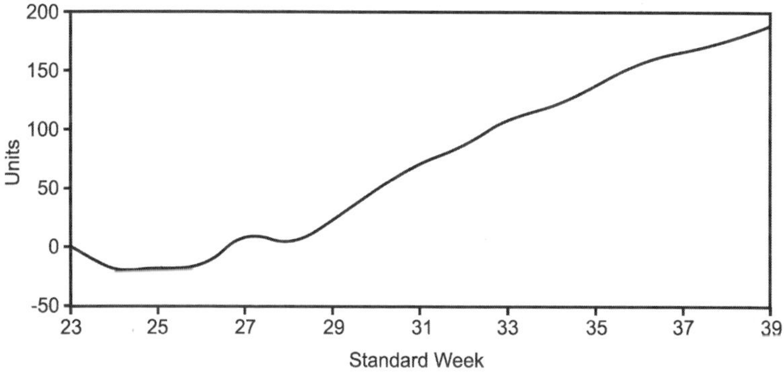

Fig. 4.9 : Weekly cumculative deviation of water deficit for Jodhpur during the year 1987

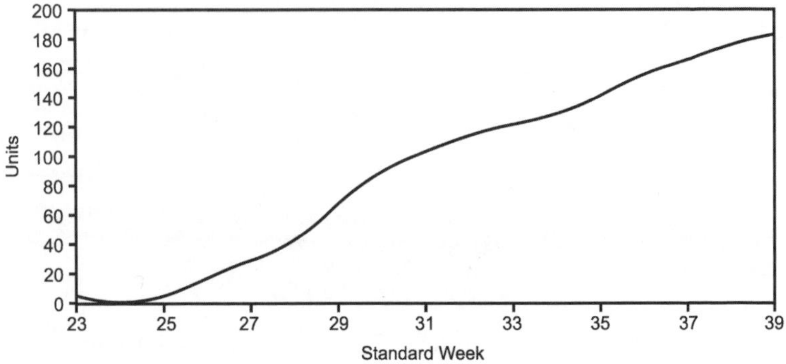

Fig. 4.10 : Weekly cumulative deviation of water deficit for Sriganganagar during the year 1987

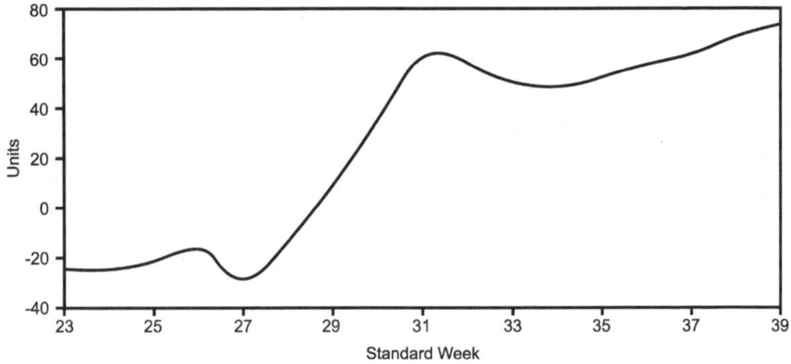

Fig. 4.11 : Weekly cumulative deviation of water deficit for Fathepur during the year 1987

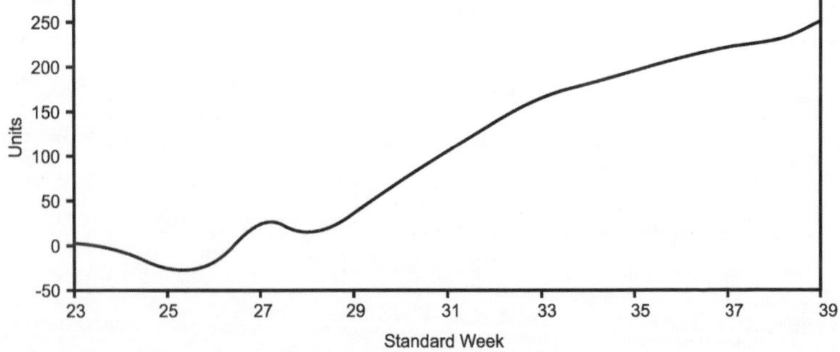

Fig. 4.12 : Weekly cumulative deviation of water deficit for Pali during the year 1987

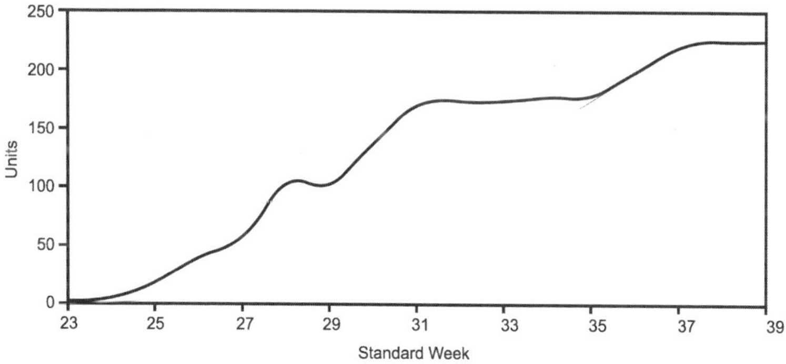

Fig. 4.13 : Weekly cumulative deviation of water deficit for Jaipur during the year 1987

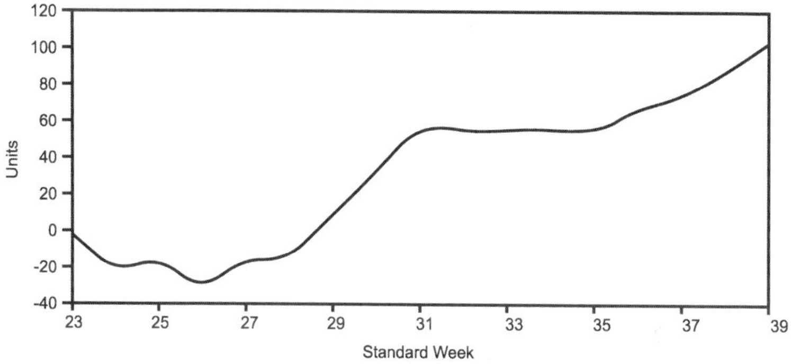

Fig. 4.14 : Weekly cumulative deviation of water deficit for Udaipur during the year 1987

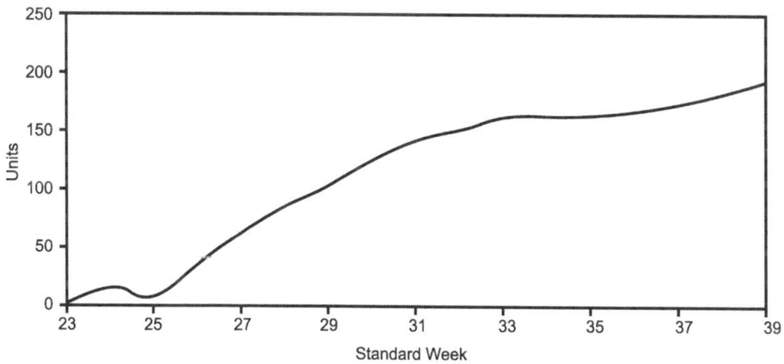

Fig. 4.15 : Weekly cumulative deviation of water deficit for Banswara during the year 1987

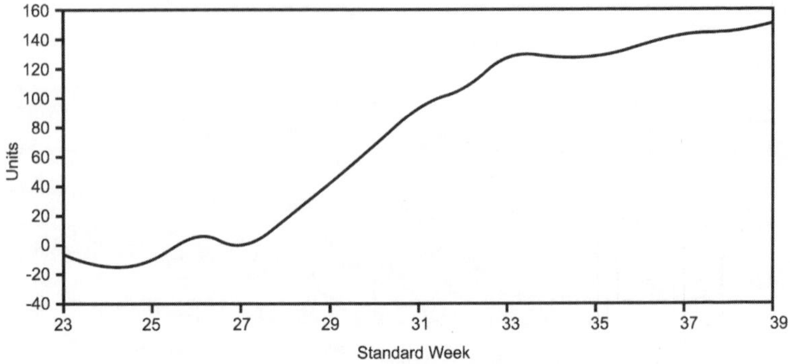

Fig. 4.16 : Weekly cumulative deviation of water deficit for Kota during the year 1987

Table 4.3 : Duration and severity indices of drought spells during the year 1987

S. No.	Station	Number of drought spell	Drought spell week	Duration (week)	Severity index	Category
1.	Jodhpur	1	26th-39th	14	14.59	Large
2.	Sriganangar	1	25th-39th	15	11.30	Large
3.	Fatehpur	2	28th-31st	4	18.79	Large
			34th-39th	6	4.12	Moderate
4.	Pali	2	26th-27th	2	21.99	Severe
			28th-39th	12	19.76	Large
5.	Jaipur	3	23rd-28th	6	16.53	Large
			29th-34th	6	12.67	Large
			35th-39th	5	13.25	Large
6.	Udaipur	2	26th-31st	6	14.02	Large
			35th-39th	5	9.47	Moderate
7.	Banswara	3	23rd-24th	2	7.67	Moderate
			25th-33rd	9	19.51	Large
			35th-39th	5	6.86	Moderate
8.	Kota	3	25th-26th	2	8.32	Moderate
			27th-33rd	7	18.24	Large
			35th-39th	5	4.37	Moderate

4.3.1.3 Spread of drought over Rajasthan during the severe drought year 1987

Droughts, usually do not descend all of a sudden without warning, but they are usually the result of culmination of a set of a weather sequences that require extended periods of time to develop. Droughts have their origin and decay. It is of utmost importance to planners to have knowledge of the drought vulnerability, and pattern of spread of drought in a region. This can enable them to plan timely measures to combat the intensity of drought and to mitigate the drought effect, as the reasons are known in advance once the drought spell begins. In Rajasthan, often some part or the other experiences localized drought. However, in some years the whole state experiences drought conditions. Studies reveal that even in such years the drought intensity and its time of occurrence varied from zone to zone. In order to project a pattern of drought spread in Rajasthan, one case study is presented, pertaining to the severe drought conditions that prevailed over the state during the year 1987. Diagrams depicting the spread of drought based on the departures of actual weekly water deficiency from normal values to the ratio of annual water and expressed in 1000 units are presented from 23rd to 39th week, during the main rainy season in the state (Fig. 4.17).

A perusal of Figure 4.17 clearly indicates that the zone of maximum drought intensity changed both its intensity and location in different weeks indicating a sequential pattern of drought spread and decay over the state. There were no drought conditions until 25th week. The maximum values of 'D' (water deficiency) were observed first in 26th week in Banswara region, which experiences moist sub-humid type of climate over the state in comparison to the regions experiencing semi-arid, arid and extremely arid type of climates. Also interesting is the fact that the lowest values of 'D" were observed in Jodhpur region, which records the lowest annual rainfall. It indicates that the crippling effect of drought is discernible more earlier in the region of the high rainfall than in the regions of low rainfall. Thus, during the year 1987 the water deficiency remained slightly pronounced in the southern region of Rajasthan except at Udaipur region during 23rd-26th week *i.e.,* even before the monsoon onset, with elongation direction of the isolines in the south west direction into the Pali and Jodhpur regions. By 27th week, the drought condition in Rajasthan not only intensified (D > 40) but one definite core of severe drought can be identified in Pali. The isolines run south to north during this week also with an increase in the intensity of the drought over a major parts of the state. However, the decreasing

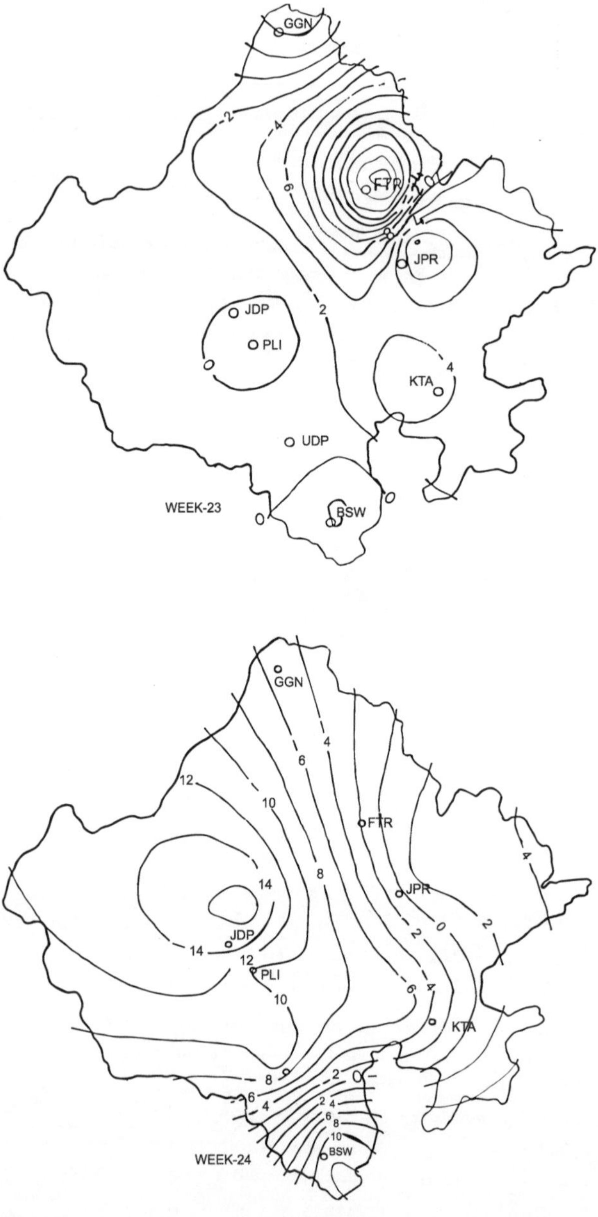

Fig. 4.17 : Weekly variation in drought spread over Rajasthan during the year 1987

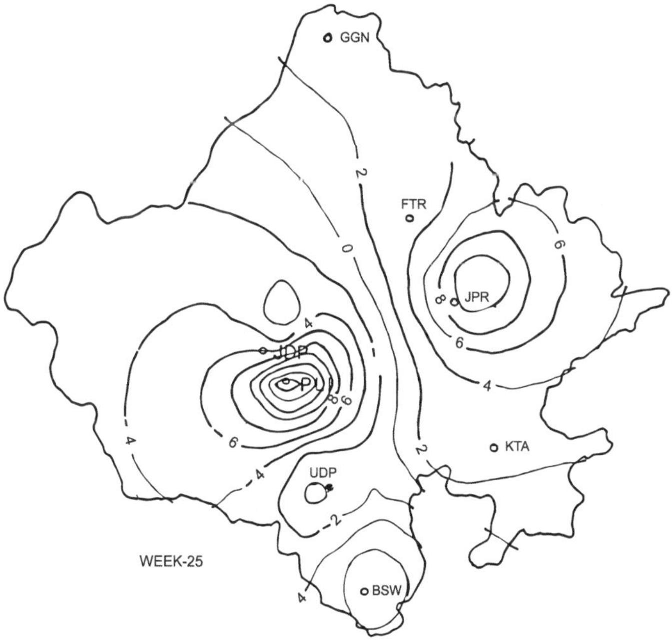
GGN
2
FTR
0
JPR
8
6
6
4
JDP
PLI
8
6
4
6
4
2
KTA
UDP
4
2
WEEK-25
BSW
4

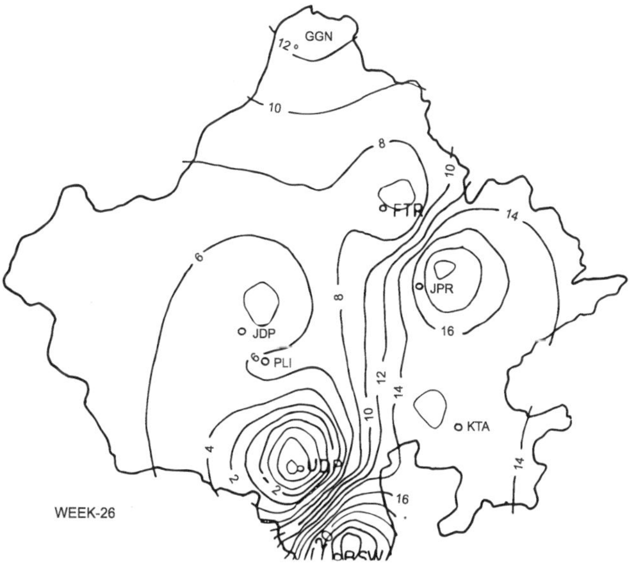
12
GGN
10
8
10
FTR
14
6
JPR
8
16
JDP
6
PLI
12
14
10
KTA
4
2
UDP
2
14
16
WEEK-26
20
BSW

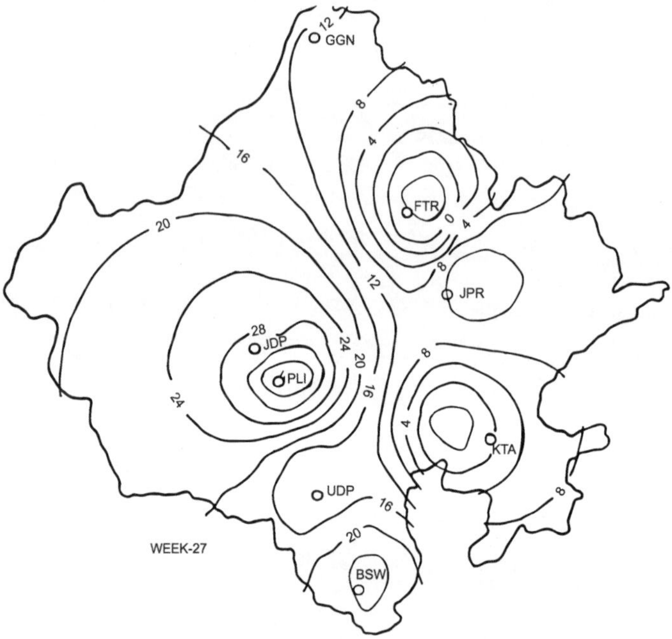

WEEK-27

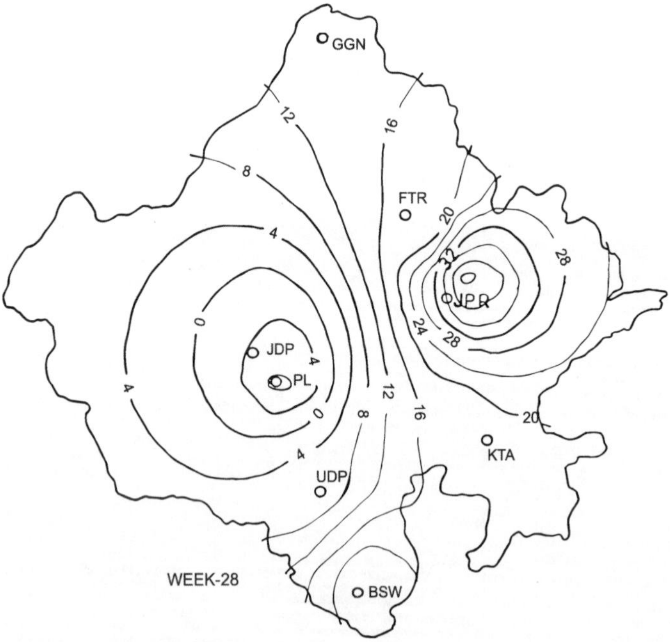

WEEK-28

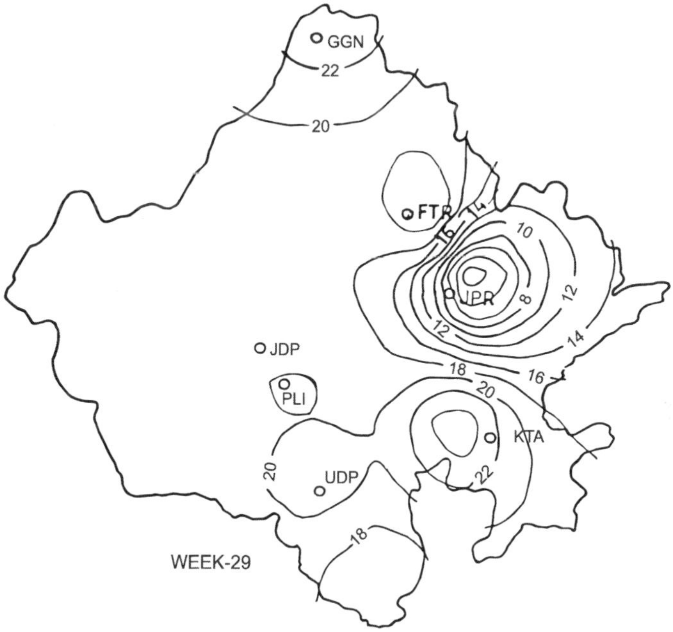
GGN
22
20
FTR
14
16
10
JPR
8
12
12
14
JDP
PLI
18
16
20
KTA
20
UDP
22
18
WEEK-29

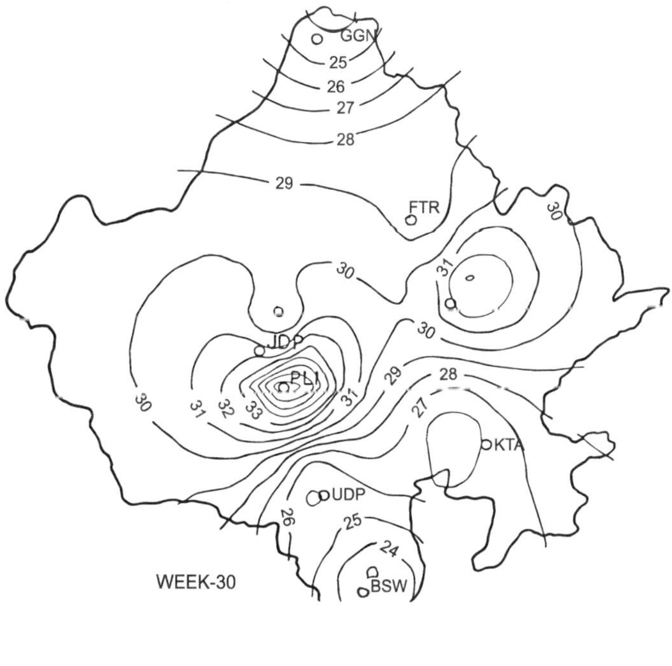
GGN
25
26
27
28
29
FTR
30
30
31
30
JDP
PLI
29
28
30
31
32
33
31
27
KTA
UDP
26
25
24
BSW
WEEK-30

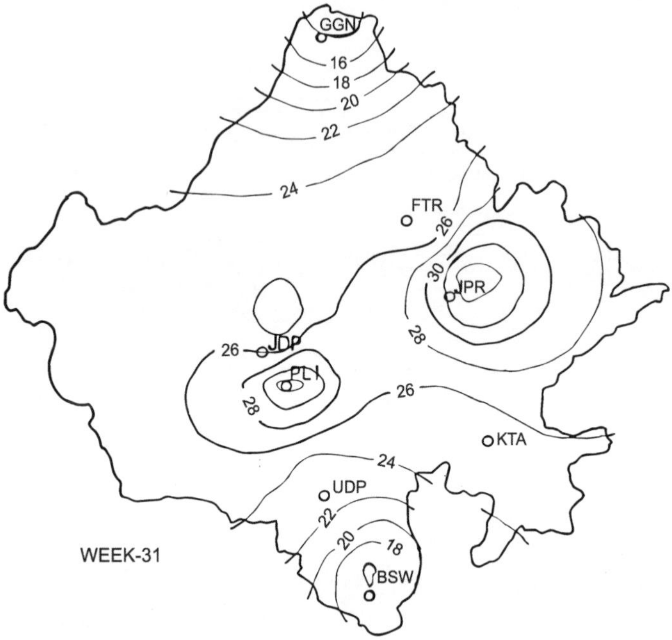

WEEK-31

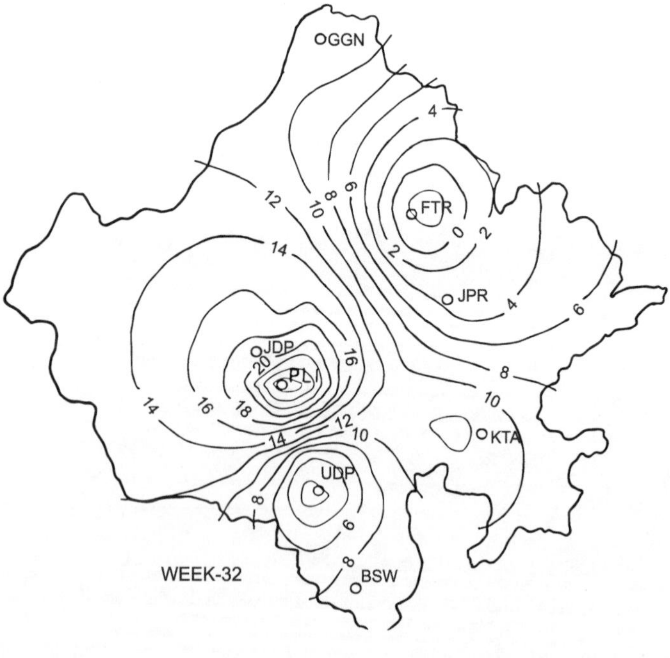

WEEK-32

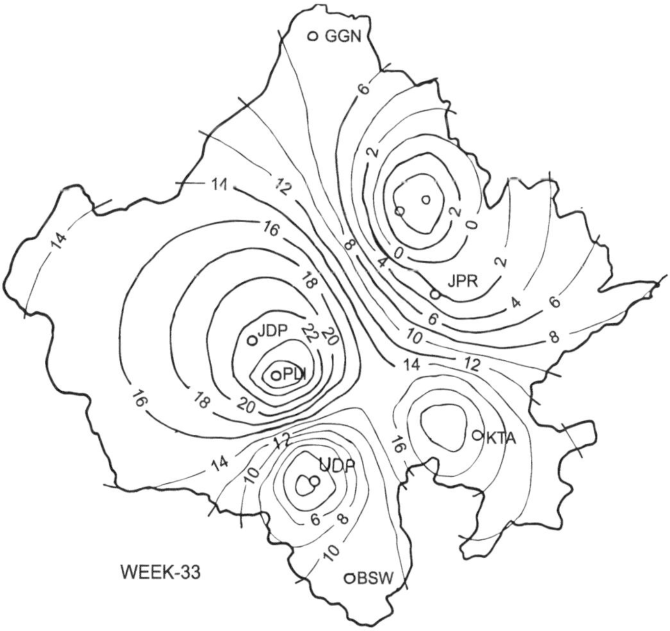

WEEK-33

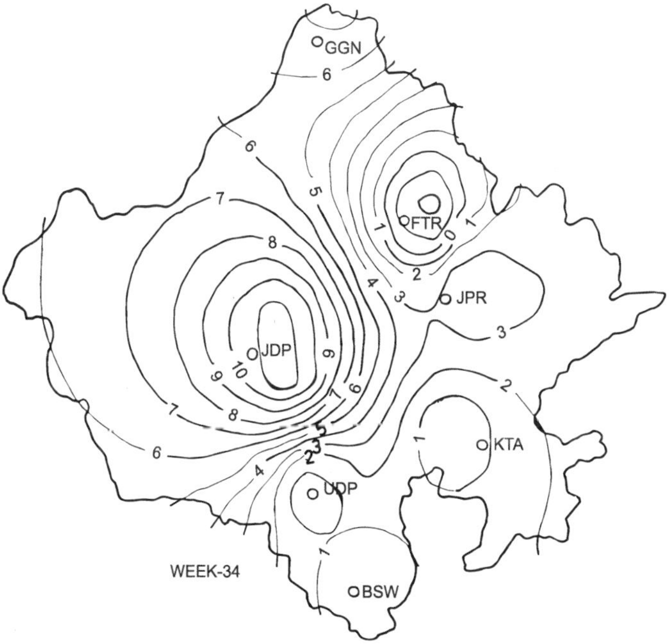

WEEK-34

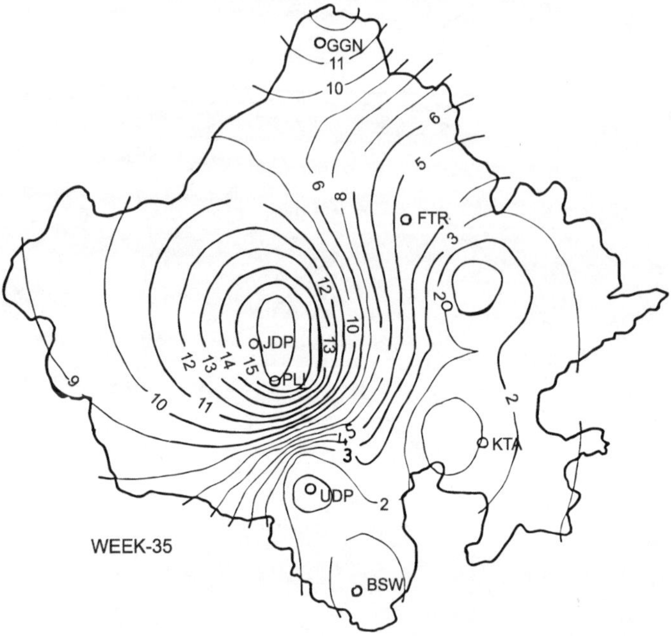
GGN
11
10
6
5
6
8
FTR
3
12
10
2
JDP
13
12
13
14
15
PLI
9
10
11
5
4
3
KTA
2
UDP
2
WEEK-35
BSW

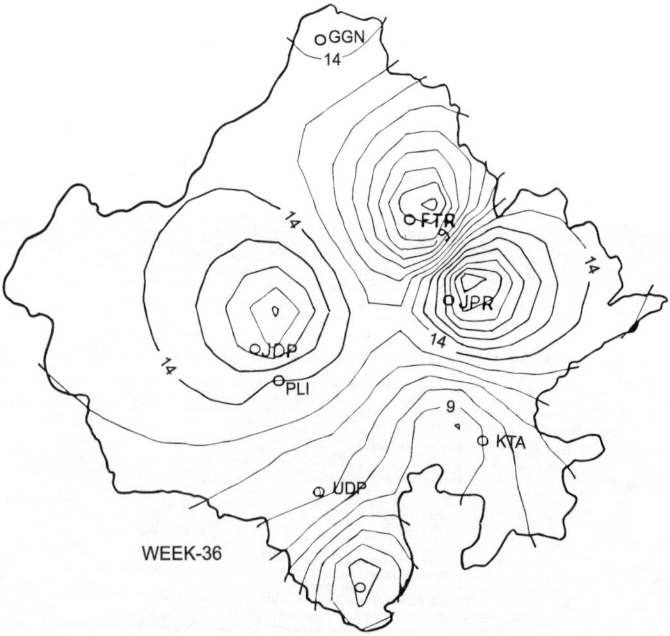
GGN
14
14
FTR
14
JPR
JDP
14
14
PLI
9
KTA
UDP
WEEK-36

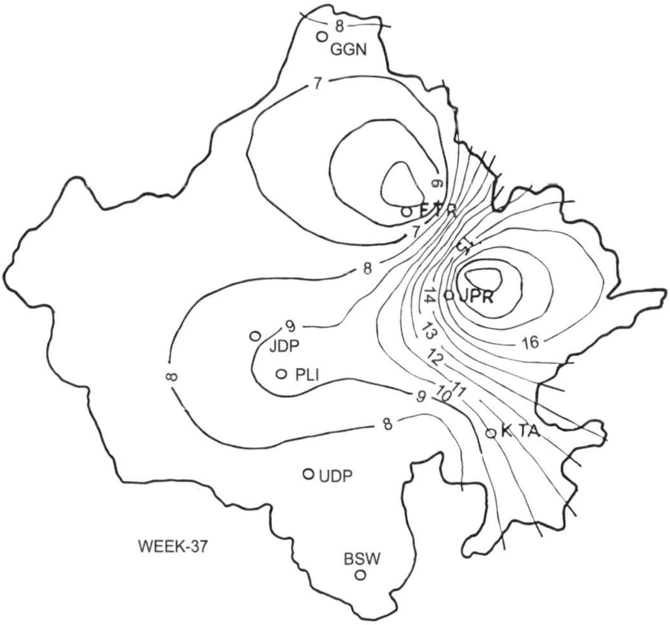
GGN
FTR
JPR
JDP
PLI
K TA
UDP
BSW
WEEK-37
8
7
9
14
13
12
11
10
16

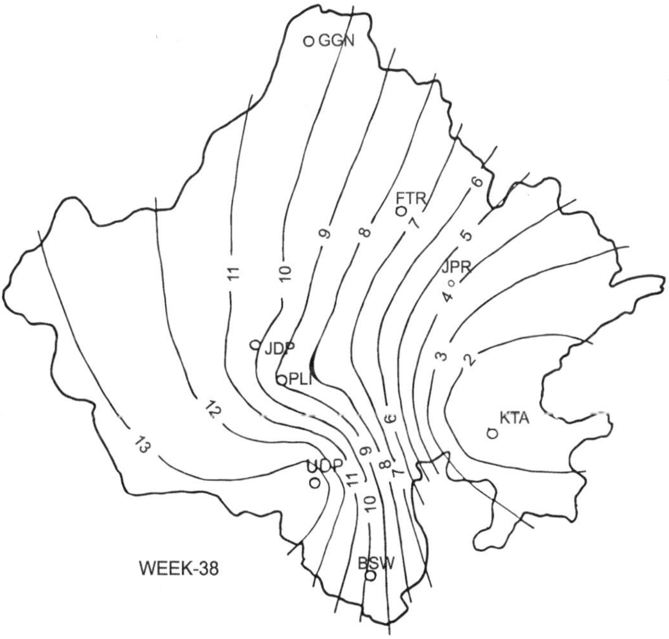
GGN
FTR
JPR
JDP
PLI
KTA
UDP
BSW
WEEK-38
13
12
11
10
9
8
7
6
5
4
3
2

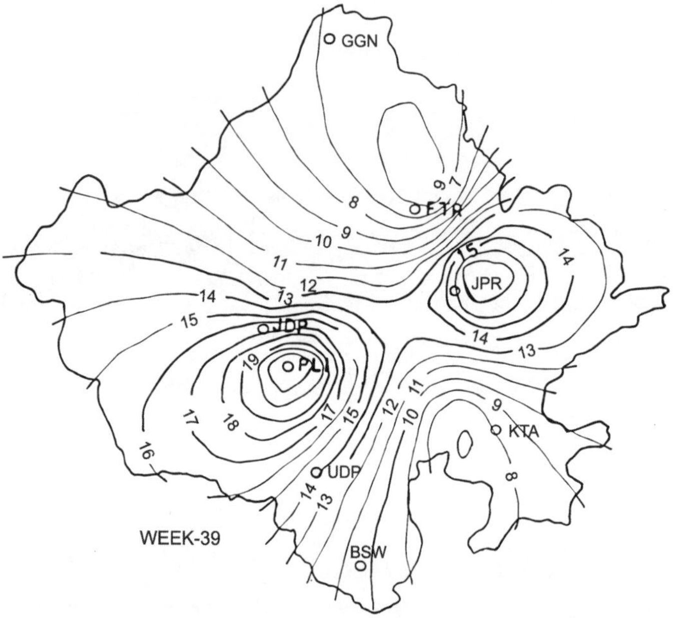

gradient continued to be in north to south and also from south to south-west direction towards the Jodhpur and Pali regions during the 28th week and lowest intensity being recorded in the Jodhpur and Pali region. Thus, the drought, which originated in Banswara region had intensified from moderate to large drought and spread north-east wards and south-east wards in Jaipur, Fatehpur and Kota respectively by 28th week. Sriganganagar, Jodhpur, Fatehpur and Udaipur region had experienced large drought conditions during this period.

The drought intensity in the Jaipur region had eased off by 29th week. During the 30th and 31st weeks during spread all over the Rajasthan the state had experienced large drought conditions and the drought had spread allover Rajasthan. Then the drought intensity had eased off in the Fatehpur region by 32nd to 34th weeks, in the Udaipur region by 32nd to 35th weeks and in the Banswara and Kota regions by 34th to 35th weeks. The large drought concentrated mainly in the Jodhpur and Pali regions during the 34th and 35th week. The drought intensity in the Jaipur region had also eased off by 35th week. By the 36th week Banswara had lowest water deficiency and isolines run from east to south and also north-east to south-east in 37th week. By the 38th week Kota had experienced lowest water deficiency and isolines run from south to north of the state. By the 39th week the water deficiency was highest at Pali region and had shifted to north-east. The drought was localised with lesser intensity around Fatehpur while over the rest of the region the drought core had merged with the seasonal drought (occurring with the rainless period after the withdrawal of the monsoon).

Thus, it clearly indicates that the occurrence and spread of drought is not a spordic event in Rajasthan state but it has a general tendency to originate in the southern region, spread in south westerly direction and dissipate with a north easterly movement. The analysis thus, clearly brings out two important facts, *viz.* that the effect of drought is more discernible and occurs early in regions receiving comparatively higher rainfall than regions of lower rainfall and secondly that they have an origin and follow a pattern of spread and decay. This type of analysis could be useful to pinpoint the origin and the gradual spread of drought into various regions of the state and would provide advance information to crop planners to mitigate the ill effects of such severe drought as well as minimizing the economic imbalances.

5. Frequency Analysis of Drought

5.1 Introduction

Frequency analysis is the most common statistical method of analyzing hydrologic data. Frequency analysis is used to predict how often certain values of a variable may occur and to assess the reliability of the precision.

Forecasts of drought events can generally reduce the damage caused by droughts. Firstly, and most obviously, warning of an event enables people to adopt drought management strategies. If sufficient lead-time is provided, vulnerable possessions may be removed from the drought region, and remedial measures can be made in advance to minimize the damage caused by drought.

Chow (1951) mentioned that the frequency analysis, if properly handled and conservatively interpreted, is a valuable hydrological tool. It is useful particularly for the economic study of any hydrologic structures, which are designed to take care of the hydrologic events.

Sen (1980a) studied the problem of regional drought and flood analysis and approached theoretically on the basis of random fields. Necessary formulations for the regional drought descriptions such as the deficit area, the total area deficit and the maximum deficit intensity have been derived.

Senapati *et al.,* (1989) analysed the annual maximum daily total rainfall data of 30 years of Bhubaneswar. The frequency curves of annual maximum rainfall data have been plotted using lognormal distribution and Gumbel distribution. The theoretical frequency curves were in close agreement with the observed data points, excepts at the highest rainfall for two distributions.

Varshney (1989) analysed weekly rainfall data of 50 years of Midnapur for the normal distribution by transforming the data into logarithmic, square roots, cube roots and fourth roots. Gamma distribution has also been applied to the original data. The goodness of fit has been checked by applying the Chi-square test. The cube and fourth roots of rainfall amounts have been found to give better approximation to the normal distribution. The frequency analysis has been done by analytical means considering fourth root normal distribution.

Bhatt *et al.,* (1996) worked out frequency analysis of one day maximum rainfall data for 24 years period (1968–1991). The observed values were estimated by Weibull's formula and expected values were estimated by extreme value type 1, log normal and log Pearson Type III distributions. The values obtained by three different probability distribution functions have been compared with observed values. The analysis indicated that log Pearson Type III distribution gives the closest fit to the observed data, hence it can be used to forecast the maximum daily rainfall.

Sharma (1997) found that the important elements of a drought phenomenon are longest duration and the largest severity for a desired return period. These elements form a basis for designing water storage systems to cope with droughts. At times, a third element, drought intensity, is also used and is defined as the ratio of severity to duration. The commonly available statistics for the causative drought variables such as annual rainfall, run-off sequences are the mean, the coefficient of variation and the leg one serial correlation coefficient and occasionally some indication of the probability distribution function of the sequences. The external values of the duration and severity were modeled using information on the aforesaid parameters. The drought severity was modeled as the product of the duration and intensity with the assumption of independence between them. An estimate of drought intensity was realized from the concept of the truncated normal distribution of the standardized form of the drought sequences in the normalized domain. A formula in terms of the extremely severity and return period was also suggested.

Kumar *et al.,* (1998) analysed rainfall data for a period of 12 years from 1979 to 1990 for Aonla in Uttar Pradesh. They used normal distribution, log normal distribution, square-root transformation and mixed distribution and compared for their ability to fit monthly rainfall values. Kolomogorov-Smirnov goodness of fit test was used. It was found that mixed probability distribution gave best fit at all probability levels.

Kumar *et al.,* (2000) analysed monthly dry days data series to compare commonly used distributions and transformations for probability analysis. The observed values were computed by Weibull's formula at different probability levels. The special values were estimated with no transformation, normal distribution with log transformation, normal distribution with log transformation, normal distribution with Smemax transformation, normal distribution with power transformation, Gumbel distribution and log-Pearson Type III distributions. The comparison was made at 0.10, 0.20, 0.50, 0.80 and 0.95 probability levels. Kolomogorov-Smirnov test and percentage deviation were used to find the goodness of fit of the distributions. The normal distribution with Smemax transformation was found to be best fit for monthly dry days data series of Pantnagar with percentage deviation of 5.66, 3.42, 2.81, 3.19 and 6.63 at 0.10, 0.20, 0.50, 0.80 and 0.95 probability levels.

5.1.1 Frequency Distribution of Water Deficit

Before undertaking frequency analysis, the adequacy of the length of record was determined from the following equation (Mockus, 1960):

$$Y = (4.3\ t \log_{10} R)^2 + 6 \qquad ...(5.1)$$

where,

Y = minimum acceptable years of record,

t = students statistical value at the 90 per cent level of significance with (Y-6) degree of freedom,

R = ratio of magnitude of 100 year event to 2-year event.

Adequacy of length of record for frequency analysis of water deficit for different agro-climatic zones of Rajasthan state were given in Appendix G. It is evident from the Appendix G that length of record of water deficit data are adequate for frequency analysis.

Frequency analysis of water deficit can be made with the help of Weibull's formula and frequency distributions. A number of frequency distributions are used for analysis of hydrological data. However, most common distributions are (i) Gumbel distribution (ii) Log normal distribution, and (ii) Log Pearson Type III distribution. These three distributions were used for the present study in the analysis of water deficit data for all the stations.

The values of water deficit at 2, 5, 10, 25, 50 and 100 years return period were worked out. The probability curves were also developed for different amount of water deficits.

5.1.1.1 Gumbel Distribution

Gumbel distribution results from any initial distribution of exponential type, which converts to an exponential function, as x increase. The probability density [P(x)] of the distribution is,

$$P(x) = \exp\left[-e^{-\left(\frac{a+x}{c}\right)}\right] \qquad ...(5.2)$$

where,

$a = \gamma c - \bar{x}$

$c = \frac{\sqrt{6}}{\pi} 6$

x = variable

γ = 0.57721 is Euler's constant

$\bar{x}$ = mean and

σ = standard deviation

5.1.1.2 Log Normal distribution

Log normal distribution is a transformed normal distribution in which the variable is replaced by logarithmic value. The probability density [P(x)] of the distribution is:

$$P_o(x) = \frac{1}{\sigma_y e^y \sqrt{2\pi}} e^{-(y-\mu_y)^2/2\sigma_y^2} \qquad ...(5.3)$$

where,

y = ln x,

x = a variable

μ_y = mean of y, and

σ_y = standard deviation of y.

This is a skewed distribution of unlimited range in both directions.

5.1.1.3 Log Pearson Type III distribution

The probability density function [P(x)] of log Pearson type III distribution is given by

$$P(x) = \frac{1}{\alpha x \Gamma(\beta)} \left(\frac{\ln x - \gamma}{\alpha}\right)^{\beta-1} e^{-(\ln x - y)/\alpha} \qquad ...(5.4)$$

where,

α = scale parameter

β = shape parameter
γ = location parameter
x = variable
n = number of observations.

5.1.1.4 Selection of frequency distribution function

To extrapolate the data series for an event and to minimize the statistical errors that may occur from observed frequencies, the theoretical probability density function gives better estimates.

It becomes necessary to test the goodness of fit of a probability distribution in order to extrapolate a data series. The chi-square test was carried out to test the goodness of fit of the theoretical probability distributions:

Chi-square test is one of the tests to check the goodness of fit of probability distribution ttlnction with the observed one. The test statistics (χ^2) was calculated from the following relationship:

$$\chi^2 = \sum_{i=1}^{N} (O_i-E_i)^2/E_i \qquad ...(5.5)$$

where,

N = number of years,
O_i = the observed values in the ith year, and
E^i = the expected values in the ith year.

5.2 Discussion

5.2.1 Frequency Analysis of Water Deficit Under Different Agro-climatic Zones of Rajasthan

The water deficit is the amount of water by which the crop fails to meet its water requirements from rainfall and stored soil moisture. The value of water deficit is required for computing the depth of irrigation for supplementary irrigations to rainfed crops during the dry spell periods. Frequency analysis of water deficit can be made by Weibull's method and probability distributions.

5.2.1.1 Frequency analysis of water deficit by Weibull's method

The frequency analysis of weekly water deficit was performed by using Weibull's method. In this method return periods were determined by Equation 2.1 for all the stations under study. Relationships between water deficit and return periods are shown in Figs. 5.1 to 5.8 for different stations.

It can be seen from Fig. 5.1 that at Jodhpur the mean weekly water deficits have been observed between 2-5 years return period from 28th to 38th week during the rainy season and during the rest of the duration the mean weekly water deficits have been observed at less than 2 years return period. Figure 5.2 indicates that at Sriganganagar the mean weekly water deficit have been observed between 2.5 years return periods from 29th to 33rd week and from 39th week during the rainy season. During the rest of the duration the mean weekly water deficits have been observed either at less than 2 years return period or at 2 years return period.

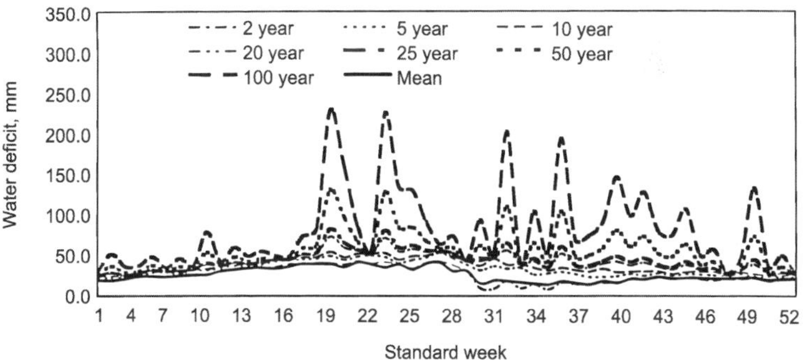

Fig. 5.1 : Frequency distribution of weekly water deficit at different return periods at Jodhpur (1981-1995)

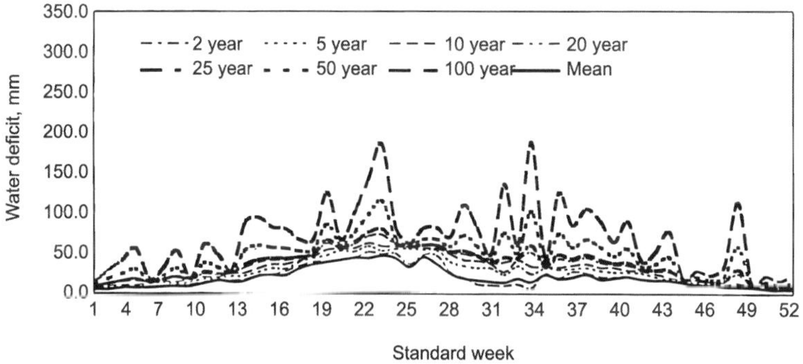

Fig. 5.2 : Frequency distribution of weekly water deficit at different return periods at Sriganganagar (1981-1995)

Figure 5.3 shows that the mean weekly water deficit have been observed between 2-5 years return periods from 28th to 33rd week and from 36th to 37th week during the rainy season of Fatehpur, while during the rest of the duration the mean weekly water deficit have been observed either at less than 2 years return period or at 2 years return period. Similarly, it can be seen from Figure 5.4 that at Pali the mean weekly water deficit have been observed between 2-5 years return period from 28th to 32nd weeks and from 34th to 37th week during the rainy season and during the rest of the duration the mean weekly water deficit have been observed either at less than 24 years return period or at 2 years return period.

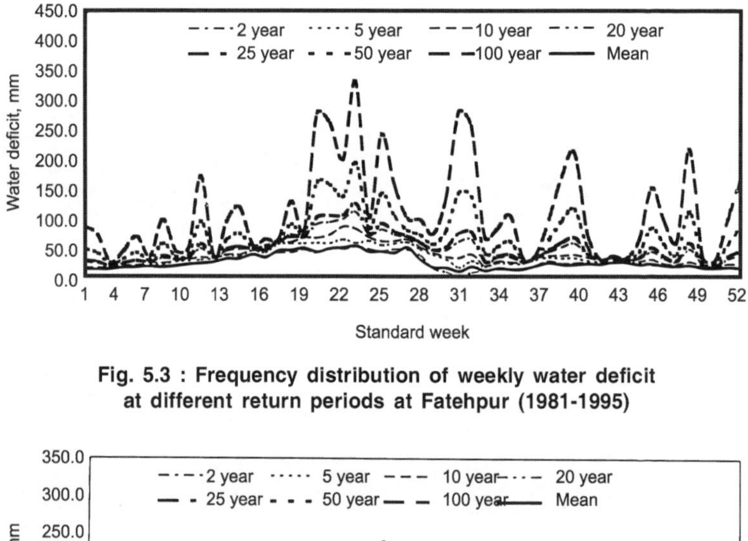

Fig. 5.3 : Frequency distribution of weekly water deficit at different return periods at Fatehpur (1981-1995)

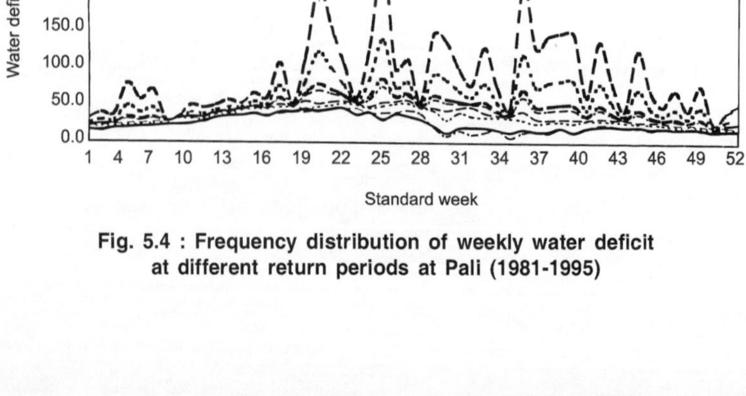

Fig. 5.4 : Frequency distribution of weekly water deficit at different return periods at Pali (1981-1995)

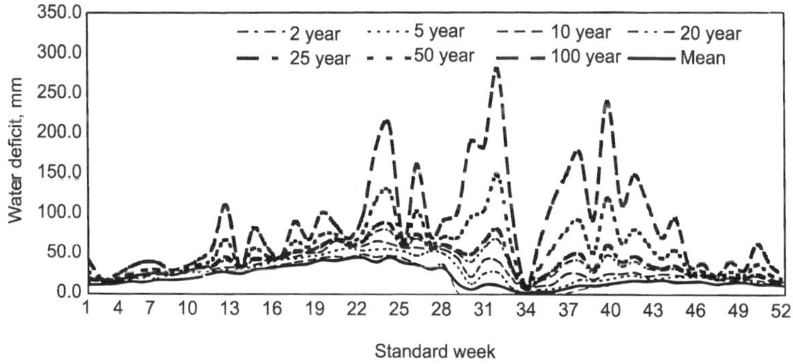

Fig. 5.5 : Frequency distribution of weekly water deficit at different return periods at Jaipur (1981-1995)

It can be seen from Fig. 5.5 that at Jaipur the mean weekly water deficit have been observed between 2-5 years return period from 28th to 38th week and in 41st week and during the rest of the duration it has been observed either at less than 2 years return period or at 2 years return period. Figure 5.6 shows that at Udaipur the mean weekly water deficit has been observed between 2-5 years return period from 27th to 40th week, while during the rest of weeks it has been observed either at less than 2 years return period or at 2 years return period. Similarly, at Kota the mean weekly water deficits have been observed between 2-5 years return period form 27th to 32th week during the rainy season, while during the rest of the duration the values have been observed either at less than 2 years return period or 2 years return period (Fig 5.8).

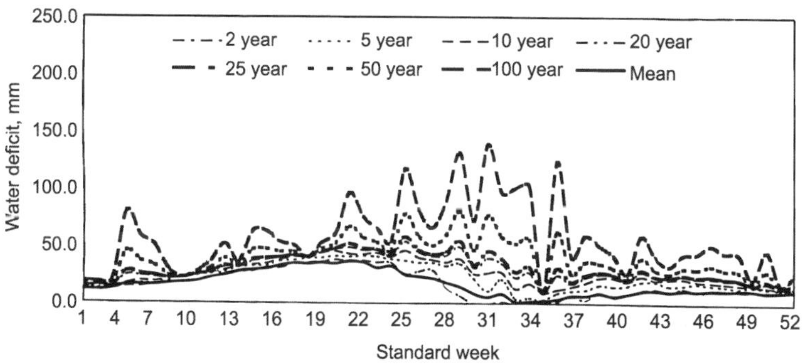

Fig. 5.6 : Frequency distribution of weekly water deficit at different return periods at Udaipur (1981-1995)

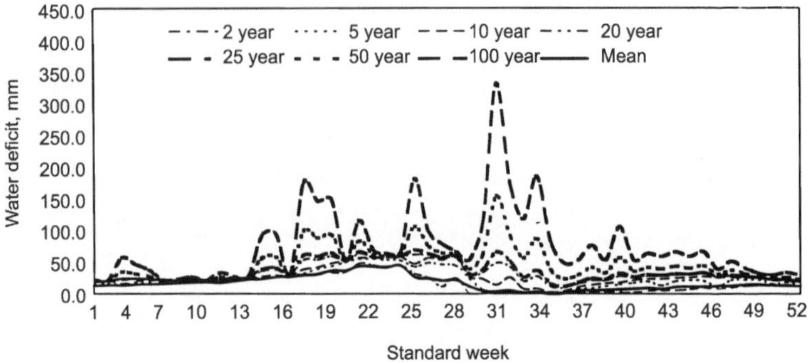

Fig. 5.7 : Frequency distribution of weekly water deficit at different return periods at Banswara (1981-1995)

It can be seen from Figs. 5.1 to 5.8 that the mean weekly water deficits have been observed between 2-5 years return period for a maximum duration of 24 weeks at Banswara. Thus, it is interesting to note that the mean weekly water deficits during the rainy season in Rajasthan have been observed between 2-5 years return period, while during the rest of the duration the mean weekly water deficits have been observed either at less than 2 years return period or at 2 years return period.

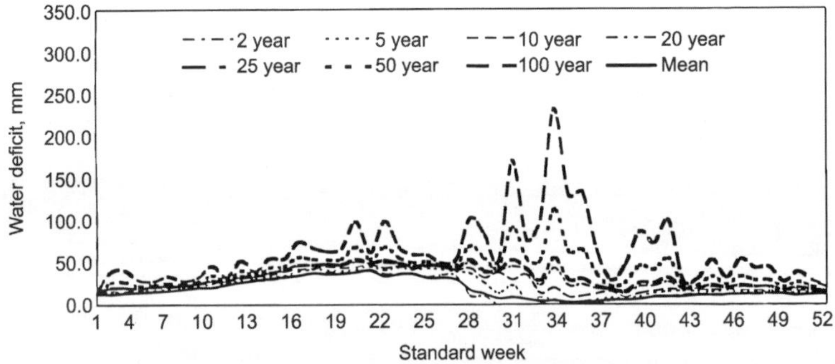

Fig. 5.8 : Frequency distribution of weekly water deficit at different return periods at Kota (1981-1995)

5.2.1.2 Frequency analysis of water deficit by different probability distributions

Based on the statistical parameters, the expected weekly water deficit amounts corresponding to different return periods were calculated with the help of Equations 5.2 to 5.4 for Gumbel's distributions, log normal and log Pearson Type-III, respectively for different stations under study. Comparison of frequency curves for various probability distribution of weekly water deficit are shown in Figures 5.9 to 5.16 for different stations.The best theoretical probability distribution was adjudged by comparing the sum of the Chi-square values for each distribution corresponding to all return periods. The observed water deficits were computed by using the Weibull's method. The Chi-square values for log Pearson type III, Log normal and Gumbel's distributions are given in Table 5.1 for different stations.

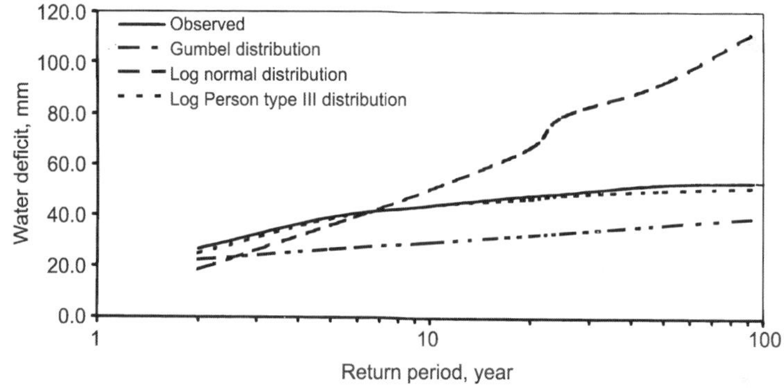

Fig. 5.9 : Comparison of frequency curves of various distributions of weekly water deficit at Jodhpur (1981-1995)

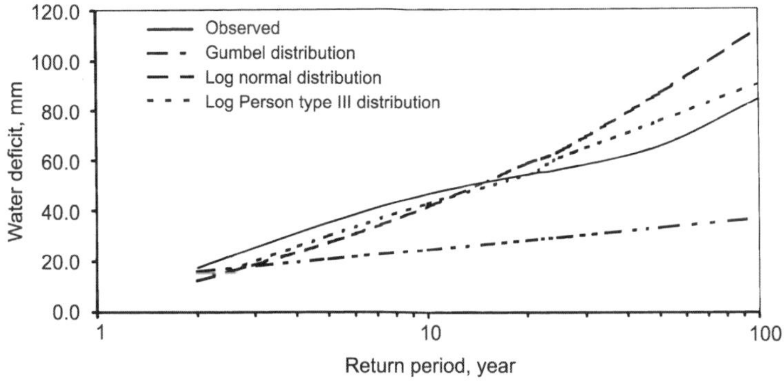

Fig. 5.10 : Comparison of frequency curves of various distributions of weekly water deficit at Sriganganagar (1981-1995)

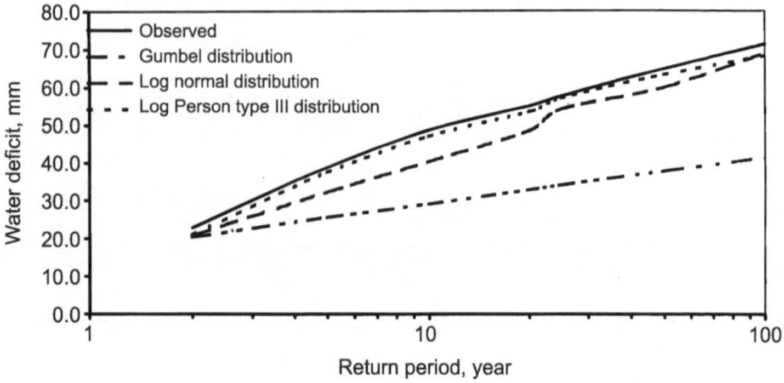

Fig. 5.11 : Comparison of frequency curves of various distributions of weekly water deficit at Fatehpur (1981-1995)

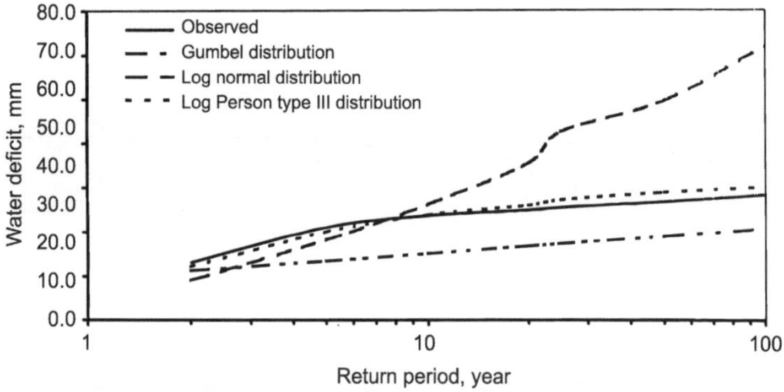

Fig. 5.12 : Comparison of frequency curves of various distributions of weekly water deficit at Pali (1981-1995)

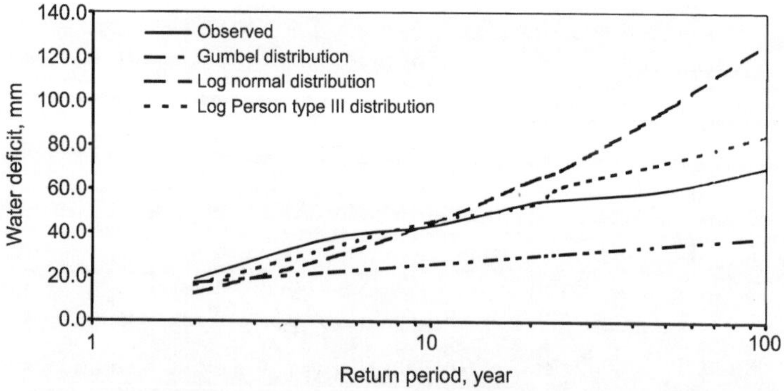

Fig. 5.13 : Comparison of frequency curves of various distributions of weekly water deficit at Jaipur (1981-1995)

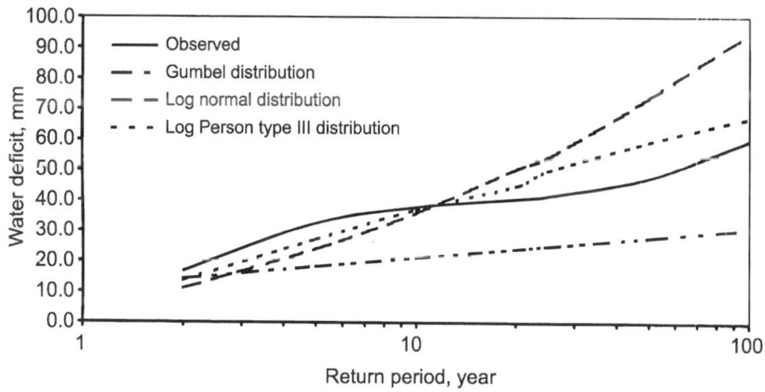

Fig. 5.14 : Comparison of frequency curves of various distributions of weekly water deficit at Udaipur (1981-1995)

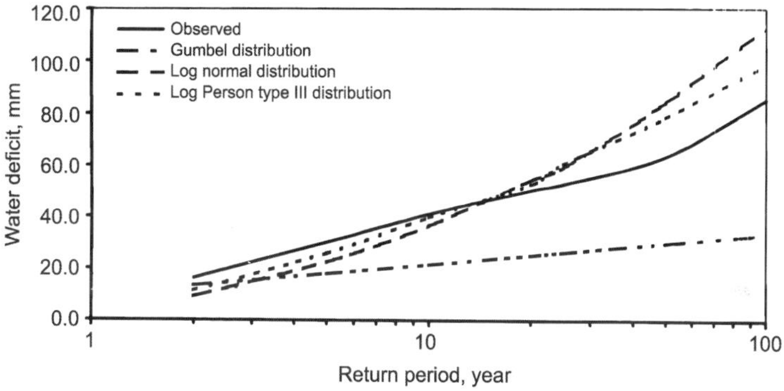

Fig. 5.15 : Comparison of frequency curves of various distributions of weekly water deficit at Banswara (1981-1995)

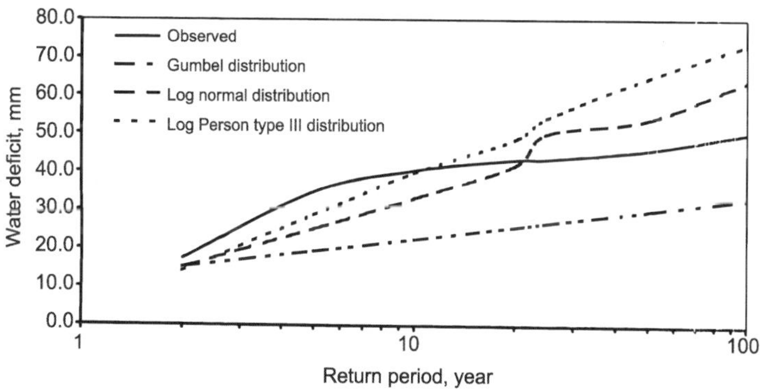

Fig. 5.16 : Comparison of frequency curves of various distributions of weekly water deficit at Kota (1981-1995)

It can be seen from Table 5.1 that the computed Chi-square values are less than the tabulated Chi-square value at 5 per cent probability level at Jodhpur, Sriganganagar, Pali, Udaipur and Banswara for Log Pearson Type III distribution, while the combined Chi-squared values are less than tabulated Chi-squared value at 1% probability level at Jaipur for Log Pearson Type III distribution. Therefore, Log Pearson Type III distribution gives the best fit for the weekly water deficit values at Jodhpur, Sriganganagar, Pali, Udaipur, Banswara and Jaipur. Figures 5.9, 5.10, 5.12, 5.14, 5.15 and 5.16 also confirm the closeness of curves for log Pearson Type III distribution with that of the observed curve for these stations respectively. The combined Chi-squares values at Fatehpur are less than the tabulated values at 5% probability by log normal and log Pearson Type III distributions. Therefore, at Fatehpur, log normal and Log Pearson Type III distributions give the best fit for weekly water deficit values. Figure 5.11 also confirms the closeness of curve for log normal and Log Pearson Type III distributions with that of the observed curve at Fatehpur.

Table 5.1 : Chi-square test of goodness of fit for different probability distributions of weekly water deficit

Station	Computed Chi-square value		
	Gumbel	Log Normal	Log Pearson Type III
Jodhpur	28.880	127.670	0.355
Sriganganagar	58.392	21.170	3.940
Fatehpur	55.780	4.210	0.466
Pali	31.410	153.820	1.044
Jaipur	58.100	87.420	13.107
Udaipur	35.110	45.080	7.350
Banswara	64.850	22.330	9.128
Kota	39.56	10.450	23.710
Chi-Square DF = 6	Tabulated Chi-square value at 5% probability level		12.592
	Tabulated Chi-square value at 1% probability level		16.812

It can also be seen from Table 5.1 that the computed Chi-square values at Kota are less than the tabulated values at 5% probability level by the log normal distribution. Therefore, log normal distribution gives the best fit for the weekly water deficit values at Kota. Figure 5.16 also confirms the closeness of curve for log normal distribution with that of the observed curve at Kota.

Thus, it is interesting to note that log Pearson Type III distribution gives the best fit for the weekly water deficit at Jodhpur, Sriganganagar, Pali, Jaipur, Udaipur and Banswara, while for Kota log normal distribution gives the best fit for the weekly values of water deficit. At Fatehpur both log normal and Log Pearsion Type III distributions give the best fit for the weekly water deficit. Gumbel distribution does not appear to fit for weekly water deficit values for any of the stations under study.

6. Stochastic Modelling of Drought

6.1 Introduction

Stochastic processes treat sequences that are governed by law of chance. Time series are considered as part of stochastic processes. The word 'stochastic' means according to its Greek origin, skillful in aiming. If an individual were shooting at a target, it is likely that the density of with near the center would be greatest and least near the edge. The location of the hits would be random with respect to centre. Thus the word 'Stochastic' has come to refer to the random nature of the variable.

The purpose of a stochastic model is to represent important statistical properties of one or more time series. Indeed different types of stochastic models are often studied in terms of statistical time series they generate. Examples of these properties include: trend, seasonality, mean, variance, skewness, serial correlation, covariance, cross-correlation and long term properties such as the rescaled range and the variance function. Since the various statistical models are described in terms of these properties, the appropriate stochastic model and numerical value of the model parameters may be inferred from statistics of the observed time series. Models formulated on stochastic concept explain the extent of dependence of the present observation on the past observations.

The first step in stochastic model construction is to select suitable classes or families of models from which the most appropriate model to a given time series can be chosen by following the identification, estimation and diagnostic check stages of model development. For example, when modelling a hydrologic time series one may wish to consider the Auto Regressive Moving Average (ARMA) families of model (Box and Jenkins, 1976), the classes of non Gaussian Models suggested by Lewis (1985) and fractional different models (Hosking, 1985). Certainly if one is not aware that certain classes of models exist, one may not fit the most appropriate model to a given time series.

Stochastic processes deal with continuous or discrete state and time parameters. Discrete series occur when the random variate in the time series is continuous but for computation and analysis purpose time is considered discrete. Discrete stochastic models may be classified in many ways. Generally, it has been subjectively decided to classify them as short memory or long memory models. A stationary series may be modeled by short memory models and non stationary series by long memory models.

Short memory models of hydrologic phenomena include moving average (MA) models autoregressive (AR) models and autoregressive integrated moving average (ARIMA) models. The use of short memory models in hydrologic analysis was introduced primarily to produce synthetic sequences of flows to route through a water resource system, the idea being to test it under variety of conditions and with longer sequences of flows that historically available. Short memory models have also extensively been used for forecasting stream flows. The long memory models are specifically designed to reproduce the Hurst phenomenon (Mandelbet, 1971).

Autoregressive (AR) model has been used extensively in hydrological analysis. It is a very useful tool in the simulation of hydrological and climatological data. As with the moving average process, this model also works with deviation, Z_t from the mean of the process or sequence of events Z'_t. However, the autoregressive process expresses the deviation from the mean of the process, a finite weighted sum of previous deviation plus a random variate, a_t. Thus

$$Z_{t-\mu} = a_t + \phi_1 (Z_{t-1} - \mu) + \phi_2 (Z_{t-2} - \mu) + \ldots + \phi_p (Z_{t-p} - \mu) \qquad \ldots(6.1)$$

is an autoregressive process of order p. It contains p+2 parameters, $\phi_1 \ldots \phi_p$ and σ_a^2 that must be estimated from a given data.

Box and Jenkins (1976) have systematically discussed the time series models. Most of the recent advances in time series analysis are based on the basic work of Box and Jenkins. A comprehensive discussion on time series modelling of hydrologic variable is presented by Salas *et al.,* (1980). Many aspects of hydrologic modelling in single variate or multivariate scenario are also discussed. Many other applications and recent advances made in the field of time series analysis applied to the field of hydrology are discussed by Hipel and McLeed (1994). Thus, stochastic models can be used to predict the frequency of occurrence of a certain hydrologic variable including water deficit. A brief review related to stochastic modelling of drought is described as under:

Parthasarathy and Rakhecha (1972) have studied the nature of fluctuations in the aridity over 12 representative stations of Rajasthan and its neighborhood during past 65 years (1901-1965) The power spectrum analysis revealed the presence of significant quasibiennial oscillations and periods somewhat similar to the sunspot cycle at Delhi and Barmer.

Rao *et al.,* (1973) have analysed drought index by power spectrum method for 23 sub-divisions of India for about 60 years. Very few of the peaks were significant at 95 per cent level through persistence in case of some divisions. No definte periodicity could be recognized in the series even the oscillatory character was noted in some of them.

Sen (1980b) developed general formulation of adaptive parameters on the basis of Kalman filters. It has been successfully applied to decompose a given monthly flow sequence into periodic and stochastic components. The results showed a definite advantage of using adaptive fourier analysis over conventional.

Thapliyal (1981) showed that the leading indicator ARIMA (Autoregressive integrated moving average) models are capable of providing some accurate forecasts than those obtained from other methods. Since the leading indicator incorporates the dynamics of the system, it is felt that it may provide better forecast of meteorological elements such as rainfall during the summer monsoon over India.

Thapliyal (1982) made an attempt to develop a stochastic dynamic model that could be used for forecasting monsoon rainfall (June to September), in the larger sub-division of India. For building such a model the atmosphere has been considered as a linear dynamic system that converts various inputs into the output, say the rainfall. In this study the 500 mm mean April subtropical ridge position along long 75E has been used as input to the atmosphere. The input and output data for 38 years (1939–1976) have been utilized for developing the model which utilizes the dynamics of the atmosphere and also that of the ARIMA process to forecast the rainfall. The performance of the model has been found good during the sample and the test (1977–1980) periods. Even in rank drought and excess rainfall years the closeness of the predicted values stands out well. In terms of seven categories being used by the India Meteorological Department for describing its long range forecasts, the skill score of the model forecast for the test period has been found equal to one which is the highest that a forecast formula can have.

Chowdhury and Abhyankar (1984a) made an attempt for drought climatology of India from 1875–1980. Only "meteorological drought" *i.e.,* rainfall deficiency exceeding 25 per cent of the normal rainfall has been considered. Frequency of occurrence of drought in different meteorological sub-divisions has been obtained which has been classified into moderate and server categories and the probability of occurrence of these types in different sub-divisions computed and discussed. Depending upon the area of the country affected, drought has been further classified into different categories, *viz.,* localised, semi vast, vast extensive and clamatious. Observed frequency of each type was determined and representation of drought incidence has been obtained and used to predict its occurrence in any 10 years period. The series is also subjected to power spectrum analysis and significant peak obtained and discussed.

Walsh (1987) estimated the spatial and temporal variations in drought conditions by evaluating the crop moisture index, drought severity index, and the hydrological deficit in combination with the remote sensing vegetation indices.

Mutua (1989) used a recurssive transfer function hydrological model for the Sagana catchment of Tana river in Kenya, based on the daily stream flow data. The averages of rainfall were achieved through the delineation of the catchment into homogenous legions by principal component analysis. The generalized partial autocorrelation technique was used to identify the stochastic characteristics of the data. The identified model appeared to forecast the low flow very well, but it shows significant deviation in the forecast of high flows, mainly on the day to day basis although the general trend was well duplicated.

Gowariker *et al.,* (1991) worked out detailed analysis of south-west monsoon (June-September) rainfall over India. The regional and global antecedent signals in numerous permutation and combinations have led to conclude that a long range forecast can be made based on one, two, three or four parameters. A detailed analysis of non-linear interactions among the antecendent climatic conditions and the monsoon has led to the development of a power regression model, which is able to quantify the effect of each parameters. Based on the model, the India Meteorological Department has been issuing the operational long range forecast of monsoon rainfall over India as a whole during the past several years and these forecasts have proved to be correct.

Srinivasan and Ramesh (1993) examined the efficiency of the periodic models in preserving the basic periodic statistics as well as the over year drought characteristics of highly variable monsoon based river flow through stochastic modelling of the monthly flows measured at Krishnarajasagar reservoir.

6.2 Formulation of Stochastic Model for Water Deficit

The mathematical procedure adopted for formulation of predictive model based on stochastic component has been discussed in the following sub-sections:

6.2.1 Stochastic Behaviour of Time Series

First step in the analysis of data is to determine its various statistical parameters. Mean, variance, coefficient of variation, skewness coefficient and kurtosis of data were estimated using the following equations:

Mean of the data ($\bar{x}$) is given by:

$$\bar{x} = \frac{1}{N} \sum_{i=1}^{N} x_i \qquad ...(6.2)$$

Variance of the data (σ^2) can be expressed as:

$$\sigma^2 = \frac{1}{N-1} \sum_{i=1}^{N} (x_i - \bar{x})^2 \qquad ...(6.3)$$

Coefficient of variation (Cv) is calculated as:

$$Cv = \frac{\sigma}{\bar{x}} 100 \qquad ...(6.4)$$

Skewness characterizes the degree of symmetry of a distribution around its mean. Positive skewness indicates a distribution with an asymmetric tail extending towards more negative values. The equation for skewness is defined as:

$$\text{Skewness} = \frac{N(N-2)}{N-1} \sum_{i=1}^{N} \left(\frac{x_i - \bar{x}}{\sigma} \right)^3 \qquad ...(6.5)$$

Kurtosis characterizes the relative peaked or flatness of a distribution compared with the normal distribution. Positive kurtosis indicates a relatively peaked distribution. Negative kurtosis indicates a relatively flat distribution. Kurtosis can be expressed as:

$$\text{Kurtosis} = \left\{ \frac{N(N+1)}{(N-1)(N-2)(N-3)} \sum_{i=1}^{N} \left(\frac{x-\bar{x}}{S} \right)^4 \right\} - \frac{3(N-1)^2}{(N-2)(N-3)} \quad \text{...(6.6)}$$

The stochastic behaviour of the average weekly water deficit series was identified by the estimation of certain standard statistical parameters. The method involves analysis of serial correlation coefficient at lag one and estimation of the coefficient of variation of the historical time series.

6.2.2 Serial Correlation Coefficient

The mathematical procedure adopted for formulation of a predictive model based on stochastic component has been discussed in the following subsections.

For a time series noted by x(t) t = 1, 2, 3, ..., N) the serial correlation coefficient (γ_k) at lag k is estimated by the following equation:

$$\gamma_k = \frac{\sum_{i=1}^{N-k} x_i x_{j+k} - \frac{1}{N-k} \left(\sum_{i=1}^{N-k} x_i\right)\left(\sum_{i=1}^{N-k} x_{j+k}\right)}{\left[\sum_{i=1}^{N-k} x_i^2 - \frac{1}{N-k}\left(\sum_{i=1}^{N-k} (x_i)^2\right)\right]^{1/2} \left[\sum_{i=1}^{N-k} x_{i+k}^2 - \frac{1}{N-k}\left(\sum_{i=1}^{N-k} (x_{j+k})^2\right)\right]^{1/2}} \quad \text{...(6.7)}$$

The numerator in Equation (6.7) is called the autocovariance, and the denominator is called the variance of the time series.

6.3 Stochastic Model

The principal aim of the analysis was to obtain a reasonable model for estimating the generation process and its parameters by decomposing the original data series into its various components. Generally a time series can be decomposed into a deterministic component, which could be formulated in a manner that allowed exact prediction of its value, and a stochastic component, which is always present in the data and can not strictly be accounted for as it is made by random effects. The time series, X(t), was represented by a decomposition model of the additive type, as follows:

$$X(t) = T(t) + P(t) + 8(t) \qquad ...(6.8)$$

where,

T(t) = trend component, t = 1, 2 N,

P(t) = periodic component, and

S(t) = stochastic component, including dependent and independent parts.

To obtain the representative stochastic model of time series, identification and detection of each component of Equation (6.8) was necessary. A systematic identification and detection of each component of X(t) was done, procedures of which are described below:

6.3.1 Trend Component

The trend component describes the long smooth movement of the variable lasting over the span of observations, ignoring the short term fluctuations. The basic idea here was to study only T(t) while eliminating the effects of other components. It leads to use the total seasonal data, Z_i, for identification of T(t) so that other components were suppressed. For detecting the trend, a hypothesis of no trend was made and following statistical tests, as suggested by Kottegoda (1980), were performed:

1. Turning point test
2. Kendall's rank correlation test

These tests are described as follows:

Turning point test: In an observed sequence Z_i, i = 1,2,3,...N, a turning point, p, occurs at time i if Z_i is either greater than Z_{i-1} and Z_{i+1} or less than the two adjacent values. The expected number of p in a random series is E(p) = 2 (N–2)/3. Variance (p) can be shown as (16N–29)/90, consequently, z can be expressed as a standard measure, z = p-E (p)/Var (p)½, which is treated approximately as a standard normal deviate. This was compared with its table value, at 5% level of significance to test the trend. If the calculated value of z is within the limits, then hypothesis of no trend is accepted.

Kendall's rank correlation test: The procedure is to determine the number of times, p, in all pairs of observation that Z_j is greater' than Z_i Then $v = 4\left\{\frac{\overline{P}}{j(j-1)}\right\} - 1$, $Var(v) = \frac{2(2j+5)}{9j(j-1)}$ and a standard measure $Z = \frac{V}{\sqrt{\{Var(V)\}}}$ were calculated. This was again, compared with its table value at 5% level of significance and hypothesis of no trend is tested. If the calculated value of z is within its table value, then, it can be concluded that the trend is not present in the series.

6.3.2 Periodic Component

The periodic component concerns an oscillating movement, which is repetitive over a fixed interval of time (Kottegoda, 1980). The existence of P(t) was identified by the correlogram, a plot of autocorrelation coefficients, r_1, versus lag 1. The oscillating shape of the correlogram verifies the presence of P(t), with the seasonal period P, at the multiples of which peak of estimation can be made by a Fourier Analysis followed by the tests for significant harmonics. The correlogram of the time series clearly shows the presence of the periodic variations indicating its detection. The time series X(t) was expressed in Fourier form as follows:

$$X(T) = A_0 + \sum_{i=1}^{N} \left[A_k \text{Cos} \left(\frac{2\pi Kt}{p} \right) + B_k \text{Sin} \left(\frac{2\pi Kt}{p} \right) \right] \quad ...(6.9)$$

where

$$A_0 = \frac{1}{N} \sum_{i=1}^{N} x(t) \quad ...(6.10)$$

$$A_k = \frac{2}{N} \sum_{i=1}^{N} x(t) \text{Cos} \left(\frac{2\pi Kt}{P} \right) \quad ...(6.11)$$

$$B_k = \frac{2}{N} \sum_{i=1}^{N} x(t) \text{ Sin} \left(\frac{2\pi Kt}{P} \right) \quad ...(6.12)$$

where,

K = number of significant harmonics,
P = base period
N = number of observation points and
A_k and B_k = Fourier coefficients.

These coefficients were obtained by a least square fit of the data to the Kth harmonic components, then a least squares approximation can be given by the finite series.

$$P(t) = A_0 + \sum_{k=1}^{M} \left[A_k \text{Cos} \left(\frac{2\pi Kt}{p} \right) + B_k \text{Sin} \left(\frac{2\pi Kt}{p} \right) \right] \quad ...(6.13)$$

where, M is the number of significant harmonics (maximum, P/2). For later use, it was more convenient to use the alternate form for P (t) given as under:

$$P(t) = A_0 \sum_{i=1}^{M} D_k \text{Cos}\left(\frac{2\pi Kt}{P} - \theta_k\right) \quad ...(6.14)$$

where,

$$D_k = \sqrt{A_k^2 + B_k^2} \quad ...(6.15)$$

and

$$\theta_k = \text{Arc tan}\left(\frac{A_k}{B_k}\right) \quad ...(6.16)$$

A computer programme was used for computation of the Fourier coefficients, periodic components, amplitude and phase angle (Appendix C3).

In Equation (6.12) if $M \to \infty$, $P(t) \to X(t)$, then X(t) can be represented satisfactorily by Equation (6.11) only. However it may not be practical or desirable to allow the condition, $M \to \infty$. Thus the appropriate approach would be the selection of a value of M which contains only those harmonics which are significantly contributing towards X(t). With this as the objective, following tests were conducted to select an appropriate value of M.

1. Test of analysis of variance
2. Fourier decomposition of mean square.
3. Cumulative periodogram test

Test for analysis of variance: If the points r = 1, 2, 3, ..., P indicates the time span of periodicity, then periodic function of the periodic mean, m(r) of time series can be written by replacing P(t) and A_o by m(r) and m_o, respectively in Equation (6.9). The Fourier coefficient, α_K and β_K, (K = 1, 2, ..., P/2) can be estimated as follows:

$$m_o = \sum_{r+1}^{P} m(r)/P \quad ...(6.17)$$

$$\alpha_K = \frac{2}{p} \sum_{r+1}^{P} m(r)\,\text{Sin}\frac{2\pi Kr}{P}, \quad K = 1, 2,...,P/2 \quad ...(6.18)$$

$$\beta_K = \frac{2}{p} \sum_{r+1}^{P} m(r)\text{Cos}\frac{2\pi Kr}{P}, \quad K = 1, 2,...P/2\text{–}1 \quad ...(6.19)$$

and

$$\beta_{p/2} = \frac{1}{p}\sum_{r+1}^{P} m(r)(-1)^r \qquad ...(6.20)$$

Amplitude, $R_K = \sqrt{\alpha_k^2 + \beta_k^2}$ ...(6.21)

After determination of the Fourier coefficients, variance test was conducted for selecting the number of significant harmonics. In this test, the null hypothesis was that the variance explained by a harmonic K, which was $(N/2)(\alpha_K^2 + \beta_K^2)$ is zero. Steps were taken to test the (α_K, β_K) values for K = 8, 7, ..., 1, in order, to obtain the F-ratios, if this value of F-ratio is less than its table value at 1 per cent level of significance, the corresponding harmonic was selected otherwise test is repeated for higher harmonics until the obtained value of F-ratio was less than its table value.

Fourier decomposition of mean square : The contribution of the individual harmonics, towards the mean square, was calculated and the number of harmonics which were dominantly contributing to mean square were selected as the significant harmonics. For the zeroth and Kth harmonics, contribution is R_K^2 and for other harmonics, the average power is $2R_K^2$. The plot of the average power versus corresponding frequency indicates the number of significant harmonics.

Cumulative periodogram test: A graphical method is employed for selecting the significant harmonics in Fourier series. In this method mean weekly water deficit series was used to determine the mean squared deviation (MSD(u)) by the following equation:

$$MSD(u) = \frac{1}{p}\sum_{r=1}^{P} (u_r - \bar{u})^2 \qquad ...(6.22)$$

where,

MSD(u) = mean squared deviation

u_r = periodic estimate

$\bar{u}$ = mean of periodic estimate

$= \frac{1}{p}\sum_{r=1}^{P} u_r$

Mean square deviation MSD(j) of each harmonic j was calculated by the following expression:

$$MSD_j = \frac{1}{2}(A_j^2 + B_j^2) \qquad ...(6.23)$$

$$j = 1\ 2\ ....\ P$$

The cumulative periodogram (I) was determined by the following formula:

$$P_i = \frac{\sum_{i=1}^{P} MSD(j)}{MSD(u)} \qquad ...(6.24)$$

$$i = 1\ 2\ ...,\ p$$

where, MSD(j) and MSD(u) are as described as above.

The graph was drawn between P_i and number of harmonics for selecting the significant harmonics. The significant harmonics were selected up to the fast increase in P_i and rest of harmonics were ignored.

6.3.3 Stochastic Component

The stochastic component is constituted by various random effects, which cannot be estimated exactly. In the case of water deficit time series various climatic and soil parameters response to value of component without changing the cyclicity itself and thus add randomness to the time series of water deficit for all the stations. A stochastic model of the form of autoregressive model (AR), was used for the presentation of the time series. In this model, the current value of the process is expressed as a finite, linear aggregate of values of the process and a variate that is completely random. This model was applied to the S(t) which was treated as a random variable *i.e.* deterministic components were removed and the residual was stationary in nature. Mathematically, an autoregressive model of order p, AR(p) can be written as:

$$S(t) = \sum_{K=1}^{P} \phi_{p.K}\, S_{(t-k)} + a(t)$$

$$= \sum_{K=1}^{P} \phi_{p.1}\, S_{(t-1)} + \phi_{p.2}\, S_{(t-2)} + ... + \phi_{p.p}\, S_{(t-p)} + a(t) \qquad ...(6.25)$$

where,

$\phi_{p.K}$ = autoregressive model parameters, K = 1, 2, ..., p.

a(t) = independent random number

The fitting procedures of the AR(p) model to the water deficit series involved selection of order (p) of the model.

Selection of order of the AR(p) model: Residual variance method was used for selection of the model order. In this method, residual variance, S_z^2 (p) was calculated using following equation for different orders:

$$S_Z^2(p) = \frac{1}{N-2p-1} S(\mu, \alpha_1, \alpha_2, ..., \alpha_p) \quad ...(6.26)$$

where, $S(\mu, \alpha_1, \alpha_2, ..., \alpha_p)$ is known as the residual sum of squares and was calculated as follows:

$$S(\mu, \alpha_1, \alpha_2, ..., \alpha_p) = (N-p)(C_0 - \alpha_1 C_1 - \alpha_2 C_2, ..., \alpha_p C_p) \quad ...(6.27)$$

where,

N = number of observation points

$\alpha_1, \alpha_2 ... \alpha_p$ = parameters of the corresponding model

$C_0, C_1, ..., C$ = autocovariance function at lag p, p = 0, 1, 2, .a., p.

The minimum value of $S^2_z(p)$ suggested an approximate order of the autoregressive model to be used further.

Based on the above procedures, and order of the AR model was selected and the model was tried with time series data.

Estimation of autoregressive parameters: The parameter estimation deals with the estimation of the autoregressive parameters of Equation (6.25). These parameters can be expressed in terms of serial correlation coefficient, as Yule-Walker equations. The general recursive formulae for estimating these parameters ($\phi_{p.k}$), where suffix p and k indicate the order and the number of parameter of the order in AR(p) model, respectively may be written as:

$$\phi_{p.p} = \left[\frac{r_p - \sum_{k=1}^{p-1} (\phi_{p-1.k})(r_{p-k})}{1 - \sum_{k=1}^{p-1} (\phi_{p-1.k})(r_{p-k})} \right] \quad ...(6.28)$$

and

$$\phi_{p.k} = \phi_{p-1.k} - \phi_{p.p} \cdot \phi_{p-1.p-k}; \; k = 1, 2, 3, ...p-1 \quad ...(6.29)$$

In Equation (6.7) r_k is the autocorrelation coefficient, Auto correlation regression of the series for lag *K* and was computed, for any series Y(t) at any lag, 1, as follows:

$$r_1 = \sum_{t=1}^{N-1} [Y(t)-Y(\bar{t})][Y(t+1)-Y(\bar{t})] / \sum_{t=1}^{N} [Y(t)-Y(\bar{t})]^2 = C_1/C_0 \quad ...(6.30)$$

where,

Y(t) = mean of the series, Y(t)

N = total number of discrete values of X(t)

C_1 = autocovariance function at lag, 1, 1= 0, 1, ...p

After estimating the AR parameters $\phi_{p.k}$, S(t) was calculated by using Equation (6.25) a computer programme was used for computation of model parameters and stochastic components.

The sum of the periodic and stochastic component forms the generated value of the observed data. The difference, was termed as residuals which were tested to check the adequacy of the formulated model.

6.3.4 Diagnostic Checking of the Model

Diagnostic checking concerns the verification of the adequacy of the fitted model. Following tests were conducted to check the adequacy of the model.

Serial correlation analysis: The serial correlation coefficient of the a(t) were estimated for lags, *l* and a correlogram was obtained. The confidence limits at 95 per cent tolerance limit were drawn on the plotted correlogram to test independent normal distribution. If the correlogram is well within the tolerance limits, then it can be derived from correlogram that residuals were normally distributed with zero mean and standard deviation one. The confidence limits, t_1, for lag *l*, were estimated by the following equation, suggested by Kottegoda (1980):

$$t_1 = \frac{-t \pm 1.645\sqrt{(N-1-1)}}{N-1} \qquad ...(6.31)$$

Sum of square of analysis: This is the most commonly used method for appraisal of the formulated models and is described below as suggested by Clarke (1973). It involves computation of statistical measure, R^2, in the following way:

The sum of squares of residuals is given as:

$$A = \Sigma\,(Y_t - \hat{Y}_t)^2 \qquad ...(6.32)$$

and the sum of squares of deviations of observed values from their mean is

$$B = \Sigma\,(Y_t - \bar{Y}_t)^2 \qquad ...(6.33)$$

Then R^2 is the measure required and is given as

$$R^2 = (B - A)/B \qquad ...(6.34)$$

If all the residuals are zero i.e., $R^2 = 1$ (or 100, if R^2 is expressed as a percentage). It indicates perfect fitting of the formulated model. The goodness is better if the value of R^2 is closer to unity.

6.3.5 Validation of Stochastic Model

Validation of the developed model was made by comparison of the generated series with the observed historical series of water deficit. The variation of the generated and observed series was presented graphically with respect to time. Correlation coefficient of mean generated series and mean observed series was determined. Other statistical parameters of generated and observed series of water deficit were also computed for validation of the model.

6.4 Discussion

6.4.1 Stochastic Modelling of Water Deficit Under Different Agro-climatic Zones of Rajasthan

The mathematical procedure of analysis as described in Section 2 has been used in investigating the structure of time series of weekly water deficit. The weekly water deficit values for thirteen years (1981-93) were calculated by Thornthwaite and Mather (1955, 1957) procedures as described in chapter 3. The weekly water deficit series was tested for its stochastic behaviour. The trend of series was identified for its component. The periodic component in the series was modelled by Fourier series and the residuals from the Fourier series were modelled by using an autoregressive model. The results are presented as under:

6.4.1.1 Behaviour of water deficit time series

For testing statistical characteristics of the weekly water deficit series, thirteen years data (1981-93) were taken and mean weekly water deficit values were generated for different stations. The nature of curves for all the stations confirms the presence of dependent cyclic component and independent part in the series of weekly water deficit.

The statistical characteristics of the mean weekly water deficit time series were estimated and are presented in Tables 6.1 to 6.8. Table 6.1 indicates values of mean weekly water deficit, which vary from 13.815 mm in 34th week to 41.592 mm in the 20th week. No larger variability among the weekly values of water deficit of different years was observed. This is further confirmed by the estimated standard deviations. The standard deviation of weekly water deficit values ranged from 2.73 mm in 7th week to 19.12 mm in 22nd week during the entire year. The variation may be attributed towards the nature changes in the yearly climate.

At Sriganganagar, Fatehpur, Pali, Jaipur, Udaipur, Banswara and Kota the mean weekly water deficit values vary from 5.17 mm in 2nd week to 47.04 mm in the 22nd week, 8.469 mm in the 30th week to 51.354 mm in the 22nd week, 13.946 mm in the 29th week to 46.20 mm in the 21st the week, 1.492 mm in the 33rd week to 45.623 mm in the 23rd week, 1.354 mm in the 32nd week to 36.215 mm in the 20th week, 0.792 mm in the 34th week to 44.923 mm in the 23rd week and 2.662 mm in the 34th week to 44.923 mm in the 23rd week, and 2.662 mm in the 34th week to 40.377 mm in the 20th week respectively. No larger variability among the weekly values of weekly water deficit during different years was observed at these stations. The standard deviation values of weekly water deficit ranged form 1.527 mm in 40th week to 22.227 mm in 26th week at Sriganganagar, 2.016 mm in 50th week to 23.141 mm in 21st week at Fatehpur, 2.938 mm in 12th week to 22.744 mm in 26th week at Pali, 1.697 mm in 51st week to 21.792 mm in 26th week at Jaipur, 1.697 mm in 51st week to 21.792 mm in 26th week, 1.3121 mm in 32nd week to 16.062 mm in 26th week at Udaipur, 2.253 mm in 34th week to 22.530 mm in 22nd week at Banswara and 1.541 mm in 1st week to 19.028 mm in 27th week at Kota during the entire year. The variation may be attributed towards the nature changes in the yearly climate.

6.4.1.1.1 Coefficient of variation

The estimated values of coefficient (CV) of variation are given in Table 6.1 to 6.8 for different stations, Tables 6.1 to 6.8 indicate that the coefficient of variation is greater than zero at all the stations which shows the importance of time variability of weekly water deficit series. Since the values of coefficient of variation are significantly different than zero at all the stations, it confirmed that water deficit is mutually dependent.

Table 6.1 : Characteristics of weekly water deficit at Jodphur (1981-1993)

Week	Mean (mm)	SD (mm)	CV	Skewness	Kurtosis
1	19.162	2.979	0.155	0.158	–1.121
2	18.792	5.305	0.282	–1.751	4.801
3	20.669	3.111	0.151	–0.927	0.078
4	23.631	3.081	0.130	–1.227	1.426
5	23.846	4.953	0.208	-1.517	2.931

contd.....

Table 6.1 – contd...

Week	Mean (mm)	SD (mm)	CV	Skewness	Kurtosis
6	23.438	6.257	0.267	–2.116	5.741
7	24.654	2.733	0.111	–0.274	–0.431
8	25.708	8.033	0.312	–2.526	7.430
9	26.377	7.384	0.280	–1.376	2.458
10	30.715	3.717	0.121	–0.097	–1.239
11	32.446	4.480	0.138	–0.660	0.794
12	34.285	3.174	0.093	–0.756	0.011
13	34.800	4.274	0.123	–0.865	–0.169
14	33.623	11.512	0.342	–1.622	1.793
15	36.231	11.180	0.309	–2.449	6.637
16	39.546	5.841	0.148	0.020	–1.076
17	39.215	12.502	0.319	–2.920	9.541
18	39.477	9.058	0.229	–1.191	1.672
19	35.646	14.392	0.404	–1.349	2.094
20	41.592	9.951	0.239	–1.516	3.533
21	39.177	10.903	0.278	–0.928	0.769
22	35.323	19.123	0.541	–0.860	0.214
23	38.846	5.482	0.141	–0.127	2.998
24	32.785	14.281	0.436	–0.765	0.487
25	39.915	11.150	0.279	–1.984	4.747
26	41.131	15.816	0.385	–1.899	3.437
27	30.738	17.155	0.558	–0.687	–0.912
28	31.131	10.507	0.338	–1.192	1.820
29	15.346	15.066	0.982	0.456	–1.363
30	18.985	17.650	0.930	0.409	–1,397
31	18.285	15.698	0.859	0.437	–0.562
32	16.669	15.726	0.943	0.326	–1.633
33	15.269	13.730	0.899	0.318	–1.367
34	13.815	12.574	0.910	0.236	–1.584
35	17.854	13.353	0.748	0.449	–0.191
36	15.846	11.716	0.739	0.028	–1.231
37	14.054	10.728	0.763	0.294	–1.019
38	18.092	9.958	0.550	0.050	0.549
39	15.854	11.033	0.696	0.040	–0.415
40	20.938	7.689	0.367	–0.283	0.032
41	20.277	8.926	0.440	–0.654	1.647
42	23.192	6.024	0.260	–0.397	–0.099
43	21.338	5.608	0.263	–0.921	1.214

contd.....

Table 6.1 – contd...

Week	Mean (mm)	SD (mm)	CV	Skewness	Kurtosis
44	22.154	8.422	0.380	–1.486	3.487
45	22.115	5.656	0.256	–0.091	–0.910
46	21.231	4.979	0.235	0.880	–0.037
47	21.092	3.663	0.174	–0.623	–0.581
48	22.146	4.591	0.207	0.167	–0.993
49	22.077	4.539	0.206	1.239	3.856
50	19.292	4.691	0.243	–0.526	0.092
51	20.515	3.407	0.166	0.210	–0.697
52	20.838	4.042	0.194	–0.093	–0.846

Table 6.2 : Characteristics of weekly water deficit at Sriganganagar (1981-1993)

Week	Mean (mm)	SD (mm)	CV	Skewness	Kurtosis
1	5.238	2.521	0.481	–1.089	0.553
2	5.169	2.847	0.551	–0.037	–0.308
3	7.100	2.496	0.352	0.429	0.597
4	7.223	3.585	0.496	–1.110	1.710
5	7.523	4.111	0.547	–0.963	–0.089
6	8.154	5.652	0.693	–0.579	–1.342
7	9.038	4.838	0.535	–0.760	0.502
8	8.915	5.291	0.593	–0.605	–0.676
9	12.400	5.151	0.415	–1.172	1.495
10	16.031	3.495	0.218	–1.027	1.101
11	14.492	8.245	0.569	–1.030	–0.301
12	15.485	5.032	0.325	0.003	–1.070
13	21.569	5.535	0.257	1.346	2.224
14	23.015	8.980	0.390	–1.036	3.622
15	22.277	12.148	0.545	–1.030	0.033
16	31.400	10.880	0.346	–2.182	5.921
17	35.985	9.779	0.272	–1.600	3.046
18	39.162	13.852	0.354	–1.869	5.589
19	41.769	13.574	0.325	–1.370	1.684
20	44.308	10.939	0.247	–0.557	–1.020
21	44.108	19.128	0.434	–1.175	1.036
22	47.038	9.642	0.205	0.111	0.996
23	43.623	16.736	0.384	–1.361	0.470
24	33.300	19.270	0.579	–0.446	–0.900
25	45.008	12.591	0.280	–1.533	1.941
26	36.715	22.227	0.605	–1.061	–0.552

contd.....

Table 6.2 – contd...

Week	Mean (mm)	SD (mm)	CV	Skewness	Kurtosis
27	24.646	17.294	0.702	–0.231	–0.996
28	18.738	15.018	0.801	0.129	–0.828
29	16.515	15.326	0.928	0.553	–1.019
30	15.015	14.811	0.986	0.339	–1.477
31	13.915	10.786	0.775	0.175	–0.866
32	18.300	15.742	0.860	0.263	–1.574
33	14.392	13.100	0.910	0.958	0.464
34	22.777	10.559	0.464	–1.210	0.679
35	18.369	13.836	0.753	–0.331	–1.490
36	19.831	14.468	0.730	–0.503	–1.626
37	23.462	12.993	0.554	–0.882	–0.211
38	17.415	12.910	0.741	–0.233	–1.613
39	18.931	10.810	0.571	–0.641	–0.588
40	20.538	7.530	0.367	–0.844	0.621
41	17.869	8.086	0.453	–0.567	–0.431
42	16.438	7.034	0.428	–1.022	1.082
43	16.323	3.505	0.215	–0.079	1.703
44	13.562	2.899	0.214	–0.582	0.069
45	11.277	4.251	0.377	–1.624	3.451
46	10.646	2.105	0.198	–0.244	0.066
47	8.938	2.460	0.275	–0.363	0.713
48	8.415	4.301	0.511	1.140	5.132
49	7.931	1.527	0.192	–0.292	–0.682
50	6.292	2.488	0.395	–0.413	–0.132
51	5.538	2.173	0.392	–1.054	–0.222
52	5.923	2.406	0.406	–1.200	2.053

Table 6.3 : Characteristics of weekly water deficit at Fatehpur (1981–1993)

Week	Mean (mm)	SD (mm)	CV	Skewness	Kurtosis
1	12.977	3.406	0.262	–0.188	4.127
2	12.269	4.218	0.344	–1.927	7.084
3	12.954	4.350	0.336	–2.453	7.111
4	15.362	4.404	0.287	–1.753	3.234
5	15.992	4.167	0.261	–1.587	3.768
6	17.323	6.104	0.352	–2.211	5.455
7	15.869	9.039	0.570	–1.056	–0.141
8	17.969	8.337	0.464	–1.207	0.439
9	20.377	7.579	0.372	–1.270	2.189
10	22.615	9.948	0.440	–0.633	1.872

contd.....

Table 6.3 – contd...

Week	Mean (mm)	SD (mm)	CV	Skewness	Kurtosis
11	24.377	5.654	0.232	–0.380	–1.243
12	29.785	5.593	0.188	0.494	–0.630
13	30.131	9.110	0.302	–2.412	7.875
14	36.754	6.147	0.167	–0.187	–0.923
15	31.962	11.271	0.353	–2.031	5.383
16	43.377	9.046	0.209	0.579	–1.071
17	44.408	11.136	0.251	–0.207	–0.361
18	47.800	9.448	0.198	0.994	0.380
19	41.400	19.127	0.462	–0.178	1.479
20	46.900	19.005	0.405	–0.730	3.252
21	48.554	23.141	0.477	–0.058	1.199
22	51.354	18.246	0.355	1.321	2.464
23	41.631	21.303	0.512	–0.895	–0.228
24	41.985	14.757	0.351	–0.147	1.166
25	39.969	17.639	0.441	–0.974	1.001
26	48.554	10.111	0.208	–0.182	–0.429
27	28.477	20.106	0.706	–0.284	–1.776
28	15.146	15.787	1.042	0.596	–1.107
29	14.546	15.081	1.037	0.333	–1.727
30	8.469	14.471	1.709	2.007	4.023
31	16.092	14.935	0.928	1.230	1.360
32	9.538	7.576	0.794	–0.030	–1.800
33	13.362	10.045	0.752	–0.037	–1.473
34	11.623	10.411	0.896	0.075	–1.485
35	12.123	9.382	0.774	–0.159	–1.724
36	17.331	11.911	0.687	–0.379	–1.690
37	22.515	12.058	0.536	–0.820	–0.503
38	19.038	12.286	0.645	0.434	–0.394
39	19.508	13.802	0.708	0.248	0.102
40	20.992	8.510	0.405	–0.577	–0.281
41	19.777	5.923	0.300	–0.460	–0.145
42	23.554	5.079	0.216	–0.443	–0.195
43	23.292	3.422	0.147	–0.691	–0.471
44	19.208	6.710	0.349	–2.078	5.787
45	20.377	5.483	0.269	2.201	5.707
46	17.346	3.802	0.219	0.315	1.982
47	16.054	3.609	0.225	–0.215	–0.791
48	17.446	6.196	0.355	2.868	9.345
49	13.662	3.188	0.233	–0.541	–0.644
50	13.169	2.016	0.153	0.065	–1.378
51	14.662	3.786	0.258	2.082	5.098
52	12.962	5.487	0.423	1.951	5.399

Table 6.4 : Characteristics of weekly water deficit at Pali (1981-1993)

Week	Mean (mm)	SD (mm)	CV	Skewness	Kurtosis
1	16.738	5.747	0.343	–2.270	6.255
2	16.562	5.041	0.304	–0.790	–0.156
3	19.338	3.135	0.162	–0.315	–0.883
4	20.354	5.181	0.255	–1.138	2.621
5	21.415	4.723	0.221	–1.588	3.191
6	22.208	4.282	0.193	–0.474	0.991
7	23.638	4.059	0.172	–0.972	–0.089
8	24.569	6.940	0.282	–1.291	2.244
9	25.362	6.888	0.272	–1.917	4.926
10	29.023	3.199	0.110	–0.092	–1.364
11	33.685	3.646	0.108	0.212	–1.079
12	36.262	2.938	0.081	–0.499	–0.214
13	37.100	4.749	0.128	0.023	–1.135
14	39.138	6.182	0.158	–1.421	2.971
15	35.938	12.266	0.341	–2.369	6.451
16	41.569	7.009	0.169	–0.048	0.549
17	41.731	6.440	0.154	–1.354	2.479
18	44.654	5.910	0.132	1.566	2.183
19	42.362	12.645	0.299	–2.189	7.623
20	45.385	7.174	0.158	0.372	0.676
21	46.200	8.395	0.182	–1.445	3.840
22	38.038	17.766	0.467	–1.771	2.027
23	41.154	5.603	0.136	0.799	0.777
24	31.608	18.787	0.594	0.132	–0.674
25	39.185	13.803	0.352	–0.441	–1.154
26	36.400	22.744	0.625	–0.873	–1.065
27	31.585	18.430	0.584	–0.831	–0.950
28	23.354	20.919	0.896	0.017	–1.930
29	13.946	17.229	1.235	1.201	0.108
30	20.500	18.059	0.881	0.335	–1.180
31	20.977	17.605	0.839	0.216	–1.164
32	20.615	15.222	0.738	0.045	–1.265
33	17.931	15.604	0.870	–0.075	–1.750
34	14.608	14.504	0.993	0.612	–1.358
35	18.546	16.150	0.871	0.798	0.218
36	16.246	12.961	0.798	0.069	–1.138
37	16.008	12.732	0.795	0.124	–1.121
38	20.538	10.155	0.494	–0.140	0.154
39	15.269	13.322	0.872	0.196	–1.143
40	20.608	10.048	0.488	–0.633	–0.368
41	22.531	10.435	0.463	–0.327	–0.871
42	23.200	8.301	0.358	–0.585	–0.677

contd.....

Table 6.4 – contd...

Week	Mean (mm)	SD (mm)	CV	Skewness	Kurtosis
43	21.846	7.500	0.343	–0.244	–1.441
44	22.323	6.648	0.298	–0.054	0.026
45	20.138	6.725	0.334	–0.009	–0.868
46	20.546	5.377	0.262	0.178	–0.905
47	19.892	4.777	0.240	–0.113	–0.486
48	19.985	4.209	0.211	0.486	–0.197
49	19.846	4.118	0.207	0.302	–0.091
50	16.846	5.063	0.301	–1.599	2.437
51	17.700	4.425	0.250	–1.378	2.518
52	18.100	3.629	0.200	–0.284	–1.057

Table 6.5 : Characteristics of weekly water deficit at Jaipur (1981–1993)

Week	Mean (mm)	SD (mm)	CV	Skewness	Kurtosis
1	11.408	4.273	0.375	–1.515	3.791
2	11.654	4.379	0.376	–1.693	3.479
3	12.715	2.517	0.198	–0.741	–0.480
4	15.131	3.062	0.202	–0.875	0.610
5	14.546	5.540	0.381	–1.191	1.172
6	17.415	4.191	0.241	–0.198	–0.112
7	17.469	5.743	0.329	–2.276	6.626
8	18.692	7.041	0.377	–1.453	2.840
9	20.808	9.143	0.439	–0.986	0.714
10	25.762	4.411	0.171	–0.744	–0.004
11	27.300	4.725	0.173	0.815	1.463
12	25.292	8.914	0.352	–1.767	3.690
13	29.931	6.815	0.228	–2.062	6.334
14	31.623	7.625	0.241	–0.961	–0.173
15	34.000	11.754	0.346	–2.214	6.037
16	35.877	12.551	0.350	–2.137	5.604
17	37.323	13.301	0.356	–1.698	3.485
18	41.931	7.794	0.186	–0.215	–0.015
19	41.154	16.294	0.396	–0.441	–0.844
20	44.500	14.747	0.331	–1.862	4.320
21	45.415	15.337	0.338	–2.366	6.986
22	39.662	16.101	0.406	–1.439	1.894
23	45.623	9.687	0.212	–0.337	–0.939
24	38.985	14.251	0.366	–0.618	–0.357
25	36.554	19.206	0.525	–0.901	0.515
26	30.769	21.792	0.708	–0.210	–1.524

contd.....

Table 6.5 – contd...

Week	Mean (mm)	SD (mm)	CV	Skewness	Kurtosis
27	32.292	18.399	0.570	–0.411	–0.775
28	13.515	18.402	1.362	1.137	–0.397
29	6.038	10.099	1.672	1.883	3.330
30	11.738	17.695	1.507	1.496	1.110
31	11.054	18.561	1.679	1.672	1.934
32	4.931	8.078	1.638	1.810	2.631
33	1.492	2.229	1.494	1.483	0.949
34	3.254	4.404	1.353	1.337	1.180
35	3.377	5.729	1.697	2.076	4.125
36	7.331	10.293	1.404	1.718	2.622
37	10.823	10.298	0.952	1.098	1.192
38	10.669	6.230	0.584	0.554	–0.283
39	12.462	10.345	0.830	1.441	3.038
40	14.938	7.279	0.487	0.944	0.281
41	16.715	8.741	0.523	0.172	1.273
42	16.523	5.991	0.363	–0.456	2.873
43	18.100	5.558	0.307	–0.406	1.283
44	15.469	8.684	0.561	–0.651	–0.056
45	15.731	6.248	0.397	–0.925	1.145
46	16.077	4.907	0.305	0.300	–1.336
47	12.777	5.014	0.392	–0.654	0.063
48	14.685	3.265	0.222	–0.241	–0.989
49	15.062	3.102	0.206	–0.057	–1.615
50	12.746	4.776	0.375	–1.290	4.366
51	12.892	1.697	0.132	0.318	0.279
52	11.223	4.336	0.386	–1.784	3.042

Table 6.6 : Characteristics of weekly water deficit at Udaipur (1981–1993)

Week	Mean (mm)	SD (mm)	CV	Skewness	Kurtosis
1	11.354	1.475	0.130	–0.148	–1.512
2	11.000	3.689	0.335	–2.376	7.329
3	12.162	1.312	0.108	0.090	–0.135
4	15.746	7.926	0.503	3.196	11.080
5	15.754	8.414	0.534	2.709	8.762
6	15.554	2.202	0.142	1.195	1.719
7	17.108	2.039	0.119	0.603	–0.678
8	18.015	2.647	0.147	–0.175	–0.256
9	19.031	4.393	0.231	–1.464	3.718
10	22.300	2.458	0.110	–0.021	–0.892

contd.....

Table 6.6 – contd...

Week	Mean (mm)	SD (mm)	CV	Skewness	Kurtosis
11	24.500	2.746	0.112	0.814	–0.130
12	27.400	2.867	0.105	–0.186	–1.545
13	28.777	3.430	0.119	0.245	–1.007
14	29.831	5.744	0.193	–2.655	8.582
15	32.500	4.011	0.123	–0.194	–1.179
16	34.385	5.322	0.155	–1.254	1.973
17	34.008	6.107	0.180	–0.954	–0.220
18	34.954	5.835	0.167	–0.402	0.090
19	34.146	11.394	0.334	–2.528	7.425
20	36.215	6.357	0.176	–1.486	4.072
21	35.046	6.037	0.172	–0.629	–0.195
22	30.531	11.859	0.388	–1.797	3.104
23	32.123	10.135	0.315	–3.031	10.095
24	24.815	12.394	0.499	0.039	1.234
25	23.138	15.773	0.682	–0.669	–1.295
26	21.546	16.062	0.745	–0.561	–1.627
27	17.823	14.875	0.835	–0.025	–1.687
28	14.085	16.612	1.179	0.772	–1.072
29	8.223	11.132	1.354	0.948	–0.809
30	5.438	10.805	1.987	2.189	3.957
31	8.177	11.457	1.401	1.213	–0.035
32	1.354	3.130	2.312	2.664	7.257
33	1.823	4.442	2.436	2.858	8.303
34	2.146	3.095	1.442	1.532	1.677
35	3.685	3.889	1.056	0.343	–1.852
36	6.623	5.565	0.840	0.128	–1.284
37	6.608	5.974	0.904	0.787	–0.818
38	8.785	7.838	0.892	0.377	–0.741
39	6.354	7.556	1.189	1.062	0.530
40	8.023	6.345	0.791	0.141	–1.250
41	10.846	7.731	0.713	0.322	–0.580
42	11.815	6.524	0.552	–0.017	–1.265
43	10.738	5.830	0.543	–0.184	–0.387
44	12.008	6.155	0.513	–0.232	0.069
45	11.215	6.011	0.536	–0.041	–0.551
46	11.892	4.586	0.386	0.303	–0.073
47	11.508	4.241	0.369	0.263	–0.565
48	11.954	3.451	0.289	0.411	–0.417
49	11.646	2.786	0.239	0.241	–1.002
50	10.285	4.093	0.398	–1.017	2.854
51	10.908	2.240	0.205	–0.448	–0.988
52	11.185	1.634	0.146	–0.312	–0.198

Table 6.7 : Characteristics of weekly water deficit at Banswara (1981–1993)

Week	Mean (mm)	SD (mm)	CV	Skewness	Kurtosis
1	13.015	3.856	0.296	0.137	–0.332
2	13.477	4.362	0.324	–0.250	–0.199
3	13.708	3.293	0.240	–0.762	1.926
4	14.946	4.512	0.302	–0.867	0.237
5	15.069	3.947	0.262	–0.053	–0.400
6	16.023	3.350	0.209	–0.216	0.854
7	16.531	2.470	0.149	–0.074	0.324
8	17.454	2.504	0.143	0.368	0.800
9	18.085	2.471	0.137	–0.668	0.469
10	17.392	4.936	0.284	–2.855	9.289
11	20.346	2.437	0.120	–0.249	–0.329
12	22.931	2.380	0.104	0.151	–0.959
13	23.746	3.942	0.166	0.037	3.354
14	26.669	3.366	0.126	0.756	2.643
15	27.054	2.488	0.092	–0.717	0.642
16	29.731	7.251	0.244	1.172	3.214
17	31.115	10.214	0.328	–0.672	1.644
18	36.754	8.422	0.229	0.995	0.078
19	37.092	14.381	0.388	–1.304	2.772
20	43.962	12.062	0.274	0.154	–2.013
21	43.738	10.155	0.232	0.111	–2.003
22	42.585	9.270	0.218	–0.171	1.769
23	44.923	11.056	0.246	–1.623	4.790
24	26.915	17.079	0.635	–0.515	–1.334
25	24.669	22.530	0.913	0.099	–1.583
26	21.185	21.358	1.008	0.247	–1.905
27	22.154	19.680	0.888	0.330	–0.907
28	11.715	12.492	1.066	0.337	–1.944
29	4.715	9.933	2.107	2.056	3.231
30	3.415	9.881	2.893	3.369	11.615
31	4.802	11.529	2.357	2.409	5.038
32	1.800	4.423	2.457	2.922	8.874
33	2.569	5.679	2.210	2.976	9.532
34	0.792	2.253	2.844	3.461	12.196
35	1.431	2.299	1.607	2.315	5.625
36	5.062	4.407	0.871	0.586	–0.118
37	5.431	6.400	1.179	0.921	–0.415
38	5.977	6.804	1.138	0.998	–0.257
39	6.808	8.270	1.215	0.995	–0.259

contd.....

Table 6.7 – contd...

Week	Mean (mm)	SD (mm)	CV	Skewness	Kurtosis
40	7.977	8.706	1.091	0.955	–0.677
41	8.769	9.538	1.088	1.037	–0.489
42	7.585	8.858	1.168	1.347	0.767
43	10.785	9.233	0.856	0.959	–0.507
44	11.531	9.390	0.814	0.806	–0.740
45	11.400	9.751	0.855	1.025	–0.015
46	12.492	8.390	0.672	0.916	–0.391
47	12.131	9.132	0.753	0.539	–0.533
48	14.085	7.606	0.540	0.615	–0.479
49	14.462	7.045	0.487	0.502	–0.686
50	12.969	5.027	0.388	0.397	0.025
51	12.946	5.122	0.396	0.493	–0.422
52	12.300	3.438	0.280	–0.481	–0.046

Table 6.8 : Characteristics of weekly water deficit at Kota (1981–1993)

Week	Mean (mm)	SD (mm)	CV	Skewness	Kurtosis
1	13.238	1.541	0.116	–0.075	0.960
2	12.731	3.767	0.296	–1.163	1.743
3	13.969	1.728	0.124	0.633	0.461
4	14.600	4.733	0.324	–2.785	8.687
5	15.800	4.118	0.261	–2.222	6.787
6	16.338	5.247	0.321	–2.829	9.190
7	17.815	4.327	0.243	–1.967	3.721
8	20.162	2.812	0.139	–0.854	1.529
9	20.377	3.882	0.190	–0.229	0.010
10	23.646	2.542	0.108	0.209	–1.250
11	26.515	3.217	0.121	0.196	–0.980
12	28.954	3.489	0.120	–0.352	–1.558
13	30.800	3.192	0.104	0.134	–0.640
14	33.285	3.392	0.102	0.537	–0.193
15	34.754	6.795	0.196	–1.227	2.441
16	37.469	4.499	0.120	0.032	–1.464
17	36.508	11.653	0.319	–2.907	9.490
18	36.708	5.284	0.144	–0.382	–0.104
19	38.562	7.520	0.195	–2.100	6.039
20	40.377	5.011	0.124	–0.899	1.789
21	33.731	10.676	0.317	–1.647	3.460
22	36.000	5.428	0.151	–0.116	–0.324
23	36.262	5.436	0.150	–1.024	0.864
24	32.008	9.493	0.297	0.033	–1.108
25	31.923	13.285	0.416	–1.577	2.006

contd.....

Table 6.8 – contd...

Week	Mean (mm)	SD (mm)	CV	Skewness	Kurtosis
26	29.869	14.125	0.473	–1.671	1.778
27	17.523	19.028	1.086	0.441	–1.787
28	12.992	12.616	0.971	0.511	–1.206
29	7.131	14.437	2.025	1.904	2.242
30	8.454	13.416	1.587	1.406	0.751
31	6.669	13.494	2.023	2.060	2.927
32	4.238	6.031	1.423	1.525	1.997
33	4.608	9.814	2.130	2.618	7.102
34	2.662	4.773	1.793	2.650	7.758
35	3.008	5.064	1.684	2.590	7.416
36	3.600	5.301	1.472	1.558	1.758
37	5.362	5.287	0.986	1.190	0.614
38	5.146	3.166	0.615	–0.002	0.181
39	6.492	5.203	0.801	0.593	0.702
40	9.046	5.622	0.621	–0.145	–0.043
41	9.169	7.190	0.784	0.225	–0.521
42	9.631	5.826	0.605	–0.125	–1.034
43	11.292	4.442	0.393	–0.304	–1.097
44	11.115	5.003	0.450	–0.662	0.867
45	11.754	4.937	0.420	–0.917	1.531
46	12.400	3.502	0.282	–0.187	1.001
47	12.754	3.237	0.254	–0.246	0.020
48	13.062	2.934	0.225	–0.260	–0.320
49	12.915	2.338	0.181	–0.440	–0.455
50	10.977	5.371	0.489	–1.365	1.353
51	12.369	2.322	0.188	–0.026	–0.887
52	12.300	2.236	0.182	–1.313	2.290

6.4.1.1.2 Serial correlation coefficient

The lag one serial correlation coefficient (SCC) of observed series was calculated by using Equation (6.7) for different stations. The values of SCC were found to vary from 0.598 at Jodhpur to 0.807 at Kota as shown in Table 6.9. The respective confidence limits were estimated as –0.2 to 0.2 by using the relationships presented in Subsection 3.13.3.4. The value of lag one SCC lies outside the range of confidence limits and is significantly different from zero at all the stations. This again confirms that weekly water deficit is mutually dependent.

From the analysis of CV and SCC it is confirmed that water deficit process is time variant and not an independent one. The mutual dependence of the observed water deficit series was confirmed by the correlograms (Figures 6.1to 6.8) for all the eight stations under study.

Table 6.9 : Lag one serial correlation coefficient of observed weekly water deficit at different stations (confidence limit = –0.2 to 0.2)

Station	Lag one serial correlation coefficient
Jodhpur	0.598
Sriganganagar	0.723
Fatehpur	0.692
Pali	0.627
Jaipur	0.773
Udaipur	0.731
Banswara	0.793
Kota	0.807

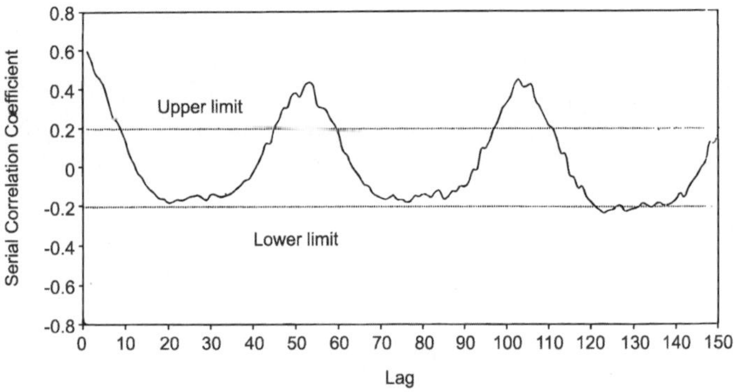

Fig. 6.1 : Correlogram of observed weekly water deficit at Jodhpur (1981–1993)

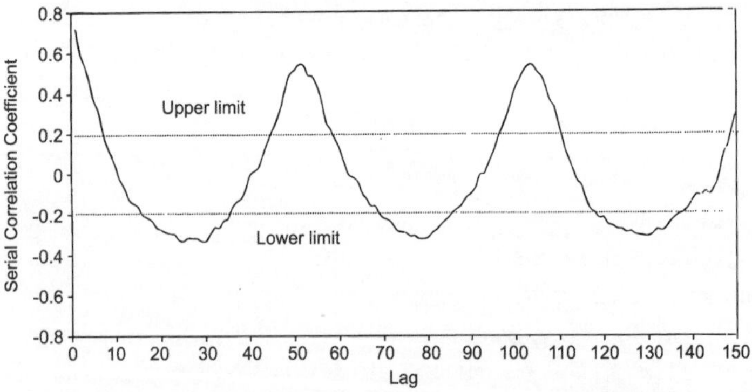

Fig. 6.2 : Correlogram of observed weekly water deficit at Sriganganagar (1981–1993)

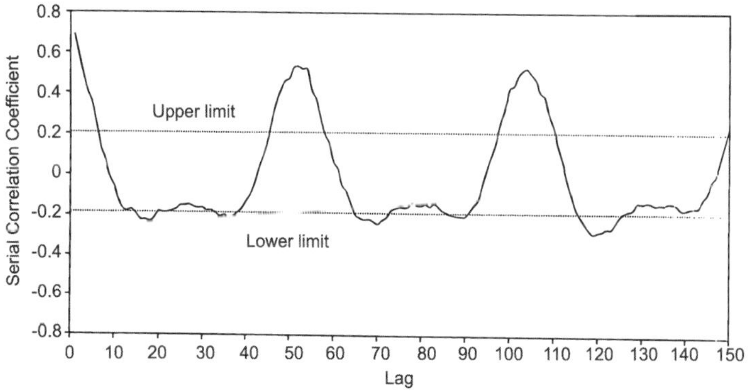

Fig. 6.3 : Correlogram of observed weekly water deficit at Fatehpur (1981–1993)

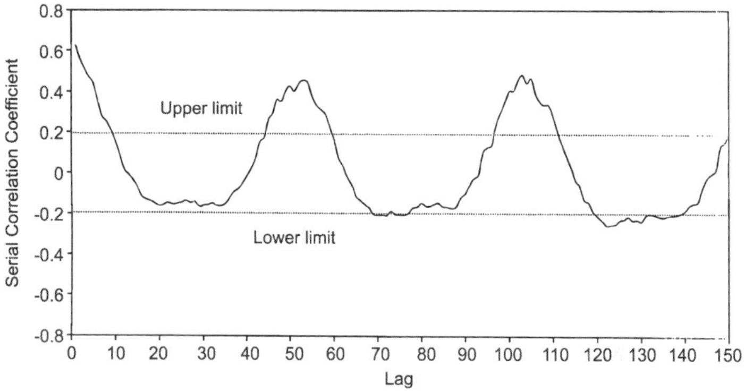

Fig. 6.4 : Correlogram of observed weekly water deficit at Pali (1981–1993)

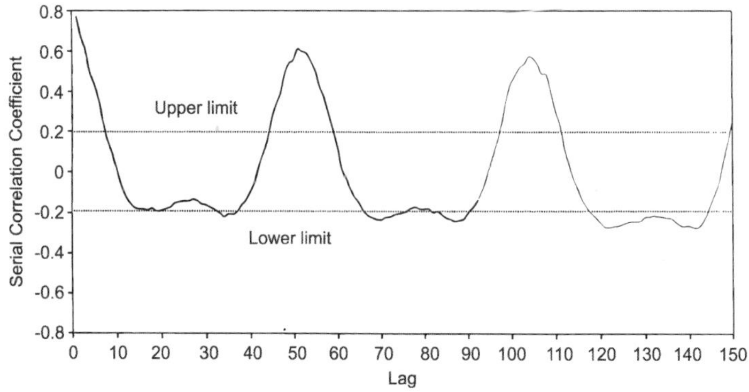

Fig. 6.5 : Correlogram of observed weekly water deficit at Jaipur (1981–1993)

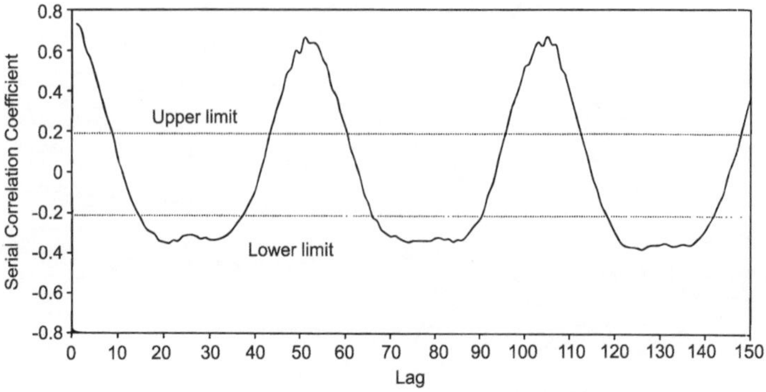

Fig. 6.6 : Correlogram of observed weekly water deficit at Udaipur (1981–1993)

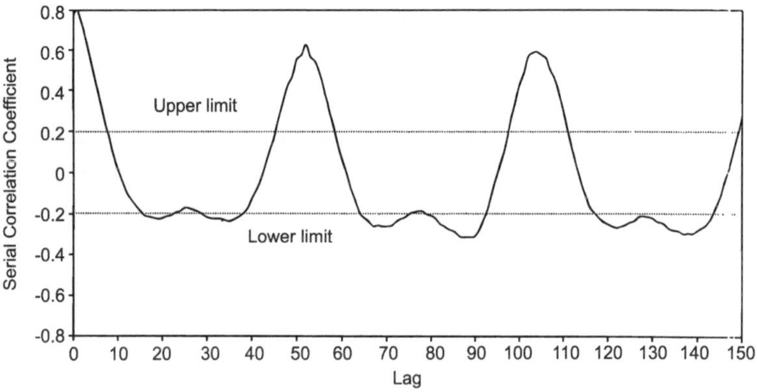

Fig. 6.7 : Correlogram of observed weekly water deficit at Banswara (1981–1993)

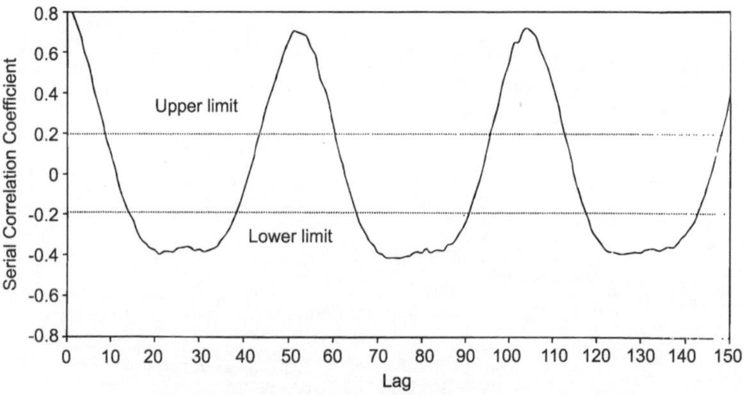

Fig. 6.8 : Correlogram of observed weekly water deficit at Kota (1981–1993)

4.1.2 Trend Component

For identification of trend component annual water deficit series was used (Kottegoda, 1980). The annual water deficit series was obtained by transforming the 13 years weekly water deficit series as given in Table 6.10. For detection of trend the hypothesis of no trend was made and value of test statistic (z) was calculated by turning point (TP) test and Kendall's rank correlation (KRC) Test as described in Section 3.1

Table 6.10 : Annual series of water deficit (mm) during the period from 1981 to 1993

Year	Jodh-pur	Srigan-ganagar	Fatehpur	Pali	Jaipur	Udaipur	Banswara	Kota
1981	1465.5	1223.1	1153.1	1661.7	1019.5	842.8	724.1	108.77
1982	1223.2	921.1	1167.2	1347.5	718.8	741.1	768.2	778.1
1983	1110.5	655.1	116.4	1114.2	688.3	801.2	895.6	844.9
1984	1471.3	991.5	1246.5	1470.9	1313.8	898.3	697.1	980.8
1985	1512.6	990.2	1243.0	1598.4	1224.0	899.8	903.2	909.2
1986	1500.6	979.8	1120.4	1558.6	1268.7	995.1	1011.8	962.9
1987	1678.9	1119.9	1423.7	1771.5	1471.3	1046.7	1267.3	1219.7
1988	1508.9	1125.3	1343.5	1390.5	1226.9	998.5	793.9	897.9
1989	1483.8	1047.3	1473.5	1262.0	1158.9	817.5	779.6	1022.3
1990	1060.0	1091.9	1449.5	935.1	826.2	719.0	886.9	822.2
1991	1322.0	1142.2	1228.7	1242.4	1056.3	881.5	1167.2	979.6
1992	1155.3	844.3	1108.7	1081.4	1031.6	838.4	924.6	950.6
1993	1085.7	1058.6	1214.8	1307.3	967.6	874.2	722.4	869.6

The calculated values of test statistics (Z_{Cal}) are presented in Table 6.11 for different stations. The estimated values of test statistics (Z_{Cal}) obtained for turning point test and for Kendall's rank correlation test was within the range at 5 per cent and 1 per cent level of significance. The value of Z_{Cal} were lower than tabulated values for both the tests at all the stations. Hence, the hypothesis of no trend was accepted.

Table 6.11 : Test statistics (z) values for trend identification for different stations

Station	Values of test statistics (z)	
	TP Test	KRC Test
Jodhpur	–0.945	–2.445
Sriganganagar	–0.240	–0.733
Fatehpur	–0.473	–0.733
Pali	–0.240	–1.834
Jaipur	–0.945	–0.367
Udaipur	–1.650	0.611
Banswara	–1.650	0.856
Kota	0.473	–0.245

Further, during the turning point test the number of turning points ranged from 5 at Udaipur to 8 at Fatehpur and Kota. This indicates that the water deficit series is random at all the stations.

From the above analysis it is confirmed that the trend component in water deficit time series is absent at all the stations and the observed series may be treated as trend free series.

Table 6.12 : Fourier decomposition of periodic component of mean weekly water deficit at Jodhpur (1981–1993)

Order	Alpha	Beta	Amplitude	Explained Variance	Cumulative Explained Variance
1	–4.782	8.592	9.833	66.908	66.908
2	0.141	–5.357	5.359	19.876	86.783
3	–1.825	0.638	1.933	2.587	89.371
4	1.035	–0.116	1.042	0.751	90.122
5	–0.888	0.285	0.933	0.602	90.724
6	0.877	–0.184	0.896	0.556	91.28
7	–1.296	–0.319	1.334	1.232	92.512
8	0.471	–0.151	0.495	0.169	92.682
9	–0.371	–0.325	0.494	0.169	92.85
10	1.081	0.266	1.113	0.857	93.707
11	–0.734	–0.078	0.738	0.377	94.085
12	0.527	0.062	0.53	0.195	94.279

contd.....

Table 6.12 – contd....

Order	Alpha	Beta	Amplitude	Explained Variance	Cumulative Explained Variance
13	0.405	-0.079	0.413	0.118	94.398
14	-0.044	0.32	0.323	0.072	94.47
15	0.513	0.004	0.513	0.182	94.652
16	0.241	-0.15	0.284	0.056	94.708
17	-0.257	0.256	0.362	0.091	94.798
18	0.24	-0.302	0.386	0.103	94.901
19	-0.247	0.345	0.424	0.124	95.026
20	0.111	-1.031	1.037	0.745	95.771
21	0.019	1.261	1.261	1.101	96.871
22	0.294	-0.908	0.954	0.63	97.501
23	-0.281	0.171	0.329	0.075	97.576
24	-0.101	-0.14	0.173	0.021	97.597
25	-0.322	0.265	0.417	0.121	97.717
26	0.713	0.051	0.715	0.354	98.071

6.4.1.3 Periodic Component

To confirm the presence of periodic component in weekly water deficit series correlograms were drawn for all the stations. The correlograms of the observed series for different stations are shown in Figures 6.1 to 6.8. The resulting oscillating shape of the correlograms confirms the presence of periodic component in weekly water deficit series for all the stations.

Further, the correlograms have peaks at lags equal to 52 and other multiples of it. The time span of periodicity was taken as 52 for use in the harmonic analysis of periodic component.

Table 6.13 : Fourier decomposition of periodic component of mean weekly water deficit at Sriganganagar (1981–1993)

Order	Alpha	Beta	Amplitude	Explained Variance	Cumulative Explained Variance
1	-13.353	4.255	14.015	65.878	65.878
2	-0.722	-7.472	7.507	18.902	84.780
3	0.460	4.910	4.931	8.156	92.937

contd.....

Table 6.13 – contd....

Order	Alpha	Beta	Amplitude	Explained Variance	Cumulative Explained Variance
4	–0.279	–1.892	1.912	1.227	94.163
5	–0.246	0.656	0.701	0.165	94.328
6	0.795	–0.210	0.822	0.226	94.554
7	–1.009	0.144	1.019	0.348	94.903
8	0.251	–0.262	0.363	0.044	94.947
9	–0.449	0.099	0.460	0.071	95.018
10	0.563	–0.123	0.577	0.112	95.129
11	–0.538	–0.040	0.540	0.098	95.227
12	0.876	–0.016	0.876	0.257	95.484
13	–0.697	0.700	0.988	0.327	95.811
14	0.399	–0.744	0.844	0.239	96.050
15	–0.359	–0.081	0.368	0.046	96.096
16	0.656	–0.103	0.664	0.148	96.244
17	–0.028	1.279	1.280	0.549	96.793
18	0.276	–1.124	1.158	0.450	97.243
19	–0.325	0.277	0.427	0.061	97.304
20	0.060	0.099	0.116	0.004	97.308
21	0.472	0.047	0.474	0.075	97.384
22	–0.718	–0.374	0.809	0.220	97.603
23	0.630	0.630	0.892	0.267	97.870
24	–0.321	–0.311	0.447	0.067	97.937
25	–0.005	0.140	0.140	0.007	97.943
26	0.060	0.012	0.061	0.001	97.944

6.4.1.3.1 Selection of significant harmonics

For representing the periodic component of the weekly water deficit series the number of significant harmonics were determined by analysing the periodic mean weekly water deficit series. Equations (6.11) and (6.12) were used to estimate the Fourier coefficients A_k and B_k for the mean weekly water deficit series. Explained variance by harmonics, cumulative periodogram test and F-distribution were considered for selection of significant harmonics.

The amplitude, phase angle and explained variance of different harmonics alongwith cumulative explained variance are presented in Tables 6.12 to 6.19 for all the stations.

Table 6.14 : Fourier decomposition of periodic component of mean weekly water deficit at Fatehpur (1981–1993)

Order	Alpha	Beta	Amplitude	Explained Variance	Cumulative Explained Variance
1	–9.250	8.657	12.669	52.107	52.107
2	–1.304	–9.905	9.990	32.401	84.508
3	–0.900	4.120	4.217	5.773	90.281
4	1.035	–1.093	1.505	0.736	91.016
5	–1.177	0.879	1.469	0.700	91.717
6	0.897	–0.502	1.028	0.343	92.060
7	–0.710	0.202	0.739	0.177	92.237
8	0.713	–0.804	1.075	0.375	92.612
9	–1.044	–0.048	1.045	0.355	92.967
10	1.564	0.872	1.790	1.041	94.008
11	–0.998	0.681	1.208	0.474	94.481
12	1.420	–0.113	1.425	0.659	95.140
13	–0.840	0.697	1.091	0.387	95.527
14	0.530	–0.415	0.673	0.147	95.674
15	–0.385	–0.151	0.413	0.056	95.729
16	0.413	0.200	0.459	0.068	95.798
17	–0.101	–0.233	0.254	0.021	95.819
18	0.255	–0.644	0.693	0.156	95.975
19	–0.889	0.153	0.902	0.264	96.239
20	0.313	0.589	0.667	0.144	96.383
21	–0.887	–0.286	0.932	0.282	96.665
22	0.666	0.321	0.740	0.178	96.843
23	–0.160	–0.508	0.532	0.092	96.935
24	–0.191	1.034	1.052	0.359	97.294
25	–0.326	–1.208	1.252	0.509	97.802
26	0.877	0.027	0.878	0.250	98.052

It can be seen from Table 6.12 that at Jodhpur the mean and first three harmonics explain more than 89 per cent of the total variance in the mean weekly water deficit series. To confirm this a cumulative periodogram test and analysis of variance test were performed. From the cumulative periodogram of the mean weekly series (Fig. 6.9), it can be seen that three harmonics are suficient to amount more than 89 per cent of the variance of the series. This is further confirmed by analysis of variance. Values of F-ratio for different harmonics are given in Table 6.29. The calculated values are given for first three harmonics. Only the values of estimated

Table 6.15 : Fourier decomposition of periodic component of mean weekly water deficit at Pali (1981–1993)

Order	Alpha	Beta	Amplitude	Explained Variance	Cumulative Explained Variance
1	-6.543	9.205	11.293	68.232	68.232
2	-1.574	-6.316	6.509	22.668	90.900
3	-0.641	0.980	1.171	0.734	91.634
4	-0.024	-0.229	0.231	0.028	91.662
5	0.099	0.651	0.658	0.232	91.894
6	0.203	-0.275	0.342	0.063	91.957
7	-1.283	-0.342	1.328	0.944	92.900
8	0.961	-0.412	1.045	0.584	93.485
9	-0.808	0.266	0.851	0.387	93.872
10	1.188	0.501	1.289	0.889	94.761
11	-0.847	-0.489	0.978	0.512	95.272
12	0.609	0.132	0.623	0.207	95.480
13	-0.037	-0.136	0.141	0.011	95.490
14	0.183	0.269	0.325	0.056	95.547
15	0.588	-0.269	0.646	0.223	95.770
16	-0.010	0.078	0.078	0.003	95.773
17	-0.312	0.082	0.323	0.056	95.829
18	0.411	0.090	0.421	0.095	95.924
19	0.392	-0.065	0.398	0.085	96.009
20	-0.171	-0.580	0.604	0.195	96.204
21	-0.129	0.991	0.999	0.534	96.738
22	0.152	-1.019	1.031	0 568	97.307
23	0.226	0.538	0.584	0.182	97.489
24	-0.786	-0.117	0.794	0.337	97.826
25	0.225	0.204	0.304	0.049	97.876
26	0.121	0.029	0.125	0.008	97.884

F-ratio for variation explained by 4, 5, 6, ..., 26 harmonics are low than the tabulated values of F-distribution at 1 per cent level and 5 per cent level of significance. This means that only first three harmonics are highly significant and other harmonics are not significant and therefore, can be ignored.

Table 6.16 : Fourier decomposition of periodic component of mean weekly water deficit at Jaipur (1981–1993)

Order	Alpha	Beta	Amplitude	Explained Variance	Cumulative Explained Variance
1	-7.020	11.332	13.330	58.458	58.458
2	-0.036	-9.619	9.619	30.440	88.898
3	-2.218	3.319	3.991	5.241	94.139
4	0.807	-1.005	1.289	0.547	94.686
5	-0.076	0.201	0.215	0.015	94.701
6	0.073	-0.432	0.438	0.063	94.764
7	-0.507	0.297	0.587	0.114	94.877
8	0.214	-0.750	0.780	0.200	95.078
9	-0.859	0.168	0.876	0.252	95.330
10	1.014	-0.017	1.015	0.339	95.669
11	-0.507	-0.343	0.612	0.123	95.792
12	0.085	0.894	0.898	0.265	96.057
13	-0.338	-0.743	0.816	0.219	96.276
14	0.042	0.686	0.687	0.155	96.432
15	0.562	-0.858	1.026	0.346	96.778
16	-0.197	1.066	1.084	0.387	97.165
17	0.506	-0.426	0.661	0.144	97.309
18	-0.127	0.075	0.148	0.007	97.316
19	-0.127	-0.385	0.406	0.054	97.370
20	-0.044	-0.151	0.158	0.008	97.378
21	-0.120	0.381	0.400	0.053	97.431
22	-0.761	-0.323	0.827	0.225	97.655
23	0.523	0.189	0.556	0.102	97.757
24	-0.385	-0.287	0.480	0.076	97.833
25	0.504	0.053	0.507	0.084	97.918
26	-0.487	-0.005	0.487	0.078	97.996

Similarly, it can be seen that at Sriganganagar, Fatehpur, Jaipur and Banswara the mean and first three harmonics explain more than 92, 90, 94 and 93 per cent, of the total variance in the mean weekly water deficit series respectively. In order to confirm this a cumulative periodogram test and analysis of variance test were performed. From the cumulative periodograms at Sriganganagar (Fig. 6.10), Fatehpur (Fig. 6.11), Jaipur (Fig. 6.13) and Banswara (Fig. 6.15) three harmonics were found to be sufficient to amount more than 92, 90, 94 and 93 per cent of the variance of the series respectively.

This was further confirmed by analysis of variance test. Values of F-ratio for different harmonics at Sriganganagar, Fatehpur, Jaipur and Banswara are given in Tables 6.29, 6.30, 6.32 and 6.34 respectively. The calculated values are given for first harmonics only. Tables 6.29 to 6.32 show that the values of estimated F-ratio for variation explained by 4, 5, 6, ..., 26 harmonics are low than the tabulated values of F-distribution at 1 per cent level and 5 per cent level of significance. This reflects that only first three harmonics are highly significant and other harmonics are not significant and therefore, can be ignored at Sriganganagar, Fatehpur, Jaipur and Banswara.

Table 6.17 : Fourier decomposition of periodic component of mean weekly water deficit at Udaipur (1981–1993)

Order	Alpha	Beta	Amplitude	Explained Variance	Cumulative Explained Variance
1	-4.080	11.542	12.242	72.001	72.001
2	-1.305	-6.840	6.964	23.300	95.301
3	-0.359	1.558	1.599	1.229	96.530
4	0.302	-0.126	0.328	0.052	96.581
5	-0.438	0.076	0.445	0.095	96.676
6	-0.207	0.257	0.330	0.052	96.729
7	-0.006	-0.454	0.454	0.099	96.828
8	-0.153	-0.169	0.228	0.025	96.853
9	-0.162	0.018	0.163	0.013	96.866
10	0.672	0.223	0.708	0.241	97.106
11	-0.119	-0.273	0.297	0.043	97.149
12	0.073	0.122	0.142	0.010	97.159
13	0.230	-0.155	0.277	0.037	97.195
14	0.115	0.558	0.570	0.156	97.351
15	0.314	-0.198	0.371	0.066	97.418
16	-0.092	0.084	0.125	0.007	97.425
17	-0.134	0.163	0.211	0.021	97.447
18	0.054	-0.016	0.057	0.002	97.448
19	-0.274	0.231	0.358	0.062	97.510
20	0.372	-0.592	0.699	0.235	97.744
21	-0.387	0.264	0.469	0.105	97.850
22	0.141	-0.120	0.185	0.016	97.866
23	0.008	-0.096	0.097	0.004	97.871
24	-0.356	0.085	0.366	0.064	97.935
25	0.288	0.065	0.295	0.042	97.977
26	-0.051	0.013	0.053	0.001	97.979

Table 6.18 : Fourier decomposition of periodic component of mean weekly water deficit at Banswara (1981–1993)

Order	Alpha	Beta	Amplitude	Explained Variance	Cumulative Explained Variance
1	-4.845	11.704	12.667	58.649	58.649
2	1.582	-8.821	8.961	29.354	88.003
3	-0.058	3.888	3.888	5.526	93.529
4	-0.725	-1.891	2.025	1.499	95.028
5	0.482	0.398	0.625	0.143	95.171
6	-0.907	0.144	0.918	0.308	95.479
7	0.303	-0.619	0.689	0.174	95.652
8	0.146	0.497	0.518	0.098	95.751
9	-0.422	-0.529	0.677	0.167	95.918
10	0.078	0.851	0.854	0.267	96.185
11	-0.323	-0.810	0.872	0.278	96.463
12	0.153	0.894	0.907	0.300	96.763
13	0.135	-0.528	0.544	0.108	96.872
14	0..255	0.499	0.560	0.115	96.986
15	-0.279	-0.413	0.498	0.091	97.077
16	0.055	0.453	0.457	0.076	97.153
17	-0.063	-0.251	0.258	0.024	97.178
18	0.339	-0.112	0.357	0.047	97.224
19	-0.117	0.349	0.368	0.049	97.274
20	-0.028	-0.635	0.635	0.148	97.421
21	-0.233	0.436	0.494	0.089	97.511
22	-0.369	-0.365	0.519	0.098	97.609
23	0.619	0.193	0.649	0.154	97.763
24	-0.578	0.092	0.586	0.125	97.888
25	0.448	-0.168	0.478	0.084	97.972
26	-0.379	0.000	0.379	0.053	98.024

At Pali, Udaipur and Kota it can also be seen from Tables 4.54, 4.56 and 4.58 that the first two harmonics explain more than 90, 95, and 94 per cent of the total variance in the mean weekly water series respectively. From the cumulative periodogram of the mean weekly series, it can be seen that at Pali (Fig. 6.12), Udaipur (Fig. 6.14) and Kota (Fig. 6.16) two harmonics are sufficient to amount more than 90, 95 and 94, of the total variance of the series respectively.

Table 6.19 : Fourier decomposition of periodic component of mean weekly water deficit at Kota (1981–1993)

Order	Alpha	Beta	Amplitude	Explained Variance	Cumulative Explained Variance
1	-5.194	13.072	14.066	73.489	73.489
2	-0.194	-7.592	7.594	21.423	94.912
3	-0.643	1.475	1.609	0.961	95.873
4	0.627	-0.640	0.896	0.298	96.171
5	-0.542	0.791	0.958	0.341	96.513
6	0.232	-0.613	0.655	0.159	96.672
7	-0.522	0.544	0.754	0.211	96.883
8	0.503	-0.341	0.608	0.137	97.020
9	-0.398	0.451	0.602	0.134	97.155
10	0.476	-0.186	0.511	0.097	97.252
11	-0.211	0.238	0.318	0.037	97.289
12	0.121	0.006	0.121	0.005	97.295
13	0.090	-0.170	0.193	0.014	97.308
14	0.460	0.259	0.528	0.104	97.412
15	-0.455	-0.185	0.492	0.090	97.502
16	0.402	0.173	0.438	0.071	97.573
17	-0.236	0.240	0.337	0.042	97.615
18	0.183	-0.549	0.578	0.124	97.739
19	0.008	0.266	0.266	0.026	97.766
20	-0.049	-0.345	0.349	0.045	97.811
21	-0.018	0.388	0.388	0.056	97.867
22	0.025	-0.159	0.161	0.010	97.876
23	-0.243	0.041	0.246	0.023	97.899
24	0.062	0.103	0.120	0.005	97.904
25	-0.371	-0.228	0.435	0.070	97.975
26	0.248	0.020	0.249	0.023	97.998

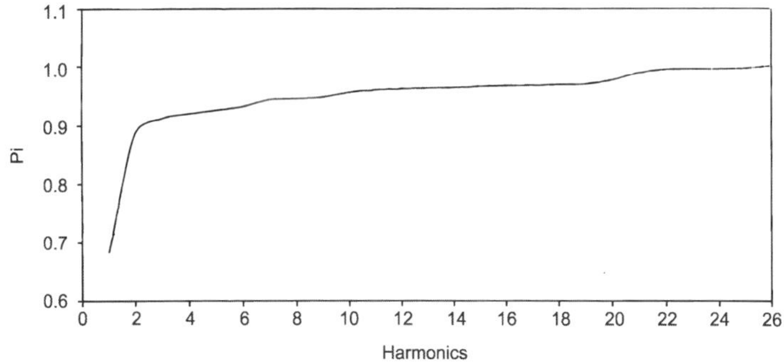

Fig. 6.9 : Cumulative periodogram for determination of significant harmonics for water deficit series at Jodhpur (1981–1993)

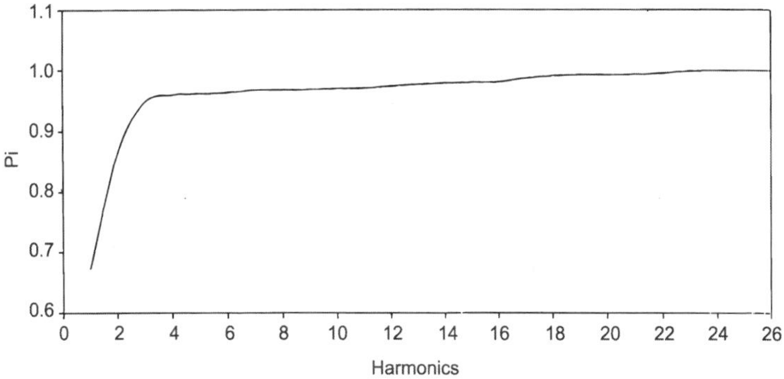

Fig. 6.10 : Cumulative periodogram for determination of significant harmonics for water deficit series at Sriganganagar (1981–1993)

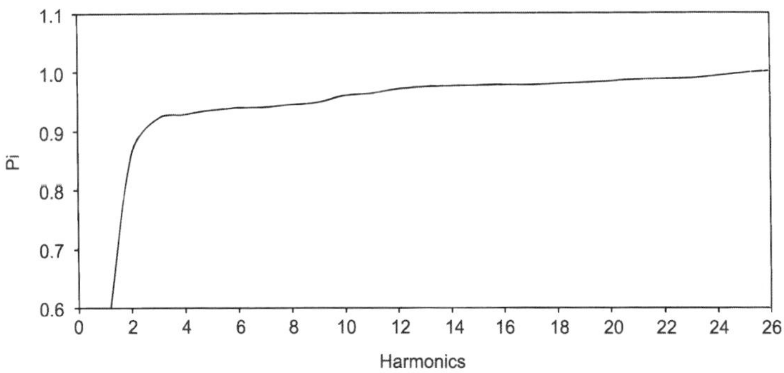

Fig. 6.11 : Cumulative periodogram for determination of significant harmonics for water deficit series at Fatehpur (1981–1993)

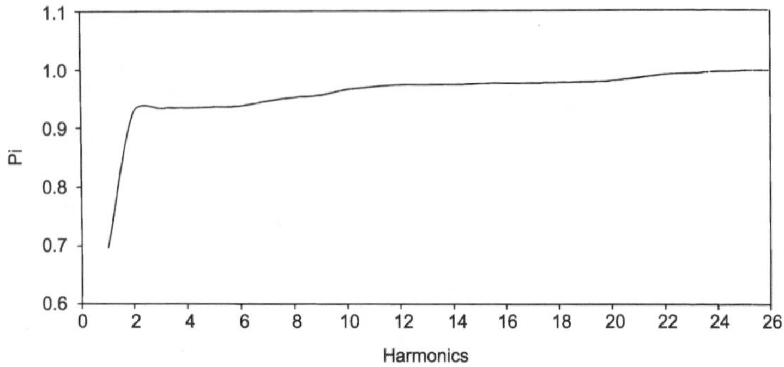

Fig. 6.12 : Cumulative periodogram for determination of significant harmonics for water deficit series at Pali (1981–1993)

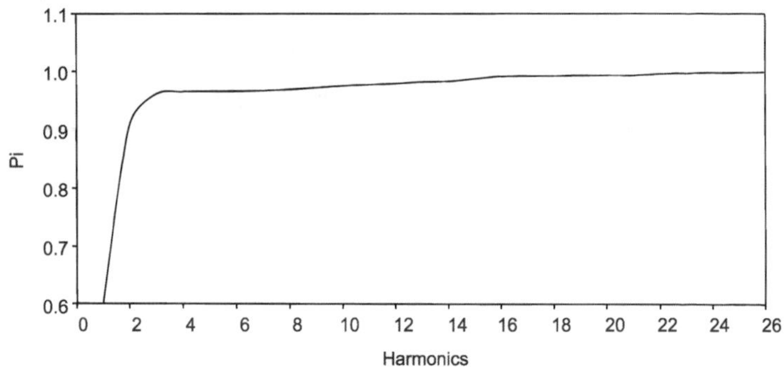

Fig. 6.13 : Cumulative periodogram for determination of significant harmonics for water deficit series at Jaipur (1981–1993)

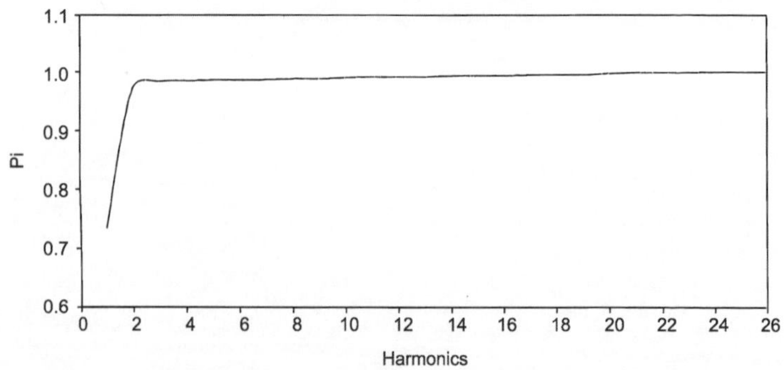

Fig. 6.14 : Cumulative periodogram for determination of significant harmonics for water deficit series at Udaipur (1981–1993)

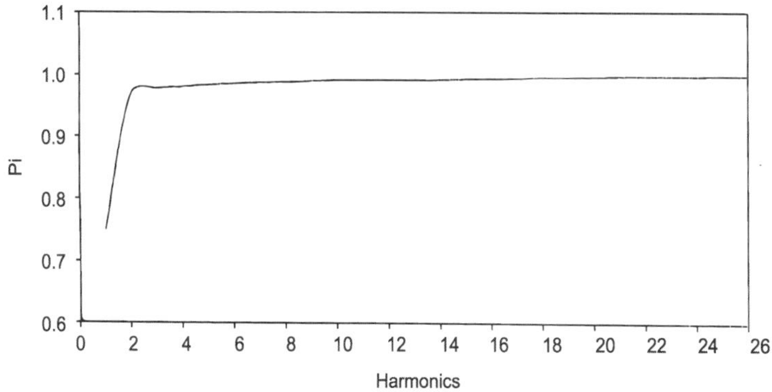

Fig. 6.15 : Cumulative periodogram for determination of significant harmonics for water deficit series at Banswara (1981–1993)

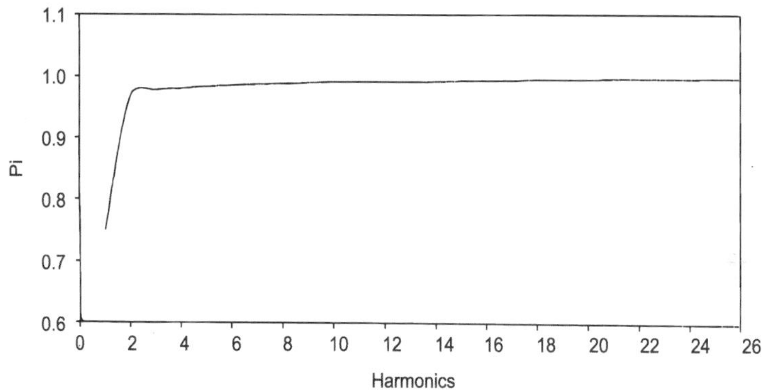

Fig. 6.16 : Cumulative periodogram for determination of significant harmonics for water deficit series at Kota (1981–1993)

Table 6.20 : Cumulative periodogram of weekly water deficit at Jodhpur (1981–1993)

Order	MSDU	MSDJ	Cum MSDJ	PI
1	70.860	48.340	48.340	0.682
2	70.860	14.360	62.700	0.885
3	70.860	1.869	64.569	0.911
4	70.860	0.543	65.112	0.919
5	70.860	0.435	65.547	0.925
6	70.860	0.401	65.949	0.931
7	70.860	0.890	66.839	0.943

contd.....

Table 6.20 – contd.....

Order	MSDU	MSDJ	Cum MSDJ	PI
8	70.860	0.122	66.961	0.945
9	-70.860	0.122	67.083	0.947
10	70.860	0.619	67.703	0.955
11	70.860	0.273	67.975	0.959
12	70.860	0.141	68.116	0.961
13	70.860	0.085	68.201	0.962
14	70.860	0.052	68.253	0.963
15	70.860	0.132	68.385	0.965
16	70.860	0.040	68.425	0.966
17	70.860	0.066	68.491	0.967
18	70.860	0.074	68.565	0.968
19	70.860	0.090	68.655	0.969
20	70.860	0.538	69.193	0.976
21	70.860	0.795	69.989	0.988
22	70.860	0.455	70.444	0.994
23	70.860	0.054	70.498	0.995
24	70.860	0.015	70.513	0.995
25	70.860	0.087	70.600	0.996
26	70.860	0.256	70.855	1.000

Table 6.21 : Cumulative periodogram of weekly water deficit at Sriganganagar (1981–1993)

Order	MSDU	MSDJ	Cum MSDJ	PI
1	146.209	98.209	98.209	0.672
2	146.209	28.178	126.387	0.864
3	146.209	12.159	138.546	0.948
4	146.209	1.829	140.375	0.960
5	146.209	0.246	140.620	0.962
6	146.209	0.338	140.958	0.964
7	146.209	0.519	141.477	0.968
8	146.209	0.066	141.543	0.968
9	146.209	0.106	141.649	0.969
10	146.209	0.166	141.815	0.970
11	146.209	0.146	141.960	0.971
12	146.209	0.384	142.344	0.974
13	146.209	0.488	142.832	0.977
14	146.209	0.356	143.188	0.979
15	146.209	0.068	143.256	0.980

contd.....

Table 6.21 – contd...

Order	MSDU	MSDJ	Cum MSDJ	PI
16	146.209	0.220	143.476	0.981
17	146.209	0.819	144.295	0.987
18	146.209	0.670	144.965	0.991
19	146.209	0.091	145.056	0.992
20	146.209	0.007	145.063	0.992
21	146.209	0.112	145.175	0.993
22	146.209	0.327	145.503	0.995
23	146.209	0.397	145.900	0.998
24	146.209	0.100	146.000	0.999
25	146.209	0.010	146.010	0.999
26	146.209	0.002	146.011	0.999

Table 6.22 : Cumulative periodogram of weekly water deficit at Fatehpur (1981–1993)

Order	MSDU	MSDJ	Cum MSDJ	PI
1	151.048	80.249	80.249	0.531
2	151.048	49.901	130.150	0.862
3	151.048	8.891	139.041	0.921
4	151.048	1.133	140.174	0.928
5	151.048	1.078	141.253	0.935
6	151.048	0.529	141.781	0.939
7	151.048	0.273	142.054	0.940
8	151.048	0.578	142.632	0.944
9	151.048	0.546	143.178	0.948
10	151.048	1.603	144.781	0.959
11	151.048	0.729	145.510	0.963
12	151.048	1.015	146.525	0.970
13	151.048	0.596	147.121	0.974
14	151.048	0.227	147.347	0.975
15	151.048	0.085	147.433	0.976
16	151.048	0.105	147.538	0.977
17	151.048	0.032	147.570	0.977
18	151.048	0.240	147.810	0.979
19	151.048	0.407	148.217	0.981
20	151.048	0.222	148.439	0.983
21	151.048	0.434	148.874	0.986
22	151.048	0.273	149.147	0.987
23	151.048	0.142	149.289	0.988
24	151.048	0.553	149.842	0.992
25	151.048	0.783	150.625	0.997
26	151.048	0.385	151.010	1.000

Table 6.23 : Cumulative periodogram of weekly water deficit at Pali (1981–1993)

Order	MSDU	MSDJ	Cum MSDJ	PI
1	91.657	63.766	63.766	0.696
2	91.657	21.184	84.950	0.927
3	91.657	0.686	85.636	0.934
4	91.657	0.027	85.662	0.935
5	91.657	0.217	85.879	0.937
6	91.657	0.058	85.937	0.938
7	91.657	0.882	86.819	0.947
8	91.657	0.546	87.366	0.953
9	91.657	0.362	87.727	0.957
10	91.657	0.831	88.558	0.966
11	91.657	0.478	89.036	0.971
12	91.657	0.194	89.230	0.974
13	91.657	0.010	89.240	0.974
14	91.657	0.053	89.293	0.974
15	91.657	0.209	89.501	0.976
16	91.657	0.003	89.505	0.977
17	91.657	0.052	89.557	0.977
18	91.657	0.089	89.645	0.978
19	91.657	0.079	89.724	0.979
20	91.657	0.183	89.907	0.981
21	91.657	0.499	90.406	0.986
22	91.657	0.531	90.937	0.992
23	91.657	0.170	91.108	0.994
24	91.657	0.315	91.423	0.997
25	91.657	0.046	91.469	0.998
26	91.657	0.008	91.477	0.998

Table 6.24 : Cumulative periodogram of weekly water deficit at Jaipur (1981–1993)

Order	MSDU	MSDJ	Cum MSDJ	PI
1	149.059	88.846	88.846	0.596
2	149.059	46.263	135.109	0.906
3	149.059	7.965	143.074	0.960
4	149.059	0.831	143.905	0.965
5	149.059	0.023	143.928	0.966
6	149.059	0.096	144.024	0.966
7	149.059	0.173	144.197	0.967
8	149.059	0.304	144.501	0.969

contd.....

Table 6.24 – contd...

Order	MSDU	MSDJ	Cum MSDJ	PI
9	149.059	0.383	144.884	0.972
10	149.059	0.515	145.399	0.975
11	149.059	0.188	145.586	0.977
12	149.059	0.403	145.990	0.979
13	149.059	0.333	146.323	0.982
14	149.059	0.236	146.559	0.983
15	149.059	0.526	147.085	0.987
16	149.059	0.588	147.673	0.991
17	149.059	0.218	147.891	0.992
18	149.059	0.011	147.902	0.992
19	149.059	0.082	147.984	0.993
20	149.059	0.012	147.997	0.993
21	149.059	0.080	148.077	0.993
22	149.059	0.342	148.418	0.996
23	149.059	0.155	148.573	0.997
24	149.059	0.115	148.688	0.998
25	149.059	0.128	148.817	0.998
26	149.059	0.119	148.936	0.999

Table 6.25 : Cumulative periodogram of weekly water deficit at Udaipur (1981–1993)

Order	MSDU	MSDJ	Cum MSDJ	PI
1	102.066	74.929	74.929	0.734
2	102.066	24.247	99.177	0.972
3	102.066	1.279	100.455	0.984
4	102.066	0.054	100.509	0.985
5	102.066	0.099	100.608	0.986
6	102.066	0.055	100.662	0.986
7	102.066	0.103	100.766	0.987
8	102.066	0.026	100.792	0.988
9	102.066	0.013	100.805	0.988
10	102.066	0.251	101.056	0.990
11	102.066	0.044	101.100	0.991
12	102.066	0.010	101.110	0.991
13	102.066	0.038	101.148	0.991
14	102.066	0.162	101.311	0.993
15	102.066	0.069	101.380	0.993
16	102.066	0.008	101.387	0.993

contd.....

Table 6.25 – contd...

Order	MSDU	MSDJ	Cum MSDJ	PI
17	102.066	0.022	101.410	0.994
18	102.066	0.002	101.411	0.994
19	102.066	0.064	101.475	0.994
20	102.066	0.244	101.720	0.997
21	102.066	0.110	101.829	0.998
22	102.066	0.017	101.847	0.998
23	102.066	0.005	101.851	0.998
24	102.066	0.067	101.918	0.999
25	102.066	0.044	101.962	0.999
26	102.066	0.001	101.963	0.999

Table 6.26 : Cumulative periodogram of weekly water deficit at Banswara (1981–1993)

Order	MSDU	MSDJ	Cum MSDJ	PI
1	134.161	80.226	80.226	0.598
2	134.161	40.154	120.380	0.897
3	134.161	7.559	127.940	0.954
4	134.161	2.050	129.990	0.969
5	134.161	0.195	130.185	0.970
6	134.161	0.422	130.607	0.974
7	134.161	0.237	130.844	0.975
8	134.161	0.134	130.979	0.976
9	134.161	0.229	131.208	0.978
10	134.161	0.365	131.573	0.981
11	134.161	0.380	131.953	0.984
12	134.161	0.411	132.364	0.987
13	134.161	0.148	132.512	0.988
14	134.161	0.157	132.669	0.989
15	134.161	0.124	132.793	0.990
16	134.161	0.104	132.897	0.991
17	134.161	0.033	132.931	0.991
18	134.161	0.064	132.994	0.991
19	134.161	0.068	133.062	0.992
20	134.161	0.202	133.264	0.993
21	134.161	0.122	133.386	0.994
22	134.161	0.135	133.521	0.995
23	134.161	0.210	133.731	0.997
24	134.161	0.171	133.903	0.998
25	134.161	0.114	134.017	0.999
26	134.161	0.072	134.089	0.999

Table 6.27 : Cumulative periodogram of weekly water deficit at Kota (1981–1993)

Order	MSDU	MSDJ	Cum MSDJ	PI
1	132.021	98.924	98.924	0.749
2	132.021	28.837	127.761	0.968
3	132.021	1.294	129.055	0.978
4	132.021	0.401	129.457	0.981
5	132.021	0.459	129.916	0.984
6	132.021	0.215	130.130	0.986
7	132.021	0.284	130.414	0.988
8	132.021	0.185	130.599	0.989
9	132.021	0.181	130.780	0.991
10	132.021	0.131	130.911	0.992
11	132.021	0.050	130.961	0.992
12	132.021	0.007	130.968	0.992
13	132.021	0.019	130.987	0.992
14	132.021	0.139	131.126	0.993
15	132.021	0.121	131.247	0.994
16	132.021	0.096	131.343	0.995
17	132.021	0.057	131.400	0.995
18	132.021	0.167	131.567	0.997
19	132.021	0.035	131.602	0.997
20	132.021	0.061	131.663	0.997
21	132.021	0.075	131.739	0.998
22	132.021	0.013	131.752	0.998
23	132.021	0.030	131.782	0.998
24	132.021	0.007	131.789	0.998
25	132.021	0.095	131.884	0.999
26	132.021	0.031	131.915	0.999

6.4.1.3.2 Parameters of periodic component

Using Equations (6.11) and (6.12) the Fourier coefficient A_k and B_k were estimated. The results are presented in Tables 6.37 to 6.44 for all the stations.

The amplitude, phase angle and explained variance for different harmonics were calculated by using Equations (6.18), (6.19) and (6.20) respectively. The Fourier decomposition for weekly water deficit series are also presented in Tables 6.37 to 6.44 for all the stations.

Table 6.28 : Analysis of variance of Fourier coefficients of water deficit series at Jodhpur (1981–1993)

S.No.	Harmonics	Degree of freedom	Sum of Squares	Mean Squares	F_{cal}	$F_{tab\ 1\%}$
1	4,5,...26	45	7163.394	159.187	0.982	0.600
2	Residual	630	102084.29	162.039		
3	3	2	1626.684	813.342	22.398	4.647
4	Residual	673	107621	159.912		
1	3,4,...26	47	8790.078	187.023	1.309	1.580
2	Residual	628	89702.471	142.838		
3	2	2	10755.13	5377.565	1.36.745	4.647
4	Residual	673	98492.549	146.349		
1	2,3,...26	49	19545.208	398..882	1.252	1.570
2	Residual	626	89702.471	143.295		
3	1	2	35411.143	17705.57	161.381	4.647
4	Residual	673	73836.536	109.713		

Table 6.29 : Analysis of variance of Fourier coefficients of water deficit series at Sriganganagar (1981–1993)

S.No.	Harmonics	Degree of freedom	Sum of Squares	Mean Squares	F_{cal}	$F_{tab\ 1\%}$
1	4,5,...26	45	5046.676	112.148	0.434	1.600
2	Residual	630	162941.03	258.637		
3	3	2	8219.545	4109.772	10.629	4.647
4	Residual	673	159768.17	237.397		
1	3,4,...26	47	13266.221	282.26	1.307	1.580
2	Residual	628	135673.15	216.04		
3	2	2	19048.337	9524.168	43.036	4.647
4	Residual	673	148939.37	221.307		
1	2,3,...26	49	32314.558	659.481	1.249	1.570
2	Residual	626	135673.15	216.73		
3	1	2	66389.208	33194.6	219.885	4.647
4	Residual	673	101598.5	150.964		

It can be seen from Table 6.36 that at Jodhpur the first three harmonics explain more than 39 per cent of the variance while rest of the harmonics explain 3.91 per cent of the variance. This further confirms that only first three harmonics are significant for weekly water deficit time series. For the first three harmonics the values of Fourier coefficients (A_1, A_2 and A_3 and B_1, B_2 and B_3)

were found to be 4.918, 0.330, –1.868 and 8.514, –5.345 and 0.563 respectively. With these coefficients and using Equation (6.9) the periodic component $P_{(t)}$ resulting from periodic process may mathematically be expressed as:

$$P^t = 25.965 - 4.918\ \mathrm{Cos}\ (2\ \pi t/p) - 5.345\ \mathrm{Sin}\ (4\ \pi t/P) + 0.33\ \mathrm{Cos}\ (4\ \pi t/p) - 5.345\ \mathrm{Sin}\ (4\ \pi t/p) -1.868\ \mathrm{Cos}\ (6\ \pi t/p) + 0.563\ \mathrm{Sin}\ (6\ \pi t/p) \quad ...(6.35)$$

Table 6.30 : Analysis of variance of Fourier coefficients of water deficit series at Fatehpur (1981–1993)

S.No.	Harmonics	Degree of freedom	Sum of Squares	Mean Squares	F_{cal}	$F_{tab\ 1\%}$
1	4,5,...26	45	8091.083	179.802	0.665	1.600
2	Residual	630	170313.14	270.338		
3	3	2	6010.279	3005.14	15.793	4.647
4	Residual	673	172393.95	256.157		
1	3,4,...26	47	14101.362	300.029	1.443	1.580
2	Residual	628	130569.79	207.914		
3	2	2	33733.074	16866.54	78.462	4.647
4	Residual	673	144671.15	214.965		
1	2,3,...26	49	47834.436	976.213	1.38	1.570
2	Residual	626	130569.79	208.578		
3	1	2	54248.632	27.124.32	147.031	4.647
4	Residual	673	124155.59	184.481		

Table 6.31 : Analysis of variance of Fourier coefficients of water deficit series at Pali (1981–1993)

S.No.	Harmonics	Degree of freedom	Sum of Squares	Mean Squares	F_{cal}	$F_{tab\ 1\%}$
1	4,5,...26	45	3948.703	87.749	0.42	1.600
2	Residual	630	131725.16	209.088		
3	3	2	463.457	231.728	9.827	4.647
4	Residual	673	135210.41	200.907		
1	3,4,...26	47	4412.159	93.876	0.504	1.580
2	Residual	628	116941.06	186.212		
3	2	2	14320.648	7160.324	39.71	4.647
4	Residual	673	121353.22	180.317		
1	2,3,...26	49	18732.807	382.302	0.482	1.570
2	Residual	626	116941.06	186.807		
3	1	2	43105.653	21552.83	156.696	4.647
4	Residual	673	92568.214	137.546		

Table 6.32 : Analysis of variance of Fourier coefficients of water deficit series at Jaipur (1981–1993)

S.No.	Harmonics	Degree of freedom	Sum of Squares	Mean Squares	F_{cal}	$F_{tab\ 1\%}$
1	4,5,...26	45	3962.242	88.05	0.341	1.600
2	Residual	630	162640.28	258.159		
3	3	2	5384.508	2692.254	8.27	4.647
4	Residual	673	161218.01	239.551		
1	3,4,...26	47	9346.749	198.867	0.991	1.580
2	Residual	628	125981.82	200.608		
3	2	2	31273.946	15636.97	77.764	4.647
4	Residual	673	135328.57	201.083		
1	2,3,...26	49	40620.695	828.994	0.948	1.570
2	Residual	626	125981.82	201.249		
3	1	2	60059.816	30029.91	189.69	4.647
4	Residual	673	106542.7	158.31		

Table 6.33 : Analysis of variance of Fourier coefficients of water deficit series at Udaipur (1981–1993)

S.No.	Harmonics	Degree of freedom	Sum of Squares	Mean Squares	F_{cal}	$F_{tab\ 1\%}$
1	4,5,...26	45	1019.353	22.652	0.139	1.600
2	Residual	630	103034.92	163.547		
3	3	2	864.275	432.138	3.324	4.647
4	Residual	673	103190	153.328		
1	3,4,...26	47	1883.628	40.077	0.293	1.580
2	Residual	628	85779.389	136.591		
3	2	2	16391.257	8195.628	62.919	4.647
4	Residual	673	87663.017	130.257		
1	2,3,...26	49	18274.885	372.957	0.281	1.570
2	Residual	626	85779.389	137.028		
3	1	2	50652.277	25326.14	319.173	4.647
4	Residual	673	53401.997	79.349		

Table 6.34 : Analysis of variance of Fourier coefficients of water deficit series at Banswara (1981–1993)

S.No.	Harmonics	Degree of freedom	Sum of Squares	Mean Squares	F_{cal}	$F_{tab\ 1\%}$
1	4,5,...26	45	4156.866	92.375	0.422	1.600
2	Residual	630	137829.57	218.777		
3	3	2	5110.095	2555.048	10.219	4.647
4	Residual	673	136876.34	203.382		
1	3,4,...26	47	9266.961	197.169	1.173	1.580
2	Residual	628	105575.3	168.114		
3	2	2	27144.175	13572.09	79.535	4.647
4	Residual	673	114842.26	170.642		
1	2,3,...26	49	36411.136	743.084	1.121	1.570
2	Residual	626	105575.3	168.651		
3	1	2	54233.032	27116.52	207.962	4.647
4	Residual	673	87753.4	130.391		

Table 6.35 : Analysis of variance of Fourier coefficients of water deficit series at Kota (1981–1993)

S.No.	Harmonics	Degree of freedom	Sum of Squares	Mean Squares	F_{cal}	$F_{tab\ 1\%}$
1	4,5,...26	45	1933.202	42.96	0.225	1.600
2	Residual	630	120411.41	191.129		
3	3	2	874.851	437.425	5.355	4.647
4	Residual	673	121469.76	180.49		
1	3,4,...26	47	2808.053	59.746	0.375	1.580
2	Residual	628	100042.83	159.304		
3	2	2	19493.724	9746.862	63.778	4.647
4	Residual	673	102850.88	152.824		
1	2,3,...26	49	22301.778	455.138	0.359	1.570
2	Residual	626	100042.83	159.813		
3	1	2	66872.747	33436.37	405.659	4.647
4	Residual	673	55471.861	82.425		

Table 6.36 : Fourier decomposition of periodic component of weekly water deficit at Jodhpur (1981-1993)

Order	A_K	B_K	Amplitude	Theta	Explained Variance	Cumulative Explained Variance
1	-4.918	8.514	9.832	-0.524	29.865	29.865
2	0.330	-5.345	5.355	-0.062	8.859	38.723
3	-1.868	0.563	1.951	-1.278	1.176	39.899
4	1.033	-0.093	1.037	-1.481	0.332	40.231
5	-0.915	0.269	0.953	-1.285	0.281	40.512
6	0.871	-0.129	0.880	-1.424	0.239	40.751
7	-1.269	-0.486	1.359	1.205	0.570	41.322
8	0.477	-0.094	0.486	-1.375	0.073	41.395
9	-0.298	-0.332	0.446	0.731	0.061	41.456
10	0.979	0.360	1.043	1.218	0.336	41.793
11	-0.709	-0.167	0.728	1.340	0.164	41.956
12	0.507	0.061	0.511	1.452	0.081	42.037
13	0.382	0.036	0.384	1.476	0.045	42.082
14	-0.111	0.338	0.356	-0.316	0.039	42.121
15	0.509	0.080	0.515	1.414	0.082	42.203
16	0.248	-0.133	0.281	-1.080	0.024	42.228
17	-0.321	0.246	0.405	-0.916	0.051	42.279
18	0.308	-0.358	0.472	-0.711	0.069	42.347
19	-0.356	0.343	0.495	-0.804	0.076	42.423
20	0.395	-1.004	1.079	-0.375	0.359	42.782
21	-0.236	1.297	1.318	-0.180	0.537	43.319
22	0.484	-0.766	0.906	-0.563	0.254	43.573
23	-0.229	0.003	0.229	-1.559	0.016	43.589
24	-0.129	-0.108	0.168	0.872	0.009	43.598
25	-0.427	0.185	0.466	-1.162	0.067	43.665
26	0.666	0.136	0.680	1.369	0.143	43.808

Table 6.37 : Fourier decomposition of periodic component of weekly water deficit at Sriganganagar (1981–1993)

Order	A_K	B_K	Amplitude	Theta	Explained Variance	Cumulative Explained Variance
1	-13.420	4.049	14.017	-1.278	39.554	39.554
2	-0.490	-7.495	7.511	0.065	11.355	50.909
3	0.255	4.954	4.961	0.051	4.954	55.863
4	-0.192	-1.976	1.985	0.097	0.793	56.657
5	-0.244	0.677	0.719	-0.346	0.104	56.761
6	0.769	-0.152	0.784	-1.376	0.124	56.885
7	-0.994	0.025	0.994	-1.545	0.199	57.084
8	0.228	-0.268	0.352	-0.705	0.025	57.109
9	-0.384	0.110	0.400	-1.291	0.032	57.141
10	0.529	-0.110	0.540	-1.366	0.059	57.200
11	-0.487	-0.084	0.494	1.401	0.049	57.249
12	0.856	0.136	0.867	1.414	0.151	57.400
13	-0.898	0.560	1.059	-1.013	0.226	57.625
14	0.513	-0.615	0.801	-0.695	0.129	57.755
15	-0.305	-0.105	0.322	1.241	0.021	57.775
16	0.680	-0.061	0.683	-1.481	0.094	57.869
17	-0.335	1.325	1.366	-0.247	0.376	58.245
18	0.599	-1.097	1.250	-0.499	0.315	58.560
19	-0.447	0.153	0.472	-1.242	0.045	58.605
20	0.139	0.140	0.197	0.782	0.008	58.612
21	0.301	0.183	0.352	1.024	0.025	58.637
22	-0.509	-0.536	0.739	0.760	0.110	58.747
23	0.268	0.704	0.753	0.364	0.114	58.861
24	-0.185	-0.394	0.435	0.440	0.038	58.900
25	0.070	0.127	0.145	0.500	0.004	58.904
26	0.023	0.140	0.142	0.165	0.004	58.908

Table 6.38 : Fourier decomposition of periodic component of weekly water deficit at Fatehpur (1981–1993)

Order	A_K	B_K	Amplitude	Theta	Explained Variance	Cumulative Explained Variance
1	-9.397	8.513	12.680	-0.835	30.415	30.415
2	-0.987	-9.926	9.975	0.099	18.823	49.238
3	-1.076	4.071	4.211	-0.258	3.355	52.593
4	1.065	-1.059	1.502	-0.788	0.427	53.020
5	-1.125	0.820	1.475	-0.981	0.411	53.431
6	0.925	-0.453	1.029	-1.116	0.200	53.632
7	-0.746	0.103	0.754	-1.433	0.107	53.739
8	0.793	-0.690	1.051	-0.855	0.209	53.948
9	-1.030	-0.198	1.049	1.381	0.208	54.156
10	1.453	1.080	1.811	0.932	0.620	54.777
11	-0.914	-0.894	1.279	0.797	0.309	55.086
12	1.460	0.162	1.469	1.460	0.408	55.494
13	-0.977	0.557	1.125	-1.053	0.239	55.734
14	0.580	-0.318	0.662	-1.070	0.083	55.816
15	-0.313	-0.174	0.359	1.064	0.024	55.841
16	0.335	0.251	0.418	0.928	0.033	55.874
17	0.003	-0.252	0.252	-0.010	0.012	55.886
18	0.385	-0.611	0.722	-0.562	0.099	55.984
19	-0.961	-0.080	0.964	1.488	0.176	56.160
20	0.269	0.649	0.702	0.394	0.093	56.254
21	-0.738	-0.654	0.986	0.846	0.184	56.438
22	0.487	0.609	0.780	0.675	0.115	56.553
23	0.048	-0.624	0.626	-0.076	0.074	56.627
24	-0.570	0.955	1.112	-0.538	0.234	56.861
25	0.231	-1.237	1.258	-0.185	0.299	57.160
26	0.748	0.482	0.890	0.999	0.150	57.310

Table 6.39 : Fourier decomposition of periodic component of weekly water deficit at Pali (1981–1993)

Order	A_K	B_K	Amplitude	Theta	Explained Variance	Cumulative Explained Variance
1	6.604	9.106	11.302	-0.634	31.775	31.775
2	-1.352	-6.357	6.499	0.210	10.508	42.283
3	-0.689	0.975	1.194	-0.615	0.355	42.638
4	-0.001	-0.261	0.261	0.004	0.017	42.655
5	0.036	0.680	0.681	0.053	0.115	42.770
6	0.219	-0.273	0.350	-0.676	0.031	42.800
7	-1.272	-0.487	1.362	1.205	0.461	43.262
8	1.027	-0.264	1.060	-1.320	0.280	43.541
9	-0.842	0.122	0.851	-1.427	0.180	43.721
10	1.106	0.710	1.314	1.000	0.430	44.151
11	-0.786	-0.657	1.024	0.875	0.261	44.412
12	0.584	0.200	0.617	1.241	0.095	44.507
13	0.014	-0.161	0.161	-0.086	0.006	44.513
14	0.102	0.315	0.331	0.314	0.027	44.541
15	0.613	-0.141	0.629	-1.345	0.098	44.639
16	0.030	0.040	0.050	0.650	0.001	44.640
17	-0.359	0.115	0.377	-1.259	0.035	44.675
18	0.391	0.077	0.398	1.376	0.039	44.714
19	0.259	0.101	0.278	1.198	0.019	44.734
20	0.146	-0.633	0.650	-0.227	0.105	44.839
21	-0.440	1.010	1.101	-0.411	0.302	45.140
22	0.463	-0.969	1.074	-0.446	0.287	45.427
23	-0.036	0.602	0.603	-0.060	0.090	45.518
24	-0.580	-0.381	0.694	0.990	0.120	45.638
25	0.064	0.362	0.368	0.174	0.034	45.671
26	0.144	-0.085	0.167	-1.036	0.007	45.678

Table 6.40 : Fourier decomposition of periodic component of weekly water deficit at Jaipur (1981–1993)

Order	A_K	B_K	Amplitude	Theta	Explained Variance	Cumulative Explained Variance
1	-7.201	11.209	13.322	-0.571	35.954	35.954
2	0.295	-9.594	9.599	-0.031	18.666	54.620
3	-2.387	3.219	4.007	-0.638	3.253	57.873
4	0.861	-1.003	1.322	-0.710	0.354	58.227
5	-0.096	0.223	0.242	-0.407	0.012	58.239
6	0.073	-0.440	0.446	-0.165	0.040	58.280
7	-0.489	0.212	0.534	-1.161	0.058	58.337
8	0.301	-0.702	0.764	-0.405	0.118	58.456
9	-0.796	0.072	0.799	-1.480	0.129	58.585
10	0.957	0.111	0.964	1.455	0.188	58.773
11	-0.446	-0.450	0.633	0.780	0.081	58.854
12	-0.078	0.880	0.883	-0.089	0.158	59.012
13	-0.198	-0.710	0.737	0.272	0.110	59.122
14	-0.155	0.636	0.654	-0.239	0.087	59.209
15	0.733	-0.672	0.994	-0.828	0.200	59.409
16	-0.362	0.946	1.012	-0.365	0.208	59.617
17	0.537	-0.248	0.592	-1.138	0.071	59.688
18	-0.069	-0.027	0.075	1.195	0.001	59.689
19	-0.112	-0.380	0.396	0.287	0.032	59.721
20	0.152	-0.185	0.239	-0.689	0.012	59.732
21	-0.176	0.335	0.379	-0.483	0.029	59.761
22	-0.629	-0.447	0.772	0.953	0.121	59.882
23	0.423	0.225	0.479	1.081	0.046	59.929
24	-0.234	-0.305	0.384	0.654	0.030	59.958
25	0.435	0.180	0.471	1.178	0.045	60.003
26	-0.513	-0.026	0.514	1.521	0.054	60.057

Table 6.41 : Fourier decomposition of periodic component of weekly water deficit at Udaipur (1981–1993)

Order	A_K	B_K	Amplitude	Theta	Explained Variance	Cumulative Explained Variance
1	-4.249	11.484	12.244	-0.354	48.629	48.629
2	-1.090	-6.870	6.956	0.157	15.695	64.324
3	-0.425	1.537	1.595	-0.270	0.825	65.149
4	0.303	-0.127	0.329	-1.173	0.035	65.184
5	-0.432	0.051	0.435	-1.452	0.061	65.245
6	-0.263	0.245	0.360	-0.822	0.042	65.287
7	0.082	-0.490	0.497	-0.165	0.080	65.367
8	-0.136	-0.131	0.189	0.804	0.012	65.379
9	-0.163	-0.005	0.163	1.542	0.009	65.388
10	0.640	0.302	0.707	1.130	0.162	65.550
11	-0.094	0.277	0.292	0.327	0.028	65.577
12	0.056	0.099	0.114	0.514	0.004	65.582
13	0.282	-0.057	0.288	-1.371	0.027	65.609
14	-0.004	0.548	0.548	-0.007	0.098	65.706
15	0.347	-0.123	0.368	-1.231	0.044	65.750
16	-0.122	0.072	0.142	-1.035	0.007	65.757
17	-0.158	0.147	0.216	-0.824	0.015	65.772
18	0.042	-0.003	0.043	-1.491	0.001	65.772
19	-0.333	0.145	0.363	-1.160	0.043	65.815
20	0.588	-0.451	0.741	-0.917	0.178	65.993
21	-0.447	0.071	0.452	-1.412	0.066	66.059
22	0.184	0.023	0.186	1.446	0.011	66.071
23	-0.048	-0.191	0.197	0.246	0.013	66.083
24	-0.256	0.070	0.265	-1.306	0.023	66.106
25	0.166	0.060	0.176	1.224	0.010	66.116
26	-0.071	0.030	0.077	-1.169	0.002	66.118

Table 6.42 : Fourier decomposition of periodic component of weekly water deficit at Banswara (1981–1993)

Order	A_K	B_K	Amplitude	Theta	Explained Variance	Cumulative Explained Variance
1	-5.030	11.616	12.658	-0.409	38.084	38.084
2	1.886	-8.775	8.975	-0.212	19.147	57.231
3	-0.257	3.906	3.914	-0.066	3.642	60.873
4	-0.609	-1.968	2.060	0.300	1.009	61.882
5	0.464	0.446	0.644	0.805	0.099	61.981
6	-0.965	0.076	0.968	-1.493	0.223	62.203
7	0.390	-0.625	0.737	-0.558	0.129	62.332
8	0.085	0.578	0.585	0.147	0.081	62.414
9	-0.342	-0.625	0.713	0.500	0.121	62.534
10	-0.040	0.910	0.911	-0.044	0.197	62.732
11	-0.168	-0.921	0.936	0.181	0.208	62.940
12	-0.004	0.951	0.951	-0.004	0.215	63.155
13	0.196	-0.533	0.568	-0.352	0.077	63.232
14	0.141	0.578	0.595	0.240	0.084	63.316
15	-0.189	-0.451	0.489	0.396	0.057	63.373
16	-0.023	0.466	0.467	-0.049	0.052	63.424
17	-0.062	-0.264	0.271	0.231	0.017	63.442
18	0.485	-0.047	0.488	-1.475	0.057	63.498
19	-0.328	0.-320	0.458	-0.797	0.050	63.548
20	0.243	-0.713	0.753	-0.328	0.135	63.683
21	-0.445	0.428	0.617	-0.805	0.091	63.774
22	-0.099	-0.537	0.546	0.182	0.071	63.845
23	0.369	0.482	0.607	0.653	0.088	63.932
24	-0.407	-0.242	0.474	1.035	0.053	63.986
25	0.332	0.231	0.404	0.964	0.039	64.025
26	-0.235	-0.400	0.464	0.531	0.051	64.076

Table 6.43 : Fourier decomposition of periodic component of weekly water deficit at Kota (1981–1993)

Order	A_K	B_K	Amplitude	Theta	Explained Variance	Cumulative Explained Variance
1	-5.398	12.987	14.065	-0.394	54.568	54.568
2	0.040	-7.575	7.575	-0.005	15.828	70.397
3	-0.700	1.439	1.600	-0.453	0.707	71.103
4	0.651	-0.623	0.901	-0.807	0.224	71.327
5	-0.581	0.756	0.954	-0.655	0.251	71.578
6	0.260	-0.608	0.661	-0.403	0.121	71.698
7	-0.570	0.478	0.744	-0.873	0.153	71.851
8	0.549	-0.269	0.611	-1.115	0.103	71.954
9	-0.428	0.432	0.608	-0.781	0.102	72.056
10	0.483	-0.156	0.508	-1.258	0.071	72.128
11	-0.231	0.212	0.314	-0.829	0.027	72.155
12	0.097	0.008	0.097	1.490	0.003	72.157
13	0.127	-0.088	0.154	-0.966	0.007	72.164
14	0.384	0.275	0.472	0.949	0.061	72.225
15	-0.405	-0.228	0.465	1.059	0.060	72.285
16	0.362	0.220	0.424	1.024	0.050	72.335
17	-0.262	0.166	0.310	-1.007	0.027	72.361
18	0.272	-0.464	0.538	-0.531	0.080	72.441
19	-0.062	0.261	0.269	-0.234	0.020	72.461
20	0.011	-0.376	0.376	-0.028	0.039	72.500
21	-0.018	0.391	0.391	-0.046	0.042	72.542
22	0.011	-0.132	0.132	-0.082	0.005	72.547
23	-0.222	-0.090	0.239	1.185	0.016	72.563
24	0.022	0.169	0.171	0.129	0.008	72.571
25	-0.264	-0.372	0.456	0.618	0.057	72.628
26	0.191	0.163	0.251	0.864	0.017	72.646

Similarly, at Sriganganagar (Table 6.37), Fatehpur (Table 6.38), Jaipur (Table 6.40) and Banswara (Table 6.43) confirms that only first three harmonics are significant for weekly water deficit time series. It can be seen that the three harmonics explain more than

55, 52, 57 and 60 per cent of the variance respectively. This further confirmed that only three harmonics are significant as rest of the harmonics explain only less than 4, 5, 3 and 4 per cent variance at Sriganganagar, Fatehpur, Jaipur and Banswara respectively. Therefore, three harmonics may be used to express the periodic component of the weekly water deficit time series.

At Sriganganagar for the first three hamonics the value of Fourier coefficients (A_1, A_2 and A_3 and B_1, B_2, and B_3) were found to be -13.420, 0.490, 0.255, 4.049, -7.495 and 4.954 respectively. With these coefficients and using Equation (6.9) the periodic component $P_{(t)}$ resulting from periodic deterministic process may mathematically be expressed as:

$$\begin{aligned} P_t = {} & 19.578 - 13.420 \text{ Cos } (2\pi t/p) + 4.049 \text{ Sin } (2\pi t/p) \\ & -0.490 \text{ Cos } (4\pi t/p) - 7.495 \text{ Sin } (4\pi t/p) \\ & + 0.255 \text{ Cos } (6\pi t/p) + 4.954 \text{ Sin } (6\pi t/p) \end{aligned} \quad \text{... (6.36)}$$

At Fatehpur for the first three harmonics the values of Fourier coefficients (A_1, A_2 and A_3 and B_1, B_2 and B_3) were found to be –9.397, –0.987, –1.076, 8.513, –9.926 and 4.071 respectively. With these coefficients and using the Equation (6.9) the periodic component P_t resulting from periodic process may mathematically be expressed as:

$$\begin{aligned} P_t = {} & 24.095 - 9.397 \text{ Cos } (2\pi t/p) + 8.513 \text{ Sin } (2\pi t/p) \\ & -0.987 \text{ Cos } (4\pi t/p) - 9.926 \text{ Sin } (4\pi t/p) \\ & -1.076 \text{ Cos } (6\pi t/p) + 4.071 \text{ Sin } (6\pi t/p) \end{aligned} \quad \text{... (6.37)}$$

At Jaipur for the first three harmonics the values of Fourier coefficients (A_1, A_2 and A_3 and B_1, B_2 and B_3) were found to be –7.201, 0.295, – 2.387, 11–209, –9.594 and 3.219 respectively. With these coefficients and using Equation (6.9) the periodic component P_t resulting from periodic process may mathematically be expressed as:

$$\begin{aligned} P_t = {} & 20.643 - 7.201 \text{ Cos } (2\pi t/p) + 11.209 \text{ Sin } (2\pi t/p) \\ & + 0.295 \text{ Cos } (4\pi t/p) - 9.594 \text{ Sin } (4\pi t/p) \\ & -2.387 \text{ Cos } (6\pi t/p) + 3.219 \text{ Sin } (6\pi t/p) \end{aligned} \quad \text{... (6.38)}$$

At Banswara for the first three harmonics the values of Fourier coefficients (A_1, A_2 and A_3 and B_1, B_2 and B_3) were found to be –5.030, 1.886, –0.257, 11.616, –8.775 and 3.906 respectively. With these coefficients and using Equation (6.9) the periodic component (P_t) resulting from periodic process may be expressed as:

$$P_t = 17.025 - 5.030 \text{ Cos } (2\pi t/p) + 11.616 \text{ Sin } (2\pi t/p) + 1.886 \text{ Cos } (4\pi t/p) - 8.775 \text{ Sin } (4\pi t/p) - 2.257 \text{ Cos } (6\pi t/p) + 3.906 \text{ Sin } (6\pi t/p) \quad \text{... (6.39)}$$

Similarly, it can be seen from Tables 6.39, 6.41 and 6.43 that two harmonics explain more than 42, 64 and 70 per cent of the variance at Pali, Udaipur and Kota respectively. This further confirms that only two harmonics are signiticant as rest of the harmonics explain only less than 4, 2 and 3 per cent at Pali. Udaipur and Kota respectively. Therefore, two harmonics may be used to express the periodic component of the weekly water deficit time series for these stations.

At Pali for the first two harmonics the values of Fourier coefficients (A_1, A_2 and B_1, B_2) were found to be –6.694, –1.352, 9.106 and 6.357 respectively. With these coefficients and using Equation (6.9) the periodic component P_t resulting from periodic process may mathematically be expressed as:

$$P_t = 26.217 - 6.694 \text{ Cos } (2\pi t/p) + 9.106 \text{ Sin } (2\pi t/p) - 1.352 \text{ Cos } (4\pi t/p) - 6.357 \text{ Sin } (4\pi t/p) \quad \text{...(6.40)}$$

At Udaipur for the first two harmonics the values of Fourier coefficients (A_1, A_2 and B_1, B_2) were found to be 4.249, –1.090. 11.484 and –6.870 respectively. With these coefficients and using Equation (3.28) the periodic component (P_t) resulting from periodic process may mathematically be expressed as:

$$P_t = 16.867 - 4.244 \text{ Cos } (2\pi t/p) + 11.484 \text{ Sin } (2\pi t/p) - 1.090 \text{ Cos } (4\pi t/p) - 6.870 \text{ Sin } (4\pi t/p) \quad \text{...(6.41)}$$

Similarly, at Kota tor the first two harmonics the values of Fourier coefficients (A_1, A_2 and B_1, B_2) were found to be -5.398, 0.040, 12.987 and –7.575 respectively. With these coefficients and using Equation (6.9) the periodic component (P_t) resulting from periodic process may mathematically be expressed as:

$$P_t = 18.251 - 5.398 \text{ Cos } (2\pi t/p) + 12.987 \text{ Sin } (2\pi t/p) - 0.040 \text{ Cos } (4\pi t/p) - 7.575 \text{ Sin } (4\pi t/p) \quad \text{...(6.42)}$$

The deterministic cyclic component (P_t) was computed by using Equations 6.9 to 6.16 for all the values of t (t_{max} = 676) for all the stations. After estimating the periodic component it was removed form original time series by subtracting the periodic component from observed time series. By this proccss new stationary series, S_t, were obtained from stochastic non-deterministic process for different stations.

6.4.1.4 Stochastic component

The presence of stochastic component was already confirmed by plotting the correlogram for all the stations (Figures 6.1 to 6.8) of the observed series and analysis of SCC and CV which have been discussed in Subsection 4.1.1.1 and 4.1.1.2.

As discussed earlier the periodic component was removed from the historical time series of water deficit. The rest of the data were analysed to obtain non-deterministic stochastic component by fitting the autoregressive stochastic models.

6.4.1.4.1 Estimation of autoregressive parameters

The autoregressive parameters were estimated by using Equations (6.25), (6.26), (6.28) and (6.29). The estimated values of autocovariance function and SCC of different lags (max - 150) are presented in Tables 6.44 to 6.51 for different stations. The values of autocovariance function given in Tables 6.44 to 6.51 indicate the linear dependence between each lag for all the stations. A study of Tables 6.44 to 6.51 reveals that the values of SCC are significantly different from zero which confirms the dependence of present value and past values at all the stations. In other words, it may be concluded that the past and present values are highly interrelated.

6.4.1.4.2 Selection of model order

Residual variance method was used to determine the order of the model which may significantly represent the non-deterministic stationary stochastic component.

Residual variance at different lags were estimated with the help of Equations (6.26) and (6.27). The estimated values for 150 lags are given in Tables 6.44 to 6.51 for different stations.

It can be seen from Table 6.44 that at Jodhpur the minimum residual variance was obtained for order nine. However, order three is selected, since no significant difference was obtained in residual variance at lag 3 and lag 9 and other lower values of different lags. Reducing the complexity of the model the 3rd order was selected to represent the stochastic component by the autoregressive model.

Table 6.45 shows that at Sriganganagar, the minimum residual variance was obtained for order three. The values of residual variance after order 3 showed no definite trend. However, at lag 64 the residual variance was lower than the value at lag 3.

Since no significant difference was obtained in residual variance. at lag 3, lag 64 and lower values of different lags. Reducing the complexity of the model the 3rd order was selected to represent the stochastic component by the autoregressive model.

Table 6.46 reveals that at Fatehpur the minimum residual variance was obtained for order two. The values of residual variance after 3rd order showed no definite trend, however, at lag 14 the residual variance was lower than the value at lag 2. Since no significant difference was obtained in residual variances at lag 2, and lag 14 and other lower values of different lags. Therefore, reducing the complexity of the model the 2nd order was selected to represent the stochastic component by the autoregressive model.

Table 6.47 depicts that at Pali the minimum residual variance was obtained for order fourteen and fifteen. However, lag 3, 14 and 15 and other lower values the residual variance have no significant difference. Therefore, reducing the complexity of the model the 3rd order was selected to represent the stochastic component by the autoregressive model.

Table 6.48 shows that the minimum residual variance was obtained at Jaipur for order three. The values of residual variance after 4th order showed no definite trend, however, at lag 9 and 10 the residual variance was obtained in residual variance at lag 3, lag 9 and 10 and other lower values of different lags. Therefore, reducing the complexity of the model the 3rd order was selected to represent the stochastic component by the autoregressive model.

Table 6.49 reveals that at Udaipur the minimum residual variance was obtained for order two. The values of residual variance after 3rd order showed no definite trend. Therefore, the 2nd order was selected to represent the stochastic component by the autoregressive model.

The Table 6.50 indicates that at Banswara the minimum residual variance was obtained or order two. The values of residual variance after 3rd order showed no definite trend, however, at lag 57, the residual variance was lower than the value at lag 2. Since no significant difference was obtained in residual variance at lag 2, lag 57 and other lower values at different lags. Therefore, reducing the complexity of the model the 2nd order was selected to represent the stochastic component by the autoregressive model.

Table 6.44 : Serial correlation coefficient, autocovariance function and residual variance for different lags of auto regressive model of water deficit at Jodhpur (1981-93)

Lag	Serial correlation coefficient	Autocovariance function	Residual variance
(1)	(2)	(3)	(4)
1	0.363	24794.009	0.871
2	0.305	20819.958	0.838
3	0.229	15622.791	0.833
4	0.226	15426.580	0.827
5	0.221	15077.241	0.822
6	0.158	10770.777	0.823
7	0.100	6808.234	0.823
8	0.142	9646.214	0.821
9	0.151	10258.648	0.819
10	0.145	9848.649	0.820
11	0.104	7103.716	0.821
12	0.109	7418.676	0.822
13	0.092	6241.163	0.823
14	0.116	7861.014	0.823
15	0.100	6784.892	0.824
16	0.100	6790.414	0.825
17	0.100	6779.983	0.826
18	0.080	5438.589	0.827
19	0.084	5648.030	0.829
20	0.056	3770.432	0.830
21	0.052	3531.032	0.831
22	0.063	4255.779	0.832
23	0.046	3121.509	0.833
24	0.050	3337.421	0.835
25	0.064	4262.062	0.836
26	0.065	4322.022	0.837
27	0.075	4898.583	0.838
28	0.063	4089.097	0.839
29	0.048	3132.525	0.841
30	0.099	6472.347	0.839
31	0.104	6762.166	0.839
32	0.088	5704.432	0.841
33	0.088	5687.210	0.842
34	0.086	5557.248	0.843
35	0.081	5246.204	0.845

contd.....

Table 6.44 – contd...

(1)	(2)	(3)	(4)
36	0.084	5405.131	0.846
37	0.073	4690.108	0.848
38	0.062	3975.223	0.849
39	0.011	695.285	0.848
40	0.000	–12.004	0.848
41	–0.006	–411.332	0.849
42	–0.013	–814.447	0.850
43	0.007	425.609	0.851
44	–0.077	–4851.133	0.846
45	–0.013	–838.793	0.847
46	–0.023	–1462.939	0.848
47	0.012	743.095	0.849
48	–0.029	–1840.956	0.850
49	0.032	2040.435	0.848
50	0.029	1832.982	0.849
51	0.014	–902.286	0.849
52	0.076	4787.541	0.843
53	0.117	7391.440	0.835
54	0.121	7643.624	0.834
55	–0.015	–923.914	0.821
56	0.013	791.195	0.822
57	0.051	3232.150	0.823
58	0.041	2598.516	0.824
59	0.064	4011.004	0.825
60	0.059	3686.572	0.824
61	0.003	206.861	0.823
62	0.050	3105.638	0.824
63	0.042	2640.893	0.826
64	0.061	3853.316	0.827
65	0.076	4773.725	0.827
66	0.080	5019.882	0.828
67	0.118	7390.772	0.827
68	0.097	6040.939	0.828
69	0.103	6326.281	0.830
70	0.102	6215.301	0.831
71	0.091	5455.685	0.832
72	0.100	5979.638	0.833
73	0.108	6397.646	0.834

contd.....

Table 6.44 – contd...

(1)	(2)	(3)	(4)
74	0.060	3550.250	0.836
75	0.052	3047.030	0.837
76	0.015	894.021	0.837
77	0.017	981.129	0.839
78	0.060	3498.390	0.839
79	0.057	3319.879	0.841
80	0.079	4583.404	0.842
81	0.080	4666.640	0.843
82	0.073	4245.874	0.845
83	0.122	7052.513	0.846
84	0.134	7690.566	0.848
85	0.070	4003.858	0.849
86	0.077	4441.346	0.851
87	0.107	6110.805	0.851
88	0.080	4566.106	0.853
89	0.088	5038.030	0.855
90	0.031	1781.299	0.853
91	0.011	623.073	0.855
92	0.046	2652.570	0.857
93	–0.029	–1660.932	0.857
94	0.065	3692.984	0.853
95	–0.017	–987.433	0.850
96	–0.036	–2049.729	0.849
97	–0.013	–721.403	0.851
98	–0.027	–1512.642	0.853
99	0.018	–1036.942	0.855
100	0.006	315.134	0.857
101	0.017	958.738	0.858
102	0.091	5164.965	0.856
103	0.109	6171.193	0.857
104	0.045	2573.639	0.859
105	0.061	3469.037	0.861
106	0.096	5423.495	0.862
107	–0.005	–271.392	0.859
108	–0.013	–752.120	0.861,
109	–0.037	–2082.374	0.864
110	0.045	–2568.451	0.864
111	–0.003	–184.031	0.867
112	–0.015	–827.521	0.869

contd.....

Table 6.44 – contd...

(1)	(2)	(3)	(4)
113	–0.048	–2699.274	0.870
114	0.037	2087.232	0.868
115	–0.058	–3273.049	0.868
116	0.008	469.329	0.871
117	–0.014	–792.510	0.873
118	0.065	3653.973	0.870
119	0.029	1649.589	0.871
120	0.030	1702.993	0.874
121	0.016	890.519	0.876
122	–0.009	–475.880	0.878
123	–0.026	–1464.176	0.880
124	–0.009	–522.824	0.883
125	–0.013	–713.039	0.885
126	0.009	502.432	0.887
127	0.001	55.257	0.890
128	–0.055	–2996.893	0.891
129	–0.048	–2619.413	0.894
130	–0.036	–1962.151	0.896
131	–0.022	–1206.174	0.898
132	0.014	772.189	0.901
133	0.014	753.552	0.903
134	0.006	298.628	0.906
135	0.038	2004.406	0.909
136	0.061	3244.429	0.911
137	0.042	2217.642	0.912
138	0.041	2148.877	0.913
139	0.044	2307.673	0.916
140	0.052	2726.606	0.919
141	0.066	–3467.652	0.922
142	–0.017	–898.465	0.924
143	0.010	545.395	0.926
144	–0.010	–549.949	0.929
145	–0.025	–1304.971	0.932
146	–0.047	–2464.613	0.934
147	–0.087	–4539.271	0.935
148	–0.009	–462.362	0.938
149	–0.104	–5446.179	0.936
150	–0.105	–5494.341	0.939

Table 6.45 : Serial correlation coefficient, autocovariance function and residual variance for different lags of auto regressive model of water deficit at Sriganganagar (1981-93)

Auto Covariance Function at lag zero (C_0) = 73801.901

Lag	Serial correlation coefficient	Autocovariance function	Residual variance
(1)	(2)	(3)	(4)
1	0.393	29004.151	0.848
2	0.187	13805.349	0.848
3	0.156	11496.381	0.843
4	0.125	9234.580	0.843
5	0.060	4436.220	0.844
6	0.087	6440.312	0.842
7	0.042	3080.898	0.843
8	0.024	1774.663	0.844
9	0.063	4684.114	0.843
10	0.060	4464.799	0.845
11	0.026	1908.560	0.846
12	0.035	2546.972	0.847
13	–0.022	–1612.587	0.845
14	–0.006	–467.769	0.846
15	0.003	187.355	0.847
16	–0.056	–4085.748	0.845
17	–0.039	–2823.203	0.846
18	–0.045	–3275.409	0.846
19	–0.079	–5737.902	0.845
20	–0.064	–4655.760	0.847
21	–0.049	–3533.673	0.848
22	–0.022	–1618.581	0.849
23	0.015	1062.013	0.849
24	–0.012	–842.994	0.849
25	–0.006	–447.427	0.850
26	0.041	2926.350	0.850
27	0.035	2451.462	0.851
28	0.017	1200.345	0.852
29	–0.038	–2689.787	0.850
30	–0.068	–4785.902	0.849
31	–0.037	–2597.414	0.851
32	–0.007	–465.894	0.852
33	–0.052	–3662.394	0.852
34	–0.024	–1671.149	0.853

contd.....

Table 6.45 – contd...

(1)	(2)	(3)	(4)
35	–0.046	–3234.661	0.852
36	0.031	2198.517	0.847
37	0.048	3333.346	0.849
38	0.023	1596.767	0.850
39	0.048	3315.474	0.847
40	0.136	9437.828	0.841
41	0.086	5997.190	0.842
42	0.017	1189.505	0.843
43	0.027	1903.506	0.844
44	0.011	770.018	0.846
45	–0.017	–1158.179	0.847
46	–0.062	–4296.199	0.841
47	–0.065	–4483.744	0.842
48	–0.003	–185.316	0.842
49	0.040	2778.256	0.844
50	–0.005	–326.454	0.844
51	0.000	–32.157	0.846
52	–0.014	–978.464	0.847
53	–0.103	–7165.627	0.842
54	–0.056	–3849.187	0.844
55	–0.055	–3791.653	0.845
56	–0.078	–5374.769	0.847
57	–0.178	–12326.210	0.833
58	–0.106	–7367.665	0.834
59	–0.053	–3705.834	0.835
60	–0.007	–456.066	0.835
61	–0.010	–713.734	0.836
62	–0.019	–1292.642	0.837
63	0.057	3974.815	0.834
64	0.111	7655.372	0.830
65	0.067	4618.147	0.832
66	0.060	4133.811	0.833
67	0.038	2596.837	0.835
68	0.080	5478.193	0.834
69	0.037	2478.757	0.836
70	0.039	2625.621	0.837
71	0.063	4233.828	0.838
72	0.061	4085.003	0.840
73	0.026	1713.718	0.841

contd.....

Table 6.45 – contd...

(1)	(2)	(3)	(4)
74	0.046	3079.620	0.842
75	0.066	4380.740	0.844
76	0.056	3715.157	0.846
77	0.068	4474.838	0.847
78	0.025	1666.196	0.848
79	0.008	525.412	0.850
80	–0.013	–836.674	0.851
81	–0.001	–74.146	0.853
82	–0.011	–715.555	0.855
83	–0.002	–128.403	0.856
84	–0.008	–547.604	0.858
85	–0.013	–843.429	0.860
86	0.001	41.819	0.862
87	–0.006	–414.254	0.864
88	0.000	–9.858	0.866
89	0.034	2186.450	0.867
90	0.018	1181.999	0.869
91	0.017	1094.214	0.871
92	0.080	5176.670	0.870
93	0.001	82.538	0.872
94	0.008	520.177	0.872
95	–0.007	–432.705	0.873
96	–0.009	–558.900	0.875
97	–0.034	–2180.071	0.878
98	–0.066	–4279.789	0.877
99	–0.040	–2597.584	0.880
100	–0.015	–996.837	0.881
101	–0.038	–2451.236	0.883
102	–0.023	–1518.399	0.885
103	–0.008	–522.418	0.886
104	–0.013	–859.509	0.886
105	–0.058	–3778.967	0.888
106	–0.034	–2193.922	0.890
107	–0.049	–3193.760	0.892
108	–0.077	–4969.849	0.891
109	0.003	–187.347	0.890
110	–0.058	–3745.284	0.889
111	–0.074	–4768.928	0.890
112	–0.065	–4195.339	0.891

contd.....

Table 6.45 – contd...

(1)	(2)	(3)	(4)
113	–0.034	–2210.651	0.892
114	–0.031	–1976.781	0.895
115	0.024	1563.433	0.897
116	–0.027	–1716.060	0.899
117	–0.062	–4017.655	0.901
118	–0.043	–2758.526	0.903
119	–0.045	–2872.674	0.905
120	0.001	58.442	0.903
121	–0.037	–2339.663	0.906
122	–0.023	–1417.649	0.907
123	–0.025	–1572.939	0.909
124	0.008	528.165	0.911
125	0.014	863.471	0.911
126	0.057	3480.066	0.913
127	0.044	2698.376	0.916
128	0.048	2933.351	0.918
129	0.040	2412.777	0.921
130	0.021	1248.040	0.923
131	0.032	1912.444	0.926
132	0.058	3412.502	0.929
133	0.012	704.088	0.931
134	0.007	380.558	0.933
135	0.030	1723.005	0.936
136	0.033	1926.648	0.939
137	0.032	1848.975	0.942
138	0.019	1097.663	0.945
139	0.032	1856.481	0.947
140	0.086	4910.451	0.947
141	0.114	6526.367	0.949
142	0.064	3635.893	0.952
143	0.048	2722.846	0.955
144	–0.070	–3960.542	0.952
145	–0.092	–5177.995	0.954
146	–0.069	–3906.572	0.957
147	–0.001	–79.022	0.956
148	–0.049	–2749.024	0.959
149	–0.028	–1569.307	0.962 –
150	–0.020	–1139.160	0.965

Table 6.46 : Serial correlation coefficient, autocovariance function and residual variance for different lags of auto regressive model of water deficit at Fatehpur (1981-93)

Auto Covariance Function at lag zero (C_0) = 84649.964

Lag	Serial correlation coefficient	Autocovariance function	Residual variance
(1)	(2)	(3)	(4)
1	0.371	31405.137	0.865
2	0.223	18826.228	0.858
3	0.111	9420.736	0.859
4	0.058	4886.351	0.860
5	0.094	7908.865	0.857
6	0.022	1861.545	0.857
7	–0.041	–3431.607	0.854
8	0.002	195.010	0.854
9	–0.021	–1784.318	0.855
10	0.023	1955.249	0.855
11	–0.005	–448.677	0.856
12	–0.010	–815.917	0.858
13	0.030	2569.858	0.858
14	0.088	7439.164	0.853
15	0.067	5635.661	0.855
16	0.006	509.491	0.854
17	–0.009	–774.878	0.855
18	–0.053	–4465.095	0.854
19	–0.040	–3335.996	0.855
20	0.000	–29.610	0.855
21	–0.027	–2231.143	0.856
22	–0.037	–3062.372	0.857
23	–0.034	–2815.505	0.858
24	–0.013	–1096.192	0.860
25	0.023	1945.557	0.861
26	0.023	1885.114	0.862
27	0.022	1848.105	0.863
28	0.000	–14.560	0.864
29	–0.008	–683.481	0.865
30	0.005	405.278	0.867
31	–0.012	–1022.440	0.868
32	0.012	954.491	0.868
33	0.007	562.515	0.870
34	–0.007	–582.489	0.871

contd.....

Table 6.46 – contd...

(1)	(2)	(3)	(4)
35	0.032	2605.671	0.871
36	0.013	1037.771	0.873
37	0.021	1686.349	0.874
38	0.040	3245.243	0.875
39	0.038	3095.449	0.876
40	0.048	3892.203	0.878
41	0.033	2637.070	0.879
42	0.034	2774.608	0.881
43	0.047	3783.376	0.881
44	0.033	2702.531	0.883
45	0.021	1703.556	0.885
46	0.086	6982.380	0.880
47	0.102	8243.541	0.880
48	0.080	6439.156	0.881
49	0.054	4333.097	0.883
50	–0.022	–1788.988	0.881
51	0.033	2645.015	0.881
52	0.017	1406.237	0.882
53	0.013	–1014.376	0.884
54	0.073	5887.202	0.879
55	–0.052	–4196.768	0.872
56	–0.027	–2142.731	0.874
57	–0.06	–4812.384	0.873
58	–0.062	–4999.968	0.875
59	0.009	756.523	0.875
60	0	8.975	0.876
61	0.078	6237.882	0.873
62	0.047	3800.727	0.874
63	0.105	8399.409	0.867
64	0.068	5479.025	0.869
65	0.028	2258.738	0.871
66	0.043	3429.333	0.872
67	0.04	3213.569	0.873
68	0.051	4108.25	0.875
69	0.019	1504.622	0.877
70	0.038	–3028.353	0.878
71	–0.021	–1659.424	0.88
72	–0.017	–1337.577	0.882
73	0.034	2635.197	0.883

contd.....

Table 6.46 – contd...

(1)	(2)	(3)	(4)
74	0.036	2816.676	0.885
75	0.033	2534.904	0.886
76	0.013	1037.06	0.887
77	0.009	701.517	0.889
78	0.034	2638.932	0.891
79	–0.003	–202.109	0.892
80	–0.015	–1117.094	0.894
81	–0.004	–311.585	0.896
82	–0.018	–1381.789	0.898
83	0.026	1992.715	0.899
84	0.004	337.31	0.901
85	–0.002	–138.459	0.903
86	0.001	64.442	0.904
87	0.003	200.743	0.906
88	0.013	958.686	0.908
89	0.009	699.842	0.91
90	–0.013	–1000.211	0.912
91	0.004	278.253	0.914
92	0.001	67.73	0.916
93	–0.037	–2780.161	0.915
94	0.023	1752.445	0.915
95	0.038	2898.97	0.915
96	0.024	1775.21	0.918
97	–0.004	–319.933	0.92
98	0.036	2697.515	0.921
99	0.033	2489.636	0.923
100	0.089	6669.675	0.923
101	–0.008	–617.834	0.921
102	0.000	–34.709	0.923
103	–0.002	–125.264	0.925
104	0.002	118.338	0.927
105	–0.038	–2865.208	0.924
106	–0.033	–2504.941	0.926
107	–0.077	–5777.840	0.926
108	0.012	902.652	0.928
109	–0.019	–1425.910	0.930
110	0.017	1271.500	0.933
111	–0.003	–204.451	0.935
112	0.022	1665.922	0.937

contd.....

Table 6.46 – contd...

(1)	(2)	(3)	(4)
113	–0.023	–1691.327	0.936
114	–0.017	–1275.987	0.938
115	–0.037	–2747.619	0.939
116	–0.019	–1388.692	0.940
117	–0.022	–1625.463	0.942
118	–0.045	–3372.604	0.945
119	–0.057	–4213.562	0.947
120	–0.043	–3148.422	0.949
121	–0.049	–3657.633	0.952
122	–0.048	–3527.511	0.953
123	–0.082	–6077.479	0.954
124	–0.055	–4018.020	0.957
125	–0.006	–462.757	0.959
126	0.011	765.262	0.962
127	0.014	1047.488	0.964
128	0.017	1243.413	0.966
129	0.063	4524.288	0.967
130	0.033	2370.139	0.969
131	0.034	2473.914	0.972
132	0.020	1442.939	0.975
133	–0.010	–706.284	0.978
134	–0.001	–88.511	0.980
135	0.006	454.023	0.983
136	0.045	3210.047	0.984
137	0.063	4425.982	0.987
138	0.073	5133.209	0.989
139	0.051	3577.280	0.991
140	0.072	5003.003	0.991
141	0.083	5823.081	0.994
142	0.079	5512.226	0.997
143	0.041	2881.159	1.001
144	0.061	4246.664	1.002
145	0.037	2555.355	1.005
146	0.049	3386.002	1.008
147	–0.025	–1757.479	1.010
148	–0.007	–519.683	1.013
149	–0.010	–703.321	1.017
150	–0.006	–435.991	1.020

Table 6.47 : Serial correlation coefficient, autocovariance function and residual variance for different lags of auto regressive model of water deficit at Pali (1981-93)

Auto Covariance Function at lag zero (C_0) = 78581.760

Lag	Serial correlation coefficient	Autocovariance function	Residual variance
(1)	(2)	(3)	(4)
1	0.364	28603.761	0.870
2	0.271	21306.252	0.849
3	0.240	18798.844	0.839
4	0.227	17841.201	0.832
5	0.248	19460.031	0.821
6	0.167	13124.380	0.823
7	0.129	10112.794	0.824
8	0.184	14423.700	0.819
9	0.211	16515.472	0.813
10	0.213	16695.313	0.810
11	0.191	14965.566	0.809
12	0.147	11551.853	0.811
13	0.174	13624.041	0.811
14	0.191	14931.960	0.810
15	0.180	14112.167	0.810
16	0.142	11102.900	0.811
17	0.147	11443.730	0.812
18	0.147	11434.790	0.813
19	0.131	10197.143	0.815
20	0.113	8846.060	0.816
21	0.117	9121.846	0.817
22	0.126	9802.388	0.818
23	0.099	7705.121	0.819
24	0.105	8147.670	0.820
25	0.114	8781.011	0.821
26	0.126	9694.504	0.822
27	0.110	8332.229	0.823
28	0.126	9570.424	0.824
29	0.089	6725.121	0.825
30	0.106	8021.734	0.826
31	0.108	8123.908	0.827
32	0.125	9458.315	0.828
33	0.101	7594.692	0.829
34	0.084	6318.483	0.830

contd.....

Table 6.47 – contd...

(1)	(2)	(3)	(4)
35	0.078	5857.517	0.832
36	0.100	7444.462	0.833
37	0.112	8396.761	0.834
38	0.082	6147.382	0.835
39	0.069	5122.988	0.837
40	0.062	4611.649	0.838
41	0.047	3518.809	0.838
42	0.051	3788.653	0.840
43	0.086	6356.813	0.840
44	0.029	2139.762	0.840
45	0.089	6552.924	0.840
46	0.049	3655.049	0.840
47	0.095	7055.839	0.840
48	0.006	471.340	0.836
49	0.038	2807.012	0.837
50	0.048	3536.367	0.838
51	–0.023	–1735.689	0.835
52	0.040	2947.721	0.835
53	0.075	5541.130	0.833
54	0.093	6876.893	0.833
55	0.021	1517.599	0.831
56	0.003	187.328	0.832
57	0.040	2973.361	0.834
58	0.088	6489.773	0.833
59	0.076	5576.370	0.834
60	0.058	4255.126	0.835
61	0.003	213.579	0.835
62	0.030	2170.528	0.836
63	0.025	1809.941	0.837
64	0.040	2908.329	0.839
65	0.010	766.162	0.841
66	0.042	3068.240	0.840
67	0.059	4309.142	0.841
68	0.071	5177.013	0.843
69	0.046	3345.198	0.844
70	0.051	3680.999	0.846
71	0.065	4663.613	0.847
72	0.057	4060.829	0.849
73	0.088	6315.608	0.849

contd.....

Table 6.47 – contd...

(1)	(2)	(3)	(4)
74	0.049	3487.158	0.851
75	0.032	2252.140	0.851
76	0.018	1242.662	0.853
77	0.022	1528.190	0.855
78	0.054	3779.581	0.856
79	0.050	3528.259	0.857
80	0.080	5590.352	0.857
81	0.061	4236.584	0.859
82	0.067	4671.113	0.861
83	0.086	5936.112	0.863
84	0.092	6343.757	0.864
85	0.063	4375.636	0.866
86	0.054	3700.090	0.868
87	0.025	1702.502	0.870
88	0.020	1359.250	0.871
89	0.055	3804.700	0.872
90	0.041	2798.306	0.874
91	0.054	3723.721	0.875
92	0.036	2472.209	0.876
93	–0.016	–1123.758	0.876
94	0.085	5864.268	0.873
95	0.039	2703.720	0.874
96	–0.036	–2475.825	0.869
97	0.052	3548.699	0.869
98	0.060	4136.280	0.868
99	0.085	5848.745	0.870
100	0.105	7192.820	0.870
101	0.052	3531.559	0.872
102	0.107	7312.874	0.872
103	0.13	8889.371	0.873
104	0.049	3368.792	0.872
105	0.104	7074.039	0.873
106	–0.001	–54.947	0.869
107	–0.019	–1320.172	0.866
108	0.011	756.308	0.868
109	0.083	5660.629	0.863
110	0.059	4012.872	0.865
111	0.031	2136.039	0.866
112	0.027	1844.529	0.867

contd.....

Table 6.47 – contd...

(1)	(2)	(3)	(4)
113	0.019	1322.849	0.867
114	0.023	1588.927	0.869
115	0.005	352.213	0.871
116	0.029	1966.250	0.873
117	–0.013	–894.979	0.874
118	0.03	2035.233	0.876
119	0.007	499.100	0.878
120	0.007	468.562	0.88
121	–0.002	–110.050	0.882
122	–0.03	–2015.798	0.884
123	–0.018	–1177.087	0.887
124	–0.018	–1195.422	0.889
125	0.005	324.989	0.892
126	0	–9.740	0.894
127	0.007	480.465	0.897
128	–0.034	–2213.662	0.899
129	–0.031	–2022.735	0.901
130	–0.05	–3253.902	0.902
131	–0.007	–456.180	0.905
132	0.001	37.799	0.907
133	0.006	383.540	0.909
134	0	–0.117	0.912
135	–0.004	–226.327	0.914
136	0.009	586.636	0.917
137	0.014	861.748	0.92
138	0.011	716.976	0.922
139	0.007	473.477	0.925
140	0.002	115.422	0.927
141	0.009	576.293	0.93
142	–0.024	–1544.698	0.933
143	–0.069	–4352.678	0.933
144	–0.028	–1759.048	0.936
145	–0.015	–934.199	0.938
146	–0.071	–4482.940	0.938
147	–0.13	–8219.142	0.928
148	–0.022	–1378.439	0.929
149	–0.08	–5077.820	0.931
150	–0.097	–6096.161	0.933

Table 6.48 : Serial correlation coefficient, autocovariance function and residual variance for different lags of auto regressive model of water deficit at Jaipur (1981-93)

Auto Covariance Function at lag zero (C_0) = 707117.897

Lag	Serial correlation coefficient	Autocovariance function	Residual variance
(1)	(2)	(3)	(4)
1	0.486	34077.298	0.766
2	0.323	22693.040	0.757
3	0.271	19048.067	0.750
4	0.164	11475.032	0.751
5	0.175	12240.114	0.746
6	0.176	12314.634	0.745
7	0.107	7491.732	0.746
8	0.106	7463.631	0.746
9	0.149	10408.962	0.742
10	0.140	9819.982	0.742
11	0.108	7563.577	0.743
12	0.095	6654.238	0.744
13	0.095	6617.901	0.744
14	0.102	7086.241	0.745
15	0.122	8478.168	0.745
16	0.142	9865.072	0.744
17	0.133	9216.695	0.745
18	0.148	10285.332	0.745
19	0.102	7040.629	0.745
20	0.094	6478.175	0.746
21	0.097	6697.612	0.747
22	0.101	7019.357	0.747
23	0.138	9568.431	0.746
24	0.144	9940.973	0.746
25	0.144	9907.234	0.747
26	0.146	10014.439	0.748
27	0.163	11093.591	0.747
28	0.166	11188.742	0.748
29	0.123	8294.479	0.748
30	0.123	8278.183	0.749
31	0.100	6711.845	0.750
32	0.083	5581.851	0.752
33	0.064	4250.966	0.752
34	0.033	2214.703	0.752

contd.....

Table 6.48 – contd...

(1)	(2)	(3)	(4)
35	0.066	4389.294	0.752
36	0.055	3687.328	0.753
37	0.059	3900.807	0.754
38	0.100	6655.546	0.754
39	0.093	6214.113	0.755
40	0.104	6892.597	0.756
41	0.104	6904.974	0.757
42	0.128	8525.832	0.756
43	0.120	7958.046	0.758
44	0.098	6495.326	0.759
45	0.150	9925.433	0.756
46	0.118	7811.722	0.757
47	0.064	4251.040	0.756
48	0.127	8377.580	0.755
49	0.113	7454.728	0.756
50	0.053	3507.634	0.755
51	0.118	7755.643	0.754
52	0.076	4987.145	0.755
53	0.086	5703.420	0.756
54	0.070	4645.433	0.755
55	0.071	4691.515	0.757
56	0.031	2028.299	0.757
57	0.070	4633.210	0.756
58	0.049	3242.051	0.756
59	0.068	4454.457	0.756
60	0.052	3406.582	0.757
61	−0.019	−1257.762	0.753
62	0.028	1815.372	0.753
63	0.040	2624.384	0.754
64	0.060	3915.362	0.755
65	0.085	5605.282	0.756
66	0.052	3384.343	0.756
67	0.038	2468.225	0.758
68	0.034	2229.453	0.759
69	0.027	1731.542	0.760
70	0.018	1152.161	0.761
71	0.038	2425.497	0.763
72	0.040	2570.740	0.764
73	0.051	3256.872	0.766

contd.....

Table 6.48 – contd...

(1)	(2)	(3)	(4)
74	0.060	3750.588	0.767
75	0.059	3657.387	0.769
76	0.072	4461.488	0.770
77	0.100	6119.748	0.769
78	0.102	6201.631	0.770
79	0.066	4007.428	0.772
80	0.068	4046.439	0.773
81	0.071	4207.429	0.775
82	0.028	1691.749	0.776
83	0.053	3151.456	0.778
84	0.024	1403.563	0.779
85	0.002	110.112	0.781
86	0.004	264.344	0.783
87	–0.008	–495.535	0.783
88	0.011	624.881	0.785
89	0.009	553.275	0.787
90	0.050	2978.096	0.788
91	0.052	3060.987	0.790
92	0.038	2260.961	0.791
93	0.045	2636.914	0.793
94	0.034	1983.825	0.795
95	0.028	1650.631	0.797
96	0.002	139.679	0.798
97	–0.026	–1537.879	0.798
98	–0.006	–342.169	0.799
99	0.049	2832.414	0.798
100	0.052	3004.258	0.800
101	–0.008	–466.451	0.798
102	–0.034	–1986.258	0.798
103	–0.073	–4238.609	0.799
104	–0.019	–1124.479	0.8
105	–0.019	–1111.733	0.802
106	–0.032	–1887.803	0.804
107	–0.049	–2859.105	0.806
108	0.095	5498.848	0.794
109	0.05	2895.591	0.795
110	0.021	1248.626	0.797
111	0.034	1955.266	0.799
112	–0.034	–1989.279	0.801

contd.....

Table 6.48 – contd...

(1)	(2)	(3)	(4)
113	–0.034	–1986.896	0.803
114	–0.032	–1880.799	0.804
115	–0.006	–362.721	0.805
116	–0.008	–469.237	0.806
117	0.007	409.824	0.808
118	0.015	879.739	0.81
119	0.017	1003.544	0.812
120	–0.006	–329.605	0.814
121	–0.047	–2652.74	0.815
122	–0.053	–2972.021	0.816
123	–0.044	–2466.411	0.818
124	–0.05	–2818.898	0.82
125	–0.042	–2285.952	0.822
126	–0.033	–1799.095	0.825
127	–0.033	–1777.175	0.827
128	–0.045	–2401.325	0.829
129	–0.043	–2326.077	0.831
130	–0.015	–780.271	0.831
131	–0.029	–1551.706	0.833
132	–0.018	–962.252	0.836
133	–0.025	–1307.715	0.838
134	–0.025	–1319.593	0.841
135	–0.019	–1010.844	0.843
136	–0.008	–421.49	0.845
137	–0.034	–1759.615	0.848
138	–0.044	–2326.817	0.85
139	–0.032	–1699.7	0.853
140	–0.013	–686.077	0.855
141	–0.038	–1983.957	0.857
142	–0.063	–3313.41	0.859
143	–0.067	–3504.788	0.861
144	–0.06	–3152.885	0.863
145	–0.041	–2160.21	0.865
146	–0.053	–2764.28	0.868
147	–0.084	–4354.522	0.87
148	–0.049	–2537.735	0.872
149	–0.043	–2223.776	0.875
150	0.002	116.165	0.878

Table 6.49 : Serial correlation coefficient, autocovariance function and residual variance for different lags of auto regressive model of water deficit at Udaipur (1981–93)

Auto Covariance Function at lag zero (C_0) = 36943.236

Lag	Serial correlation coefficient	Autocovariance function	Residual variance
(1)	(2)	(3)	(4)
1	0.266	9826.901	0.932
2	0.264	9758.513	0.893
3	0.092	3388.041	0.894
4	0.1	3694.244	0.894
5	0.068	2506.804	0.895
6	0.034	1251.057	0.896
7	0.024	875.949	0.897
8	0.056	2051.539	0.896
9	0.081	2968.195	0.895
10	0.017	628.817	0.895
11	0.052	1932.793	0.896
12	0.042	1538.716	0.896
13	0.055	2022.455	0.897
14	0.073	2685.501	0.897
15	0.073	2675.538	0.897
16	0.053	1943.87	0.899
17	0.067	2440.298	0.9
18	0.084	3087.079	0.899
19	0.021	770.105	0.899
20	0.02	719.404	0.9
21	–0.002	–73.395	0.901
22	0.034	1254.313	0.902
23	0.008	302.387	0.904
24	0.064	2320.887	0.902
25	0.075	2711.462	0.901
26	0.088	3166.068	0.902
27	0.064	2276.972	0.904
28	0.035	1264.565	0.905
29	0.058	2075.969	0.906
30	0.038	1364.047	0.908
31	0.046	1605.485	0.909
32	0.053	1853.271	0.91
33	0.064	2250.529	0.911
34	0.084	2935.921	0.911

contd.....

Table 6.49 – contd...

(1)	(2)	(3)	(4)
35	0.089	3141.176	0.911
36	0.109	3845.264	0.91
37	0.088	3079.848	0.912
38	0.063	2197.608	0.913
39	0.04	1400.005	0.915
40	–0.007	–247.706	0.914
41	0.02	700.634	0.915
42	0.008	273.409	0.917
43	0.06	2080.663	0.917
44	0.046	1609.16	0.919
45	0.051	1780.052	0.92
46	0.014	476.606	0.921
47	0.063	2195.8	0.922
48	–0.03	–1055.49	0.919
49	0.069	2389.939	0.918
50	–0.076	–2649.35	0.907
51	0.08	2790.725	0.904
52	–0.017	–591.773	0.904
53	0.016	564.489	0.906
54	0.035	1209.673	0.906
55	–0.02	–685.955	0.906
56	0.006	219.568	0.908
57	–0.097	–3382.45	0.901
58	–0.003	–103.18	0.902
59	–0.034	–1170.34	0.904
60	0.004	148.732	0.905
61	–0.01	–330.062	0.907
62	0.029	1018.345	0.908
63	0.05	1749.632	0.909
64	0.001	29.86	0.91
65	0.036	1262.795	0.912
66	–0.005	–177.387	0.913
67	0.068	2342.003	0.914
68	0.039	1336.217	0.915
69	0.063	2165.238	0.917
70	0.063	2159.605	0.918
71	0.109	3681.925	0.917
72	0.083	2800.109	0.918
73	0.054	1817.017	0.919

contd.....

Table 6.49 – contd...

(1)	(2)	(3)	(4)
74	0.021	693.582	0.921
75	0.027	923.044	0.923
76	0.015	488.624	0.924
77	0.003	93.896	0.926
78	–0.011	–378.364	0.928
79	–0.016	–522.01	0.929
80	0.038	1225.5	0.931
81	0.053	1720.947	0.931
82	0.034	1087.094	0.933
83	0.055	1782.175	0.934
84	0.008	252.432	0.936
85	0.032	1026.087	0.938
86	0.001	35.766	0.94
87	0.009	283.991	0.942
88	0.017	546.83	0.944
89	–0.003	–102.867	0.946
90	–0.06	–1921.41	0.943
91	–0.069	–2210.81	0.945
92	–0.02	–636.179	0.946
93	–0.075	–2385.59	0.947
94	–0.011	–359.457	0.949
95	–0.006	–197.043	0.951
96	–0.017	–546.655	0.951
97	0.012	373.734	0.953
98	–0.039	–1225.79	0.951
99	–0.065	–2069.9	0.952
100	–0.013	–412.485	0.954
101	–0.117	–3720.667	0.95
102	–0.05	–1586.005	0.952
103	0.005	164.856	0.952
104	–0.084	–2664.338	0.95
105	0.077	2448.660	0.946
106	0.026	807.506	0.948
107	0.13	4109.507	0.947
108	–0.033	–1035.322	0.944
109	0.043	1358.641	0.946
110	0.004	135.569	0.949
111	–0.005	–142.894	0.951
112	0.003	80.059	0.953

contd.....

Table 6.49 – contd...

(1)	(2)	(3)	(4)
113	0.014	431.184	0.953
114	0.04	1269.026	0.955
115	0.016	506.170	0.957
116	0.051	1612.285	0.958
117	0.02	635.107	0.96
118	0.059	1857.369	0.963
119	0.057	1811.494	0.964
120	0.043	1354.103	0.966
121	0.002	57.664	0.967
122	–0.013	–418.305	0.969
123	–0.007	–221.204	0.972
124	0.001	18.447	0.975
125	–0.013	–399.504	0.977
126	–0.047	–1478.933	0.98
127	–0.025	–779.059	0.983
128	–0.022	–701.263	0.986
129	–0.039	–1226.955	0.988
130	–0.053	–1643.733	0.99
131	–0.043	–1313.633	0.991
132	–0.081	–2478.367	0.989
133	–0.092	–2813.166	0.99
134	–0.078	–2349.714	0.993
135	–0.056	–1706.050	0.996
136	–0.076	–2284.489	0.997
137	–0.081	–2440.355	0.999
138	–0.044	–1318.512	1.002
139	–0.037	–1117.978	1.005
140	–0.016	–484.550	1.008
141	–0.001	–17.927	1.011
142	–0.006	–186.366	1.014
143	0	8.312	1.016
144	–0.04	–1193.287	1.018
145	0.004	124.243	1.017
146	–0.024	–709.154	1.019
147	–0.066	–1971.485	1.022
148	–0.039	–1169.696	1.025
149	–0.021	–628.195	1.029
150	–0.005	–144.812	1.032

Table 6.50 : Serial correlation coefficient, autocovariance function and residual variance for different lags of auto regressive model of water deficit at Banswara (1981–93)

Auto Covariance Function at lag zero (C_0) = 54345.388

Lag	Serial correlation coefficient	Autocovariance function	Residual variance
(1)	(2)	(3)	(4)
1	0.489	26574.895	0.704
2	0.387	21021.069	0.68
3	0.296	16051.411	0.68
4	0.236	12792.798	0.681
5	0.192	10392.385	0.681
6	0.153	8318.137	0.68
7	0.141	7617.333	0.681
8	0.167	9024.190	0.681
9	0.14	7599.632	0.682
10	0.141	7637.706	0.682
11	0.157	8506.025	0.68
12	0.141	7638.047	0.679
13	0.153	8263.315	0.676
14	0.144	7783.066	0.675
15	0.128	6903.662	0.675
16	0.097	5220.789	0.676
17	0.098	5295.294	0.675
18	0.084	4541.682	0.676
19	0.059	3171.978	0.677
20	0.059	3188.707	0.678
21	0.063	3375.548	0.679
22	0.056	3027.577	0.68
23	0.068	3646.681	0.681
24	0.067	3580.296	0.682
25	0.101	5399.659	0.683
26	0.093	4913.486	0.684
27	0.074	3863.362	0.684
28	0.067	3484.931	0.685
29	0.05	2580.386	0.686
30	0.035	1826.030	0.687
31	0.045	2291.832	0.687
32	0.06	3084.045	0.686
33	0.081	4141.209	0.684
34	0.077	3937.492	0.685

contd.....

Table 6.50 – contd...

(1)	(2)	(3)	(4)
35	0.063	3241.975	0.686
36	0.078	4005.108	0.686
37	0.082	4178.517	0.687
38	0.069	3534.719	0.688
39	0.065	3339.124	0.689
40	0.09	4603.868	0.69
41	0.072	3685.111	0.691
42	0.052	2634.751	0.69
43	0.039	1991.964	0.691
44	0.033	1668.901	0.692
45	–0.024	–1225.613	0.689
46	–0.015	–776.578	0.69
47	–0.016	–826.886	0.691
48	–0.04	–2020.278	0.692
49	–0.099	–5025.992	0.691
50	–0.022	–1124.324	0.685
51	–0.06	–3033.997	0.686
52	0.047	2375.924	0.676
53	–0.083	–4217.700	0.663
54	–0.053	–2650.668	0.664
55	–0.081	–4061.690	0.663
56	–0.108	–5459.741	0.663
57	–0.133	–6664.005	0.661
58	–0.12	–6025.744	0.662
59	–0.136	–6839.466	0.662
60	–0.114	–5720.063	0.663
61	–0.081	–4045.397	0.664
62	–0.106	–5304.892	0.664
63	–0.107	–5356.206	0.665
64	–0.133	–6626.363	0.666
65	–0.094	–4700.984	0.667
66	–0.074	–3689.698	0.667
67	–0.085	–4230.503	0.668
68	–0.046	–2276.811	0.667
69	–0.046	–2291.817	0.669
70	–0.044	–2175.049	0.67
71	–0.053	–2573.748	0.671
72	–0.026	–1258.261	0.672
73	–0.012	–591.236	0.673

contd.....

Table 6.50 – contd...

(1)	(2)	(3)	(4)
74	–0.013	–632.879	0.675
75	0.009	422.005	0.676
76	0.014	691.749	0.677
77	0.021	1015.009	0.679
78	0.018	857.019	0.68
79	–0.019	–871.011	0.681
80	–0.13	–620.258	0.683
81	–0.029	–1340.146	0.684
82	–0.068	–3144.905	0.685
83	–0.058	–2705.485	0.687
84	–0.065	–2998.188	0.688
85	–0.055	–2533.871	0.69
86	–0.069	–3176.488	0.691
87	–0.093	–4281.646	0.692
88	–0.09	–4126.621	0.693
89	–0.111	–5126.942	0.694
90	–0.138	–6338.038	0.694
91	–0.118	–5420.464	0.696
92	–0.086	–3952.112	0.697
93	–0.061	–2783.043	0.699
94	–0.012	–571.198	0.7
95	–0.012	–554.253	0.702
96	–0.038	–1721.764	0.702
97	–0.067	–3050.627	0.704
98	–0.036	–1663.326	0.705
99	–0.042	–1909.167	0.706
100	–0.03	–1359.613	0.707
101	–0.056	–2584.593	0.708
102	0.012	565.837	0.709
103	–0.028	–1285.615	0.710
104	–0.049	–2220.402	0.710
105	–0.059	–2676.304	0.712
106	–0.013	–591.971	0.713
107	–0.038	–1729.269	0.715
108	0.025	1140.783	0.716
109	0.055	2505.668	0.717
110	0.056	2523.041	0.718
111	0.018	814.666	0.718
112	–0.002	–88.579	0.718

contd.....

Table 6.50 – contd...

(1)	(2)	(3)	(4)
113	–0.020	–896.202	0.717
114	–0.030	–1332.912	0.719
115	–0.057	–2530.596	0.720
116	–0.078	–3448.498	0.722
117	–0.049	–2185.964	0.724
118	0.033	–1462.731	0.725
119	–0.035	–1542.772	0.727
120	–0.029	–1276.779	0.729
121	–0.032	–1407.340	0.731
122	–0.053	–2314.333	0.732
123	–0.038	–1652.966	0.734
124	–0.033	–1447.190	0.736
125	–0.019	–809.116	0.737
126	–0.015	–654.366	0.739
127	0.020	852.265	0.741
128	0.017	716.169	0.743
129	0.001	45.569	0.745
130	–0.012	–505.151	0.747
131	–0.048	–1990.547	0.749
132	–0.045	–1853.726	0.751
133	–0.053	–2198.157	0.753
134	–0.077	–3192.327	0.755
135	–0.056	–2295.672	0.758
136	–0.037	–1538.441	0.759
137	–0.047	–1952.073	0.762
138	–0.018	–759.736	0.764
139	–0.028	–1142.911	0.766
140	–0.001	–50.579	0.768
141	–0.001	–35.124	0.771
142	–0.012	–492.823	0.773
143	–0.013	–553.113	0.175
144	0.004	147.832	0.777
145	–0.006	–229.222	0.779
146	0.019	764.257	0.781
147	–0.030	–1217.232	0.782
148	–0.022	–903.891	0.785
149	–0.046	–1868.870	0.787
150	–0.040	–1640.158	0.789

Table 6.51 : Serial correlation coefficient, autocovariance function and residual variance for different lags of auto regressive model of water deficit at Kota (1981–93)

Auto Covariance Function at lag zero (C_0) = 36644.531

Lag	Serial correlation coefficient	Autocovariance function	Residual variance
(1)	(2)	(3)	(4)
1	0.384	14071.5	0.855
2	0.281	10303.85	0.835
3	0.203	7427.378	0.833
4	0.073	2682.606	0.831
5	0.044	1626.616	0.832
6	–0.037	–1346.03	0.83
7	0.002	84.644	0.829
8	0.024	890.035	0.829
9	0	–8.507	0.83
10	0.072	2614.721	0.828
11	0.097	3543.693	0.826
12	0.049	1780.721	0.826
13	0.075	2728.688	0.827
14	0.09	3291.307	0.826
15	0.149	5440.501	0.818
16	0.104	3781.357	0.819
17	0.071	2574.082	0.82
18	0.086	3150.557	0.821
19	0.053	1921.536	0.822
20	0.047	1717.495	0.824
21	–0.014	–500.535	0.822
22	0.018	655.817	0.823
23	0.01	356.408	0.824
24	–0.017	–609.782	0.825
25	0.035	1262.171	0.826
26	0.029	1053.298	0.827
27	0.042	1491.643	0.828
28	0.043	1528.999	0.83
29	0.002	78.302	0.829
30	0.03	1043.068	0.831
31	–0.008	–298.206	0.832
32	0.017	595.398	0.832
33	0.016	576.351	0.834
34	0.012	420.304	0.835

contd.....

Table 6.51 – contd....

(1)	(2)	(3)	(4)
35	0.041	1430.273	0.836
36	0.031	1101.262	0.837
37	0.014	488.746	0.838
38	0.00	6.731	0.839
39	0.00	–7.477	0.841
40	0.018	632.612	0.842
41	0.003	96.736	0.843
42	–0.038	–1333.07	0.842
43	–0.008	–282.257	0.843
44	–0.053	–1817.2	0.844
45	–0.039	–1335.26	0.845
46	–0.09	–3095.77	0.843
47	–0.128	–4414.48	0.839
48	–0.116	–4018.01	0.839
49	–0.032	–1112	0.834
50	0.004	149.163	0.835
51	0.066	2267.684	0.834
52	0.017	591.395	0.833
53	0.012	412.737	0.834
54	0.05	1730.178	0.834
55	–0.01	–344.7	0.835
56	0.034	1160.518	0.834
57	–0.116	–3992.33	0.821
58	–0.046	–1583.71	0.821
59	–0.028	–960.548	0.822
60	–0.089	–3063.9	0.822
61	–0.137	–4694.77	0.813
62	–0.089	–3063.47	0.814
63	–0.09	–3065.56	0.816
64	–0.044	–1506.24	0.817
65	–0.045	–1528.35	0.817
66	–0.045	–1538.87	0.818
67	–0.015	–502.68	0.819
68	–0.004	–134.755	0.819
69	0.035	1157.48	0.821
70	–0.009	–299.685	0.821
71	0.012	405.811	0.822
72	–0.027	–891.21	0.824
73	–0.027	–886.718	0.826

contd.....

Table 6.51 – contd....

(1)	(2)	(3)	(4)
74	–0.039	–1286.1	0.827
75	–0.06	–1963.6	0.829
76	–0.062	–2033.16	0.83
77	–0.072	–2335.86	0.832
78	–0.042	–1348.27	0.834
79	–0.026	–837.00	0.836
80	–0.042	–1334.43	0.838
81	0.029	937.505	0.837
82	–0.027	–854.089	0.837
83	0.018	575.002	0.837
84	0.013	405.149	0.839
85	–0.002	–55.111	0.84
86	0.054	1706.019	0.841
87	0.001	31.2	0.843
88	0.007	219.326	0.845
89	–0.018	–556.567	0.847
90	–0.048	–1507.19	0.848
91	–0.043	–1352.51	0.849
92	–0.049	–1545.66	0.851
93	–0.078	–2460.84	0.852
94	–0.08	–2521.79	0.851
95	–0.048	–1508.42	0.853
96	–0.025	–767.362	0.854
97	–9.075	–2337.94	0.855
98	–0.025	–777.501	0.855
99	0.003	96.414	0.857
100	–0.03	–923.816	– 0.859
101	0.088	2729.686	0.853
102	–0.040	–1255.233	0.849
103	0.051	1573.493	0.850
104	0.060	1855.961	0.851
105	0.021	653.670	0.853
106	0.019	590.135	0.855
107	–0.077	–2386.249	0.850
108	–0.063	–1952.249	0.852
109	–0.044	–1371.278	0.854
110	–0.042	–1316.958	0.855
111	–0.048	–1477.128	0.857

contd.....

Table 6.51 – contd....

(1)	(2)	(3)	(4)
112	–0.084	–2597.056	0.856
113	–0.039	–1197.915	0.858
114	0.012	384.132	0.860
115	–0.063	–1956.147	0.860
116	–0.008	–246.021	0.862
117	0.005	140.449	0.862
118	0.047	1461.394	0.865
119	0.007	225.118	0.866
120	0.041	1283.950	0.868
121	0.063	1952.816	0.870
122	0.049	1519.441	0.871
123	0.062	1907.339	0.873
124	0.078	2427.294	0.875
125	0.059	1807.084	0.877
126	0.035	1050.289	0.879
127	0.023	704.836	0.882
128	0.016	481.227	0.884
129	–0.011	–316.318	0.887
130	–0.042	–1262.946	0.890
131	–0.005	–146.868	0.892
132	0.003	81.687	0.895
133	0.002	49.890	0.898
134	–0.018	–527.746	0.900
135	0.021	600.894	0.903
136	0.049	1401.968	0.906
137	0.022	624.982	0.907
138	0.049	1414.771	0.909
139	0.089	2562.373	0.911
140	0.070	2004.170	0.914
141	0.057	1628.482	0.917
142	0.017	486.999	0.919
143	0.033	931.233	0.920
144	0.000	10.161	0.923
145	–0.009	–268.887	0.926
146	–0.020	–567.079	0.928
147	–0.015	–436.701	0.930
148	–0.043	–1218.959	0.933
149	–0.071	–2023.890	0.934
150	0.007	195.987	0.937

Similarly, it can also be seen from Table 6.51 that at Kota the minimum residual variance was obtained for lag sixty one. However, no significant different in residual variance was obtained at lag 3 and 61 and other lower values of different lags. Therefore, reducing the complexity of the model the 3rd order was selected to represent the stochastic component by the autoregressive model.

6.4.1.4.3 Mathematical representation of the stochastic component

The autoregression coefficients (ϕP,P) were estimated by using Equation (6.25) for orders as earlier discussed in Subsection 3.3 for different stations and are presented in Table 6.52.

Using Equations (6.28) and (6.29) the estimated autoregression coefficient of the stochastic component of the weekly water deficit time series at Jodhpur may be expressed as:

$$S_t = 0.275\ S_{t-1} + 0.176\ S_{t-2} + 0.081\ S_{t-3} + a_t \ldots \qquad \ldots(6.43)$$

Similarly, at other stations the stochastic component of weekly water deficit time series may be expressed as:

At Sriganganagar,

$$S_t = 0.374\ S_{t-1} + 0.008\ S_{t-2} + 0.083\ S_{t-3} + a_t \ldots \qquad \ldots(6.44)$$

At Fatehpur,

$$S_t = 0.335\ S_{t-1} + 0.098 S_{t-2} + a_t \ldots \qquad \ldots(6.45)$$

At Pali

$$S_t = 0.288\ S_{t-1} + 0.124 S_{t-2} + 0.116 S_{t-3} + a_t \ldots \qquad \ldots(6.46)$$

At Jaipur

$$S_t = 0.418\ S_{t-1} + 0.071\ S_{t-2} + 0.102\ S_{t-3} + a_t \ldots \qquad \ldots(6.47)$$

At Udaipur,

$$S_t = 0.211\ S_{t-1} + 0.208\ S_{t-2} + a_t \qquad \ldots(6.48)$$

At Banaswara,

$$S_t = 0.394\ S_{t-1} + 0.190\ S_{t-2} + a_t \qquad \ldots(6.49)$$

At Kota,

$$S_t = 0.314\ S_{t-1} + 0.137\ S_{t-2} + 0.062\ S_{t-3} + a_t \qquad \ldots(6.50)$$

The nondeterministic stochastic component was estimated at Jodhpur by using Equation 6.35 for all the values (t = 4 to 676). Similarly, for other stations the nondeterministic stochastic component was estimated by using Equations 6.36, 6.37, 6.38, 6.39, 6.40, 6.41 and 6.42 for Sriganganagar, Fatehpur, Pali, Jaipur,

Udaipur, Banswara and Kota respectively. The estimated values were used to generate the series and rediscussed later.

Table 6.52 : Autoregressive parameters of water deficit series for the period from 1981 to 1993 at different stations

Station	Order	$\phi p.k$	Value
Jodhpur	3rd	$\phi^{3,1}$	0.275
		$\phi^{3,2}$	0.176
		$\phi^{3,3}$	0.081
Sriganganagar	3rd	$\phi^{3,1}$	0.374
		$\phi^{3,2}$	0.008
		$\phi^{3,3}$	0.083
Fatehpur	2nd	$\phi^{2,1}$	0.335
		$\phi^{2,2}$	0.098
Pali	3rd	$\phi^{3,1}$	0.288
		$\phi^{3,2}$	0.124
		$\phi^{3,3}$	0.116
Jaipur	3rd	$\phi^{3,1}$	0.418
		$\phi^{3,2}$	0.071
		$\phi^{3,3}$	0.102
Udaipur	2nd	$\phi^{2,1}$	0.211
		$\phi^{2,2}$	0.208
Banswara	2nd	$\phi^{2,1}$	0.394
		$\phi^{2,2}$	0.190
Kota	3rd	$\phi^{3,1}$	0.314
		$\phi^{3,2}$	0.137
		$\phi^{3,3}$	0.062

6.4.1.5 Residual series of stochastic component

Residual series (a_t) is random independent part of the stochastic component. It was obtained by removing the periodic and dependent parts from the original observed series of water deficit for all the stations.

The statistical parameters were computed for observed, generated and residual series for all the stations and are presented in Table 6.53. It can be seen from Table 6.53 that mean of the residual ranges from –0.012 to + 0.011 and kurtosis ranges from –3.104 to –3.068 for all the stations. This indicates than mean is almost equal to zero and kurtosis is –3. which confirmed that the residual series follows normal distribution.

Table 6.53 : Statistical parameters of the observed, generated and residual series of water deficit at different stations

Station	Series	Mean (mm)	Standard deviation (mm)	Variance	Skewness	Kurtosis	Serial correlation coefficient at lag 1
Jodhpur	Historical	25.965	8.417	70.860	-0.070	-3.020	0.598
	Generated	25.998	8.918	79.538	0.307	-3.039	0.945
	Residual	-0.004	8.516	72.532	-1.116	-3.087	0.067
Sriganganagar	Historical	19.578	12.091	146.209	0.830	-3.020	0.723
	Generated	19.628	12.541	157.286	0.791	-3.042	0.927
	Residual	-0.000	9.664	93.409	-0.788	-3.076	0.226
Fatehpur	Historical	24.095	12.290	151.048	0.915	-3.023	0.692
	Generated	24.121	12.620	159.287	0.845	-3.037	0.934
	Residual	0.011	9.385	88.084	-0.765	-3.068	-0.037
Pali	Historical	26.217	9.574	91.657	0.077	-3.019	0.627
	Generated	26.253	10.246	104.996	0.341	-3.036	0.943
	Residual	-0.002	8.946	80.031	-0.820	-3.086	0.106
Jaipur	Historical	20.643	12.208	149.059	0.681	-3.020	0.773
	Generated	20.677	13.024	169.631	0.60	-3.042	0.936
	Residual	-0.008	8.612	74.169	-0.464	-3.104	0.211
Udaipur	Historical	16.867	10.102	102.066	0.409	-3.019	0.731
	Generated	16.880	10.278	105.652	0.534	-3.031	0.971
	Residual	-0.004	5.656	31.995	-0.157	-3.074	-0.055
Banswara	Historical	17.025	11.577	134.030	0.844	-3.021	0.793
	Generated	17.030	12.289	151.035	0.715	-3.037	0.948
	Residual	-0.012	6.347	40.293	-0.576	-3.099	-0.092
Kota	Historical	18.251	11.490	132.021	0.395	-3.018	0.807
	Generated	18.277	11.716	137.277	0.400	-3.036	0.97
	Residual	-0.012	6.463	41.782	-0.82	-3.107	-0.131

It can also be seen from Table 6.53 that the mean and variance, skewness, kurtosis and SCC at lag 1 of historical and generated series are almost same for all the stations which shows a closeness between historical and generated series.

6.4.1.6 Model structure

Since the observed weekly water deficit series was found to be trend fee series the developed model describes the periodic-stochastic behaviour of the series.

The developed model is a superposition of a harmonic deterministic process and autoregressive models of respective orders as discussed in Subsection 4.1.3.2 for different stations. The mathematical structure of the additive model described by Equation (6.8) are presented in Table 6.54.

Table 6.54 : Model structure of weekly water deficit at different stations

Station	Mean (mm)	Periodic component	Stochastic component
Jodhpur	25.085	–4.98 Cos(2πt/p) + 8.514 Sin (2πt/p) + 0.330 Cos (4πt/p) – 5.345 Sin (4πt/p) –1.868 Cos (6πt/p) + 0.563 Sin (6πt/p)	0.275 S_{t-1}+ 0.176 S_{t-2} + 0.081S_{t-3} + at
Sriganga–nagar	19.381	–13.423 Cos (2πt/p) + 4.049 Sin (2πt/p) –0.490 Cos(4πt/p) – 7.495 Sin (4πt/p) + 0.255 Cos (6πt/p) + 4.954 Sin (6πt/p)	0.374 S_{t-1} + 0.008 S_{t-2} + 0.083 s_{t-3} + at
Fatehpur	23.870	–9.397 Cos (2πt/p) + 8.513 Sin (2πt/p) –0.987 Cos (4πt/p) – 9.926 Sin (4πt/p) –1.076 Cos (6πt/p) + 4.071 Sin (6πt/p)	0.335 S_{t-1} + 0.098 S_{t-2} + at
Pali	25.681	–6.694 Cos (2πt/p) + 9.106 Sin (2πt/p) –1.352 Cos (4πt/p) – 6.357 Sin (4πt/p)	0.288 S_{t-1} + 0.124 S_{t-2} + 0.116 S_{t-3} + at
Jaipur	20.432	–7.201 Cos (2πt/p) + 11.209 Sin (2πt/p) –0.295 Cos (4πt/p) – 9.594 Sin (4πt/p) –2.387 Cos (6πt/p) + 3.219 Sin (6πt/p)	0.418 S_{t-1} + 0.071 S_{t-2} + 0.102 S_{t-3} + at
Udaipur	16.912	–4.249 Cos (2πt/p) + 11.484 Sin (2πt/p) –1.090 Cos (4πt/p) – 6.870 Sin (4πt/p)	0.211 S_{t-1} + 0.208 S_{t-2} + at
Banswara	16.607	–5.030 Cos (2πt/p) + 11.616 Sin (2πt/p) +1.886 Cos (4πt/p) – 8.775 Sin (4πt/p) –0.257 Cos (6πt/p) + 3.906 Sin (6πt/p)	0.394 S_{t-1} + 0.190 S_{t-2} + at
Kota	18.040	–5.398 Cos (2πt/p) + 12.98 Cos (2πt/p) + 0.040 Cos (4πt/p) – 7.575 Sin (4πt/p)	0.314 S_{t-1} + 0.137 S_{t-2} + 0.062 S_{t-3} + at

It can be seen from Table 6.54 that the formulated models constitute the deterministic part (periodic component) stochastic component and random component of the model where the current value of model depends of the weighted sum of the observed preceding value depending upon the order of model selected. Using the developed models the weekly water deficit series were generated for all the stations.

6.4.1.7 Diagnostic checking of formulated water deficit model

The residual obtained after fitting the formulated models were subjected to various analysis to test its adequacy for representing the weekly water deficit series.

6.4.1.7.1 Sum of squares analysis

The sum of squares of residual series were calculated by using Equations (6.32) to (6.34) and sum of squares of deviations of observed values from their mean were computed by using Equation (3.51). The value of measure (R^2) ranges from 0.948 (Jodhpur) to 0.970 at Banswara and are shown in Table 6.55, which are nearly unity. Thus, this leads to the conclusion that the developed model has a fair goodness of fit to generate the weekly water deficit series at different stations under study.

Table 6.55 : Sum of squares analysis of residual series at different stations

Station	R^2
Jodhpur	0.948
Sriganganagar	0.961
Fatehpur	0.965
Pali	0.956
Jaipur	0.965
Udaipur	0.963
Banswara	0.970
Kota	0.964

6.4.1.7.2 Serial correlation analysis

The serial correlation coefficient (SCC) for lag 1 (1 = 1, 2, ..., 150) were computed with the help of Equation (6.7) for different stations. The valves of SCC against respective lags were then plotted to obtain a correlogram. The resulting correlograms are shown in Figs. 6.17 to 6.24 for all the stations with confidence limit at 1 per cent level. The confidence limits were estimated by using Equation (6.31). A residual of these figures indicates that the correlograms are almost completely within the confidence limits at 1 per cent level at all the stations. Therefore, these may be treated as nonsignificant. Further, the correlograms show that at 1 per cent level the coefficients are within the limits. This again confirms that the residual series may be treated as nonsignificant at all the stations.

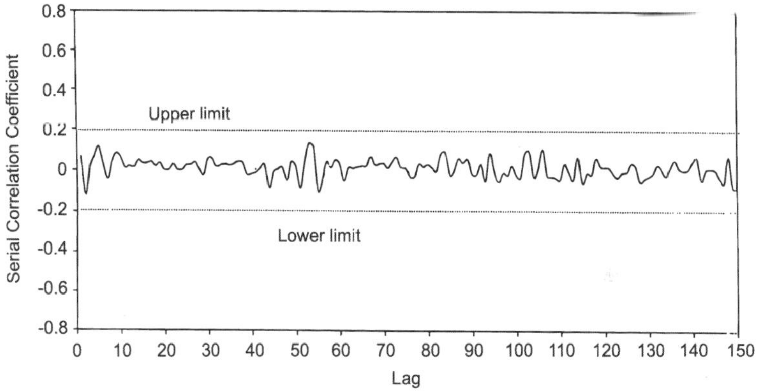

Fig. 6.17 : Correlogram of residual series of observed weekly water deficit at Jodhpur (1981–1993)

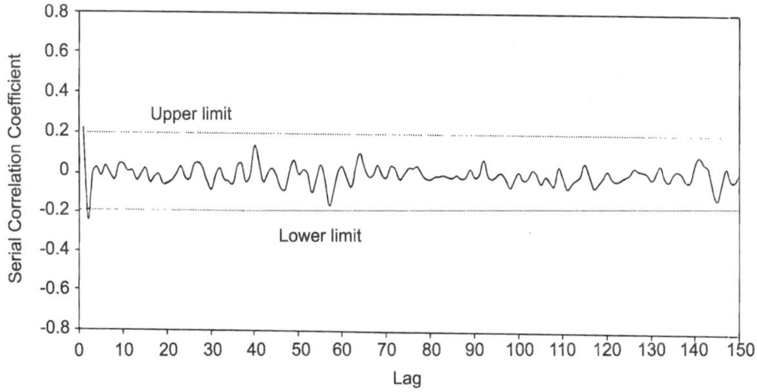

Fig. 6.18 : Correlogram of residual series of observed weekly water deficit at Sriganganagar (1981–1993)

Also, it can be seen from Table 6.56 that the mean of SCC ranges from 0.000 to 0.017 and variance ranges form 0.001 to 0.003, which indicates that mean is nearly zero and variance is equal to variance (1/150). This leads to the conclusion that the residuals are independent and normally distributed. Further it also confirms the randomness of the residuals.

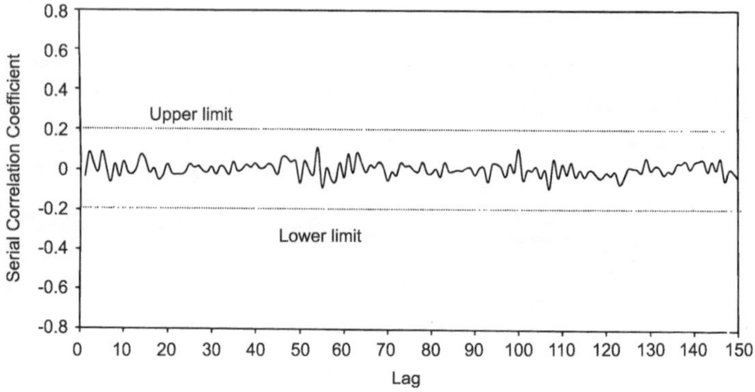

Fig. 6.19 : Correlogram of residual series of observed weekly water deficit at Fatehpur (1981–1993)

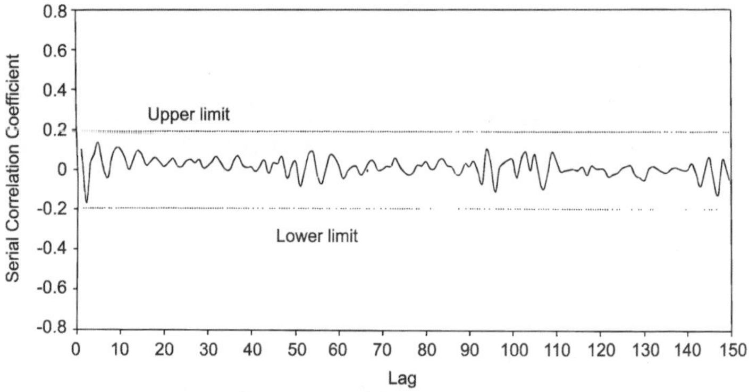

Fig. 6.20 : Correlogram of residual series of observed weekly water deficit at Pali (1981–1993)

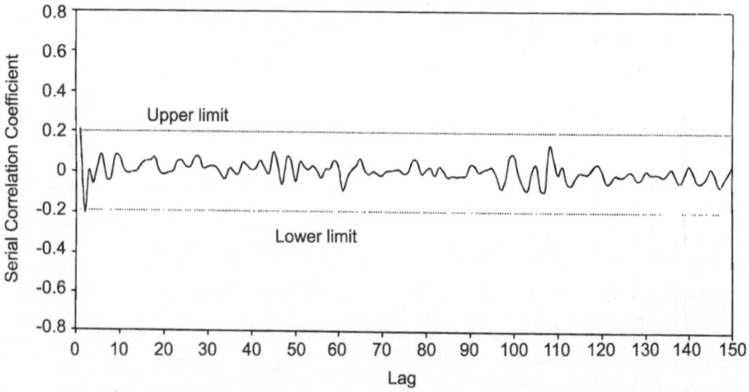

Fig. 6.21 : Correlogram of residual series of observed weekly water deficit at Jaipur (1981–1993)

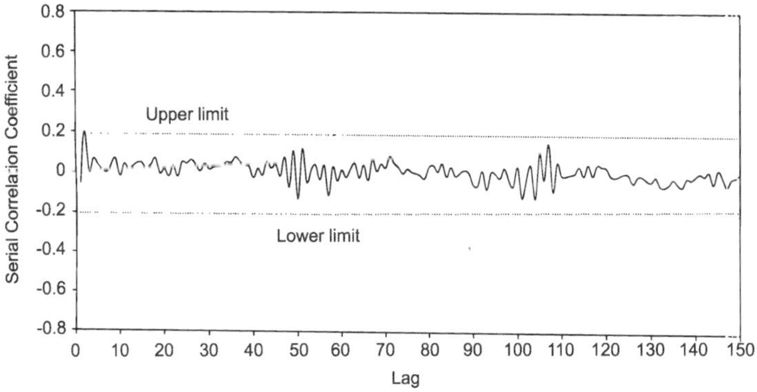

Fig. 6.22 : Correlogram of residual series of observed weekly water deficit at Udaipur (1981–1993)

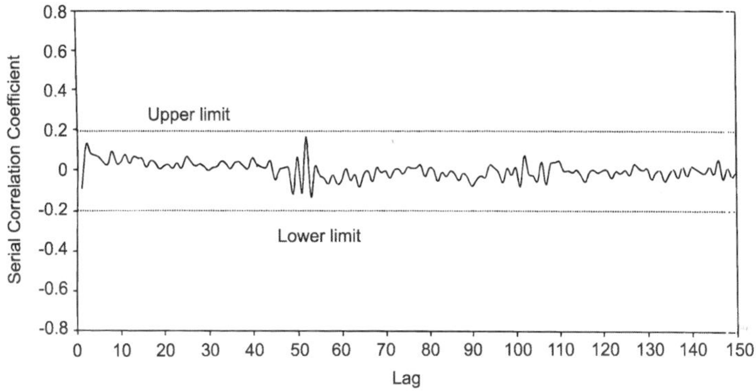

Fig. 6.23 : Correlogram of residual series of observed weekly water deficit at Banswara (1981–1993)

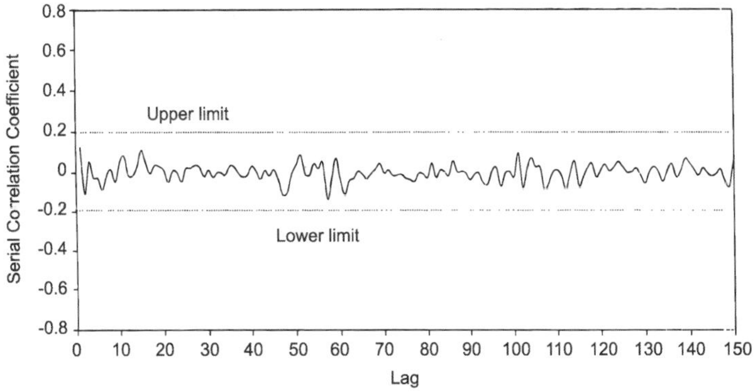

Fig. 6.24 : Correlogram of residual series of observed weekly water deficit at Kota (1981–1993)

Table 6.56 : Mean and variance of residual series at different stations

Station	Mean	Variance
Jodhpur	0.012	0.002
Sriganganagar	0.000	0.003
Fatehpur	0.005	0.001
Pali	0.017	0.002
Jaipur	0.009	0.002
Udaipur	0.009	0.002
Banswara	0.000	0.002
Kota	0.000	0.002

6.4.1.8 Validation of stochastic model of water deficit

Validation of generated water deficit series by developed stochastic models at various stations were made by comparison of generated water deficit series and observed series. The generated water deficit series and observed water deficit series are shown in Figures 6.25 to 6.32 for all the stations under study. Figures 6.25 to 6.32 depict the variation of observed weekly water deficit and generated weekly water deficit series for thirteen years (1981–1993) for all the stations. A perusal of Figures 6.33 to 6.40 indicates that there is a linear correlation between generated weekly water deficit series and observed weekly water deficit series.

This is further confirmed from Table 6.53 that mean and standard deviation of generated weekly water deficit series and observed weekly water series are almost same for all the stations under study. The values of correlation coefficient (r) between generated weekly water deficit series and observed water deficit series were worked out and are shown in Table 6.57. Table 6.57 indicates that the correlation co-efficient ranges from 0.75 at Jodhpur to 0.90 at Banswara, which further confirms the closeness

Table 6.57 : Correlation coefficient between generated and observed weekly water deficit series for thirteen years (1981–93) at different stations

Station	Correlation coefficient (r)
Jodhpur	0.75
Sriganganagar	0.79
Fatehpur	0.82
Pali	0.78
Jaipur	0.84
Udaipur	0.89
Banswara	0.90
Kota	0.88

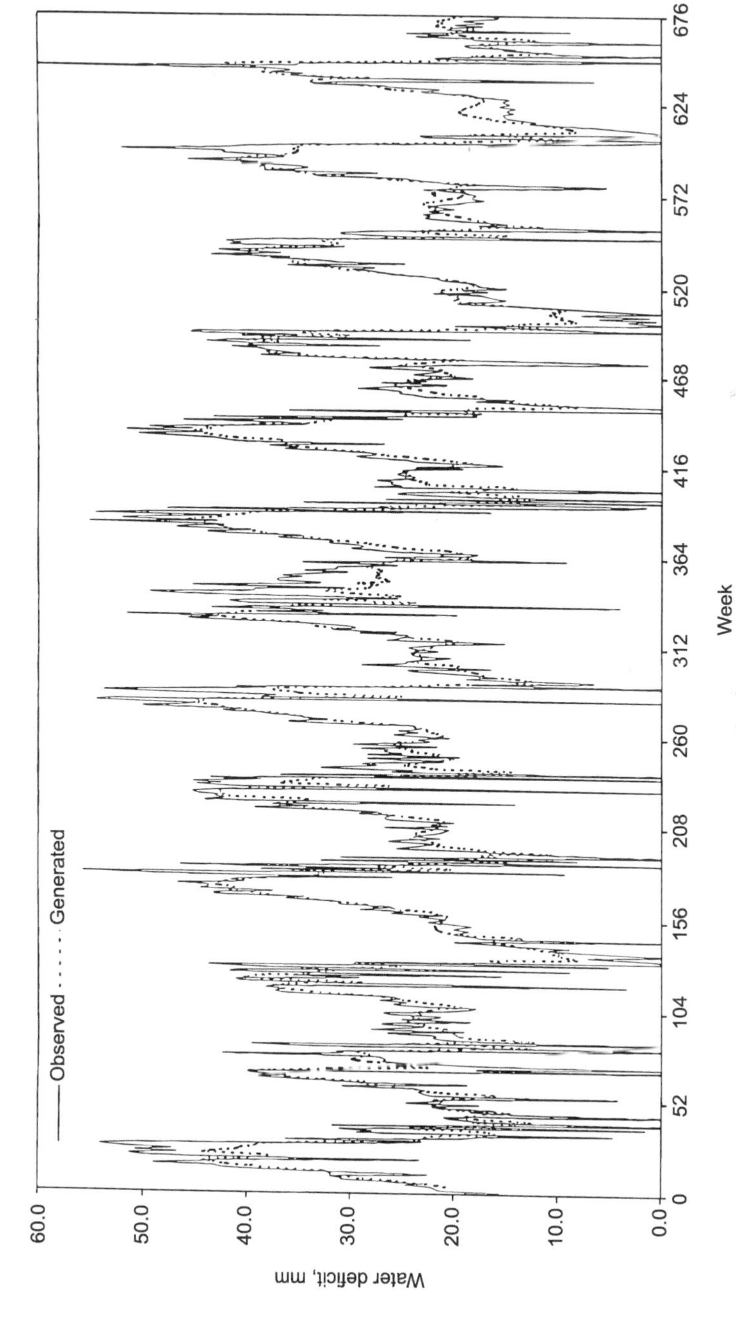

Fig. 6.25 : Variation of observed and generated weekly water deficit series at Jodhpur (1981–93)

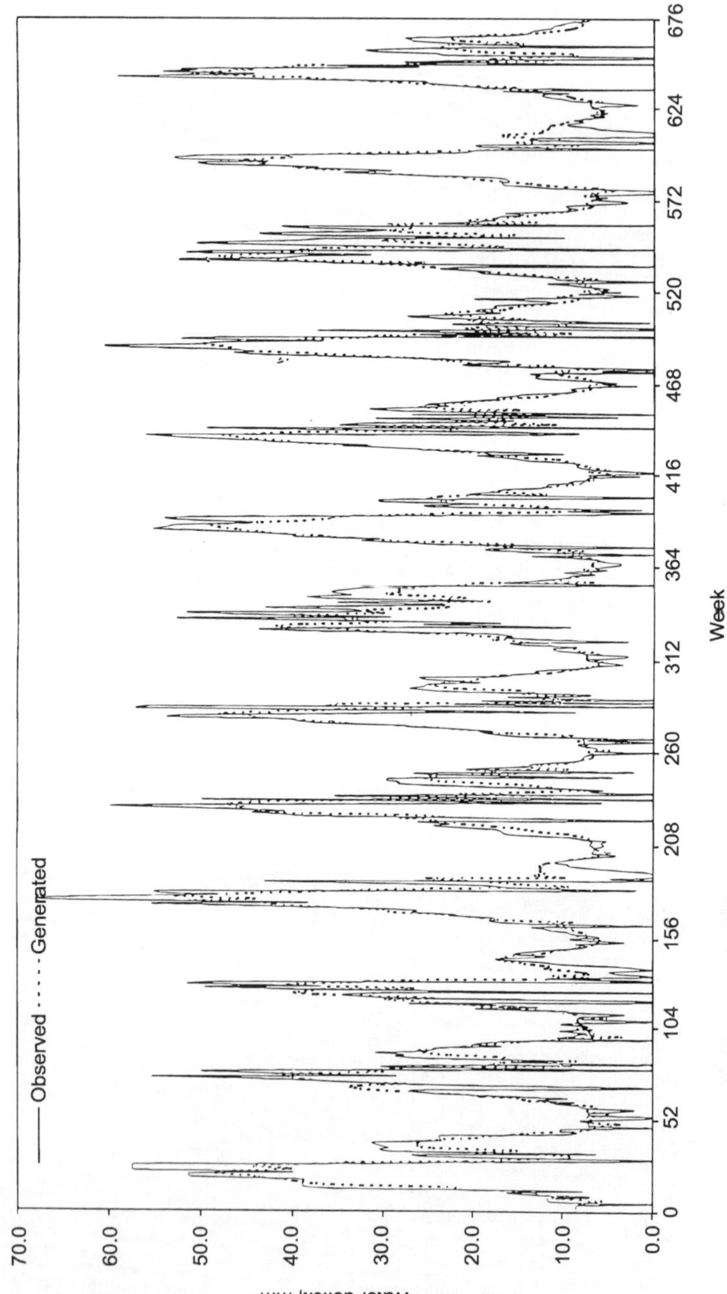

Fig. 6.26 : Variation of observed and generated weekly water deficit series at Sriganganagar (1981–93)

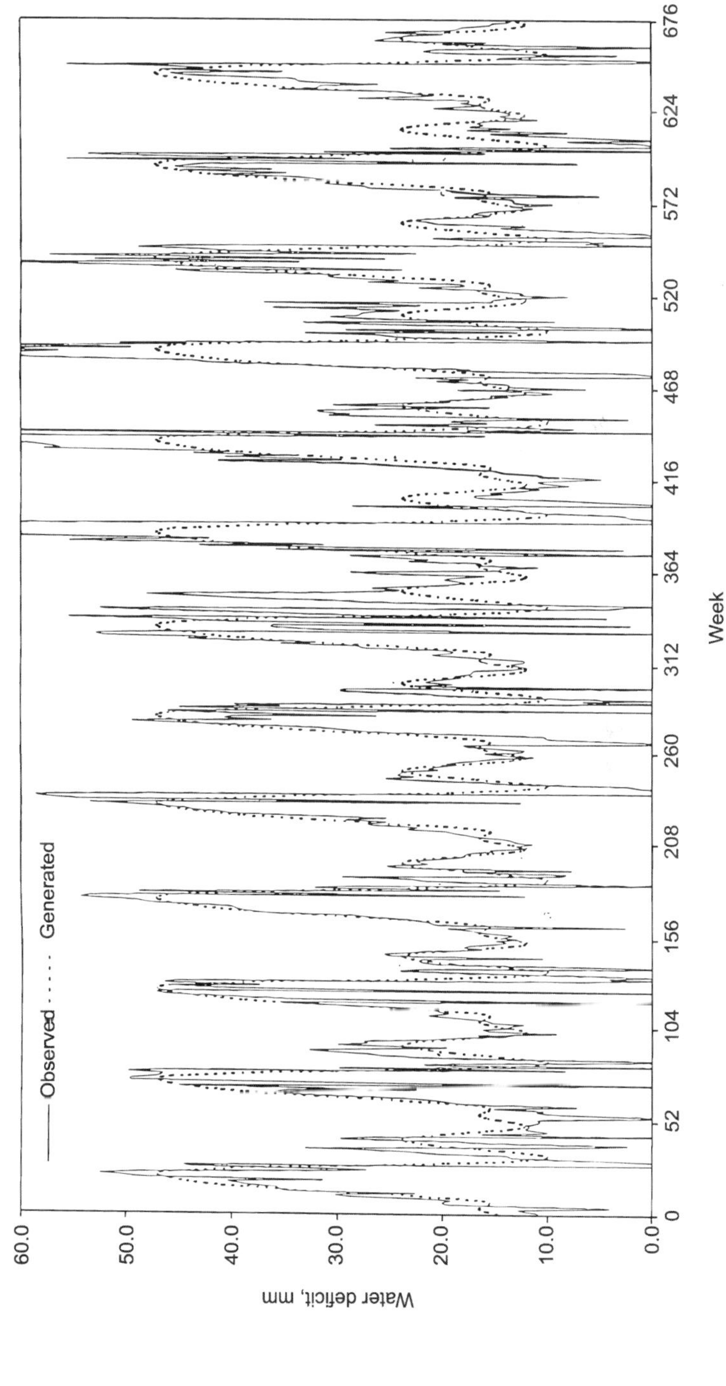

Fig. 6.27 : Variation of observed and generated weekly water deficit series at Fatehpur (1981–93)

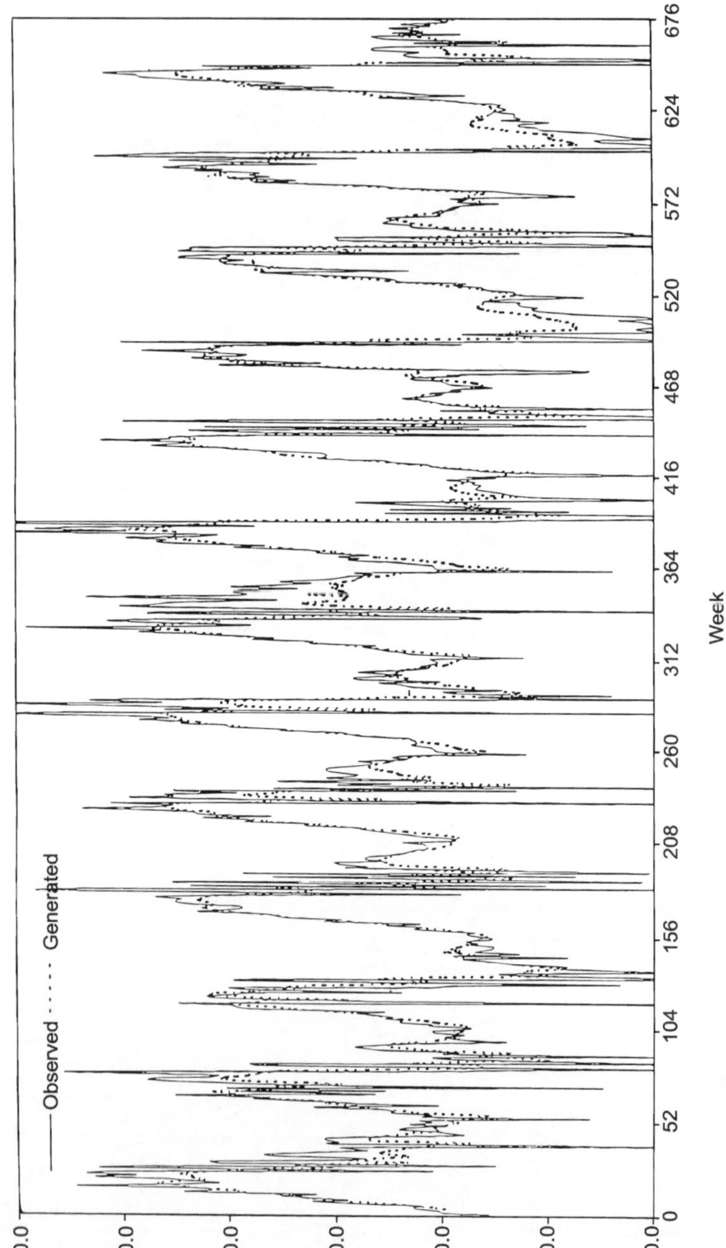

Fig. 6.28 : Variation of observed and generated weekly water deficit series at Pali (1981–93)

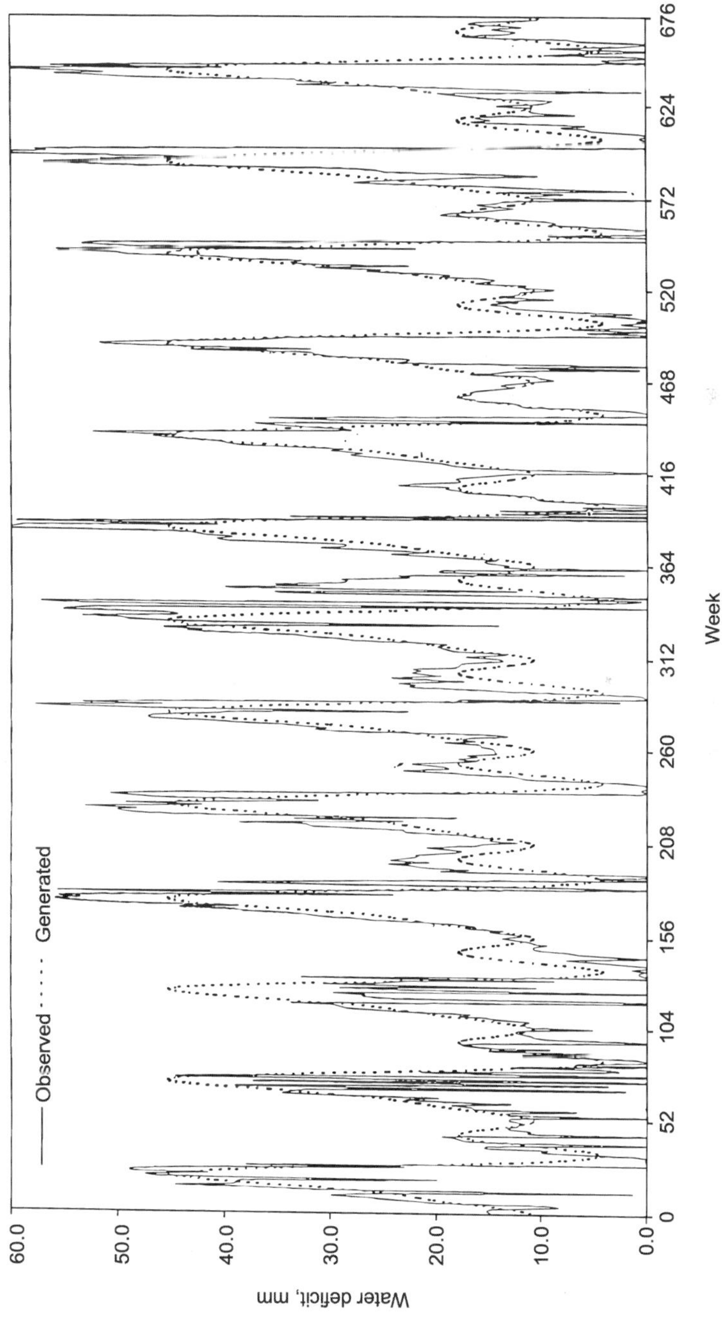

Fig. 6.29 : Variation of observed and generated weekly water deficit series at Jaipur (1981–93)

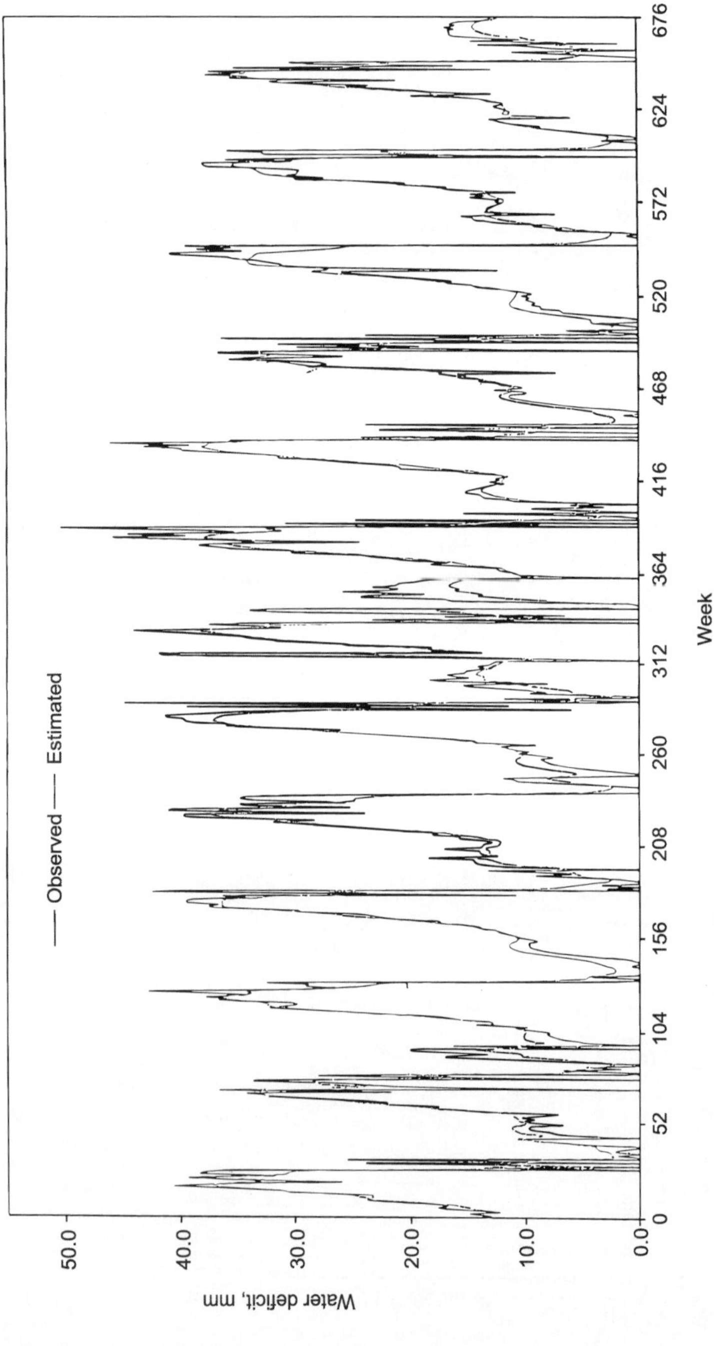

Fig. 6.30 : Variation of observed and generated weekly water deficit series at Udaipur (1981–93)

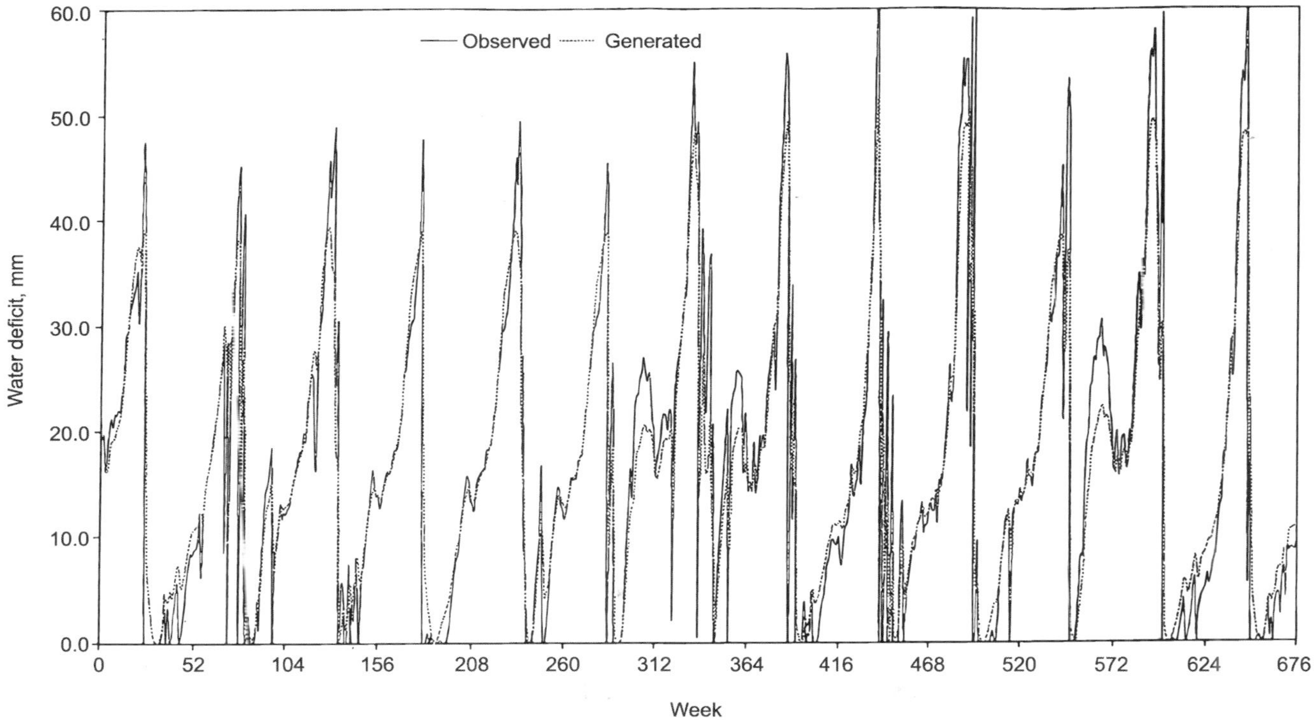

Fig. 6.31 : Variation of observed and generated weekly water deficit series at Banswara (1981–93)

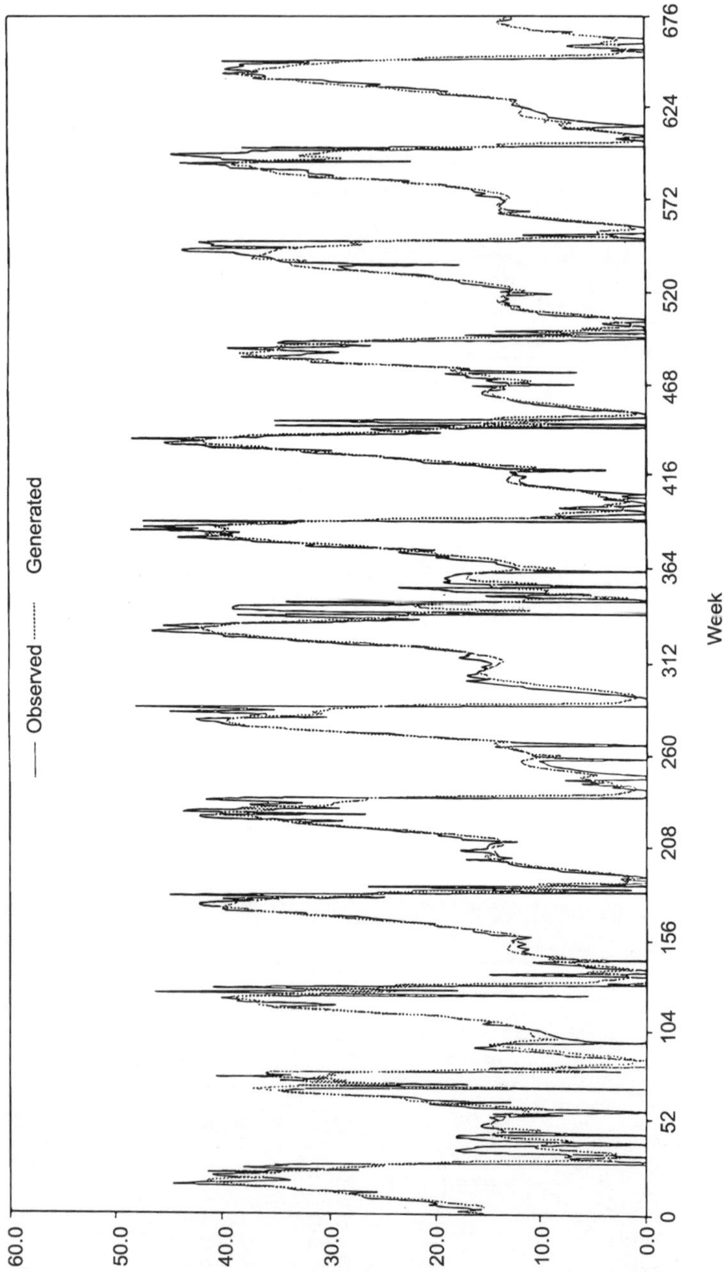

Fig. 6.32 : Variation of observed and generated weekly water deficit series at Kota (1981–93)

between generated and observed weekly water deficit series for thirteen years (1981–93) at different stations.

Validation of generated 13 years mean weekly water deficit series was made with 13 years mean observed weekly water deficit series for all the stations as shown in Figures 6.33 to 6.40. A perusal of Figures 6.33 to 6.40 indicates the closeness between the generated mean weekly water deficit series and observed mean weekly water deficit series at all the stations.

The correlation coefficient between the generated mean weekly water deficit series and observed weekly water deficit series were computed and are presented in Table 6.58. It can be seen from Table 6.58 that the correlation coefficient values were found to range from 0.947 at Jodhpur to 0.993 at Udaipur. The correlation coefficient values were tested by t test and found to be highly significant at 1 per cent level for all the stations. Thus, it also confirms the closeness between the mean generated weekly water deficit series and mean observed weekly water deficit series. Therefore, model structures presented in Table 6.54 can be used for long-term predication of weekly water deficit at respective stations.

Table 6.58 : Correlation coefficient between generated and observed mean weekly water deficit at different stations (1981–93)

Station	Correlation coefficient (r)
Jodhpur	0.947
Sriganganagar	0.972
Fatehpur	0.965
Pali	0.966
Jaipur	0.975
Udaipur	0.993
Banswara	0.987
Kota	0.986

Validation of generated 1994 and 1995 weekly water deficit series was also made with 1994 and 1995 observed weekly water deficit series (Tables 6.59 to 6.66). Other statistical parameters such as mean, standard deviation, coefficient of variation, skewness and kurtosis of observed and generated weekly water deficit for the year 1994 and 1995 at different stations were also computed (Table 6.67). It is evident from Table 6.67 that all statistical parameters at different stations show considerable variations in observed and generated weekly water deficit series of 1994 and 1995 years. It appeared that the formulated models do not adequately forecast the weekly water deficit values at all the stations under study.

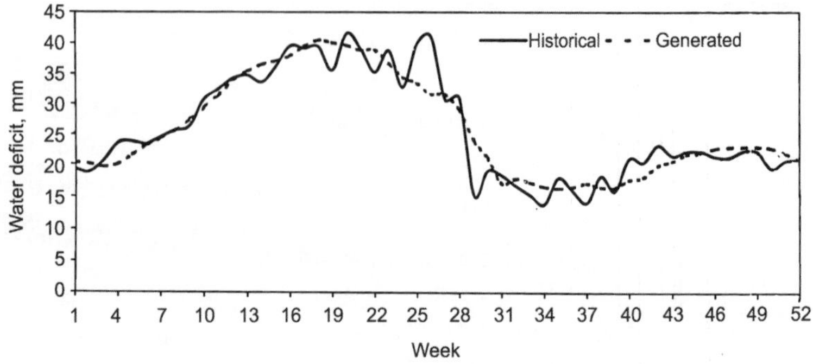

Fig. 6.33 : Variation of historical and generated mean weekly water deficit at Jodhpur (1981–1993)

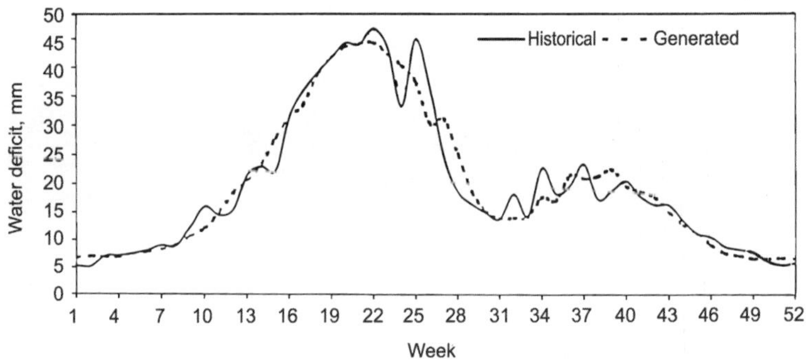

Fig. 6.34 : Variation of historical and generated mean weekly water deficit at Sriganganagar (1981–1993)

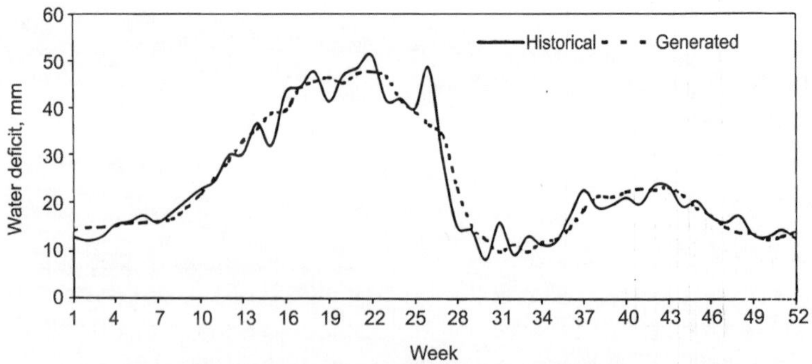

Fig. 6.35 : Variation of historical and generated mean weekly water deficit at Fatehpur (1981–1993)

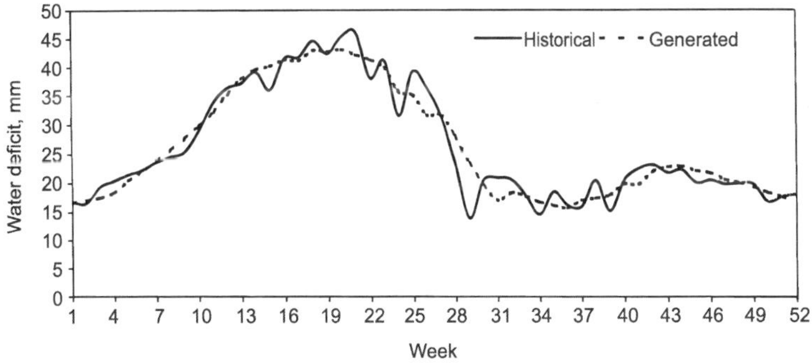

Fig. 6.36 : Variation of historical and generated mean weekly water deficit at Pali (1981–1993)

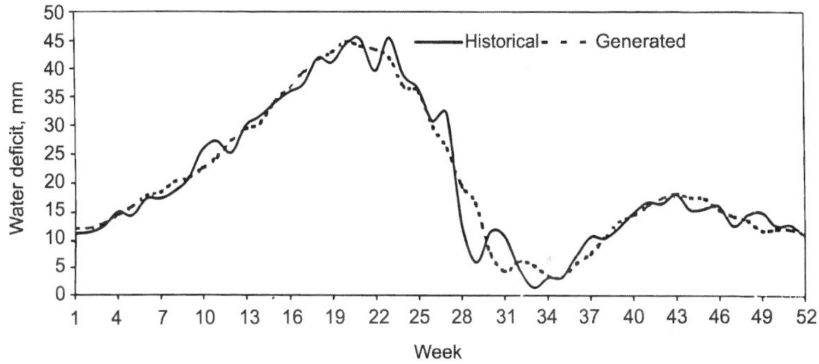

Fig. 6.37 : Variation of historical and generated mean weekly water deficit at Jaipur (1981–1993)

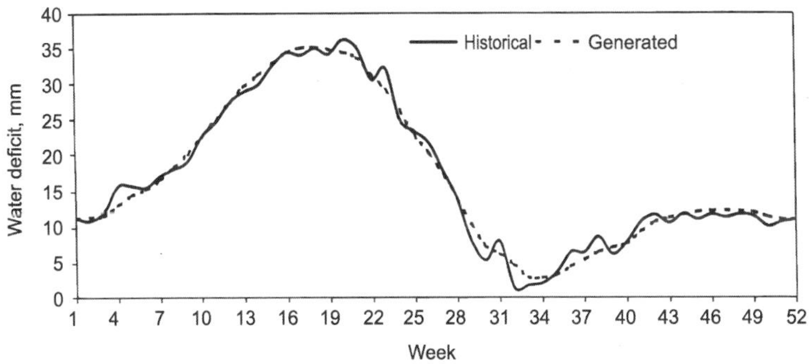

Fig. 6.38 : Variation of historical and generated mean weekly water deficit at Udaipur (1981–1993)

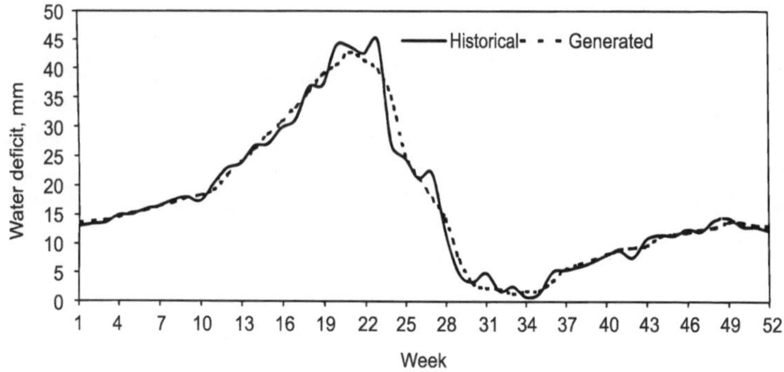

Fig. 6.39 : Variation of historical and generated mean weekly water deficit at Banswara (1981-1993)

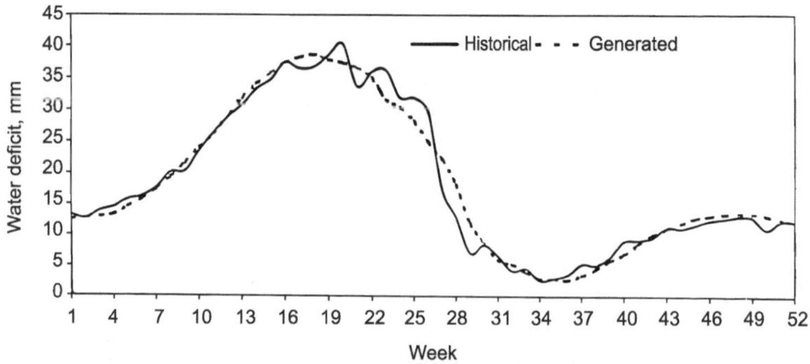

Fig. 6.40 : Variation of historical and generated mean weekly water deficit at Kota (1981–1993)

Table 6.59 : Comparison between predicted and observed weekly water deficit for the year 1994 and 1995 at Jodhpur

Week No.	Water deficit (mm) 1994		Water deficit (mm) 1995	
	Observed	Predicted	Observed	Predicted
1	14.90	14.90	13.20	13.20
2	0.00	0.00	12.60	12.60
3	13.90	13.90	10.70	10.70
4	17.90	25.47	16.10	26.96
5	17.70	25.09	17.70	28.11
6	18.70	31.77	19.10	31.40
7	18.50	36.78	18.50	37.69
8	21.30	40.02	20.10	40.86

contd.....

Table 6.59 – contd...

Week No.	Water deficit (mm) 1994 Observed	Predicted	Water deficit (mm) 1995 Observed	Predicted
9	23.50	45.02	21.90	45.15
10	26.40	49.52	23.30	49.92
11	29.20	53.78	31.30	54.06
12	31.20	58.29	27.70	58.42
13	30.70	62.46	29.50	62.63
14	26.80	66.37	32.20	66.48
15	34.40	70.05	32.90	70.12
16	11.30	73.35	31.20	73.42
17	37.20	76.29	34.90	76.34
18	66.10	78.82	38.30	78.85
19	39.40	80.83	39.20	80.86
20	36.70	82.24	43.20	82.26
21	40.20	82.93	36.60	82.95
22	42.30	82.79	38.40	82.81
23	38.50	81.76	35.70	81.77
24	34.50	79.80	31.80	79.81
25	35.10	76.98	11.40	76.99
26	18.40	73.43	36.20	73.43
27	0.00	69.35	37.90	69.35
28	12.20	64.98	22.90	64.99
29	9.40	60.61	0.00	60.61
30	0.00	56.46	0.00	56.46
31	22.20	52.73	0.60	52.73
32	3.10	49.54	9.50	49.54
33	0.00	46.94	16.70	46.94
34	2.00	44..88	0.00	44.88
35	0.00	43.27	0.00	43.27
36	0.00	42.00	13.90	42.00
37	0.00	40.96	14.60	40.96
38	1.30	40.09	15.40	40.09
39	3.60	39.37	16.20	39.37
40	8.20	38.82	17.70	38.82
41	10.80	38.52	19.70	38.52
42	12.00	38.54	14.00	38.54
43	12.90	38.96	19.20	38.96
44	14..30	39.80	17.90	39.80
45	15.40	41.01	18.00	41.01
46	14.30	42.49	16.80	42.49
47	14.90	44.08	17.00	44.08
48	12.40	45.59	17.30	45.59
49	13.00	46.82	16.20	46.82
50	10.20	47.63	15.80	47.63
51	11.90	47.93	15.30	47.93
52	12.60	47.74	16.20	47.74

Table 6.60 : Comparison between predicted and observed weekly water deficit for the year 1994 and 1995 at Sriganganagar

Week No.	Water deficit (mm) 1994		Water deficit (mm) 1995	
	Observed	Predicted	Observed	Predicted
1	8.00	8.00	6.50	6.50
2	0.00	0.00	0.00	0.00
3	1.10	1.10	0.00	0.00
4	7.20	10.87	6.40	10.21
5	12.00	8.76	7.70	8.70
6	11.10	9.20	8.10	8.78
7	9.60	13.18	1.60	12.90
8	16.80	13.37	8.30	13.32
9	22.40	14.66	10.60	14.49
10	19.10	17.87	14.90	17.75
11	24.00	20.41	15.20	20.38
12	29.30	23.90	12.00	23.83
13	29.90	28.68	21.50	28.63
14	34.80	33.73	22.90	33.72
15	35.00	39.41	27.70	39.39
16	31.60	45.68	20.00	45.66
17	42.50	51.94	32.90	51.93
18	48.00	58.02	28.60	58.01
19	38.50	63.67	40.60	63.66
20	50.50	68.43	44.70	68.42
21	53.60	72.02	41.20	72.02
22	52.40	74.24	46.90	74.24
23	39.10	74.86	54.90	74.86
24	51.30	73.85	56.00	73.84
25	47.30	71.28	0.00	71.28
26	36.00	67.37	34.30	67.37
27	7.60	62.43	40.00	62.43
28	0.00	56.89	40.10	56.89
29	11.00	51.22	0.00	51.22
30	10.60	45.89	0.00	45.89
31	0.00	41.30	9.60	41.30
32	11.50	37.76	0.00	37.76
33	3.20	35.43	7.30	35.43
34	0.00	34.32	12.50	34.32
35	11.10	34.27	8.90	34.27
36	8.40	35.00	13.80	35.00
37	0.00	36.15	16.30	36.15

contd.....

Table 6.60 – contd...

Week No.	Water deficit (mm) 1994		Water deficit (mm) 1995	
	Observed	Predicted	Observed	Predicted
38	11.30	37.31	20.80	37.31
39	14.70	38.12	20.70	38.12
40	14.90	38.28	19.90	38.28
41	15.40	37.62	18.30	37.62
42	13.90	36.10	7.10	36.10
43	13.20	33.81	11.40	33.81
44	13.90	30.93	15.90	30.93
45	11.00	27.74	13.40	27.74
46	8.70	24.51	13.50	24.51
47	11.00	21.52	10.30	21.52
48	7.20	18.96	9.20	18.96
49	7.20	16.97	8.20	16.97
50	9.30	15.57	8.10	15.57
51	7.20	14.72	7.20	14.72
52	6.20	14.30	7.70	14.30

Table 6.61 : Comparison between predicted and observed weekly water deficit for the year 1994 and 1995 at Fatehpur

Week No.	Water deficit (mm) 1994		Water deficit (mm) 1995	
	Observed	Predicted	Observed	Predicted
1	14.60	14.60	13.70	13.70
2	0.00	0.00	12.00	12.00
3	11.00	21.31	12.10	22.19
4	14.50	18.56	15.60	22.66
5	13.30	25.04	17.10	25.74
6	18.60	24.27	21.00	25.72
7	19.60	26.22	17.90	26.59
8	17.70	26.73	17.80	27.25
9	22.00	29.03	21.10	29.20
10	24.50	32.07	22.10	32.26
11	22.70	36.66	24.80	36.74
12	32.00	42.17	29.20	42.25
13	33.90	48.37	24.20	48.40
14	19.50	54.61	35.30	54.64
15	34.80	60.49	33.90	60.51
16	15.30	65.67	29.20	65.68
17	37.20	70.04	38.00	70.05

contd.....

Table 6.61 – contd...

Week No.	Water deficit (mm) 1994		Water deficit (mm) 1995	
	Observed	Predicted	Observed	Predicted
18	39.50	73.62	31.60	73.62
19	44.40	76.52	35.50	76.52
20	9.00	78.79	53.00	78.80
21	30.60	80.39	36.90	80.39
22	52.00	81.10	45.00	81.10
23	57.00	80.60	46.20	80.60
24	29.90	78.57	47.60	78.57
25	57.70	74.77	26.70	74.77
26	31.90	69.20	38.70	69.20
27	29.20	62.11	41.30	62.11
28	0.00	54.03	38.50	54.03
29	0.00	45.66	21.80	45.66
30	0.00	37.78	11.90	37.78
31	0.00	31.07	0.00	31.07
32	9.50	26.02	0.80	26.02
33	11.40	22.89	13.40	22.89
34	0.00	21.67	0.00	21.67
35	15.10	22.12	10.60	22.12
36	19.20	23.91	7.40	23.91
37	2.50	26.59	25.70	26.59
38	22.30	29.76	24.60	29.76
39	22.40	32.99	26.10	32.99
40	22.40	35.93	22.90	35.93
41	24.60	38.25	23.60	38.25
42	20.10	39.67	21.60	39.67
43	21.70	39.99	20.90	39.99
44	19.60	39.14	20.00	39.14
45	17.40	37.19	18.30	37.19
46	17.70	34.40	18.80	34.40
47	16.00	31.20	18.40	31.20
48	16.30	28.07	14.80	28.07
49	17.20	25.53	13.10	25.53
50	13.00	23.92	13.70	23.92
51	12.10	23.39	13.40	23.39
52	14.40	23.85	11.10	23.85

Table 6.62 : Comparison between predicted and observed weekly water deficit for the year 1994 and 1995 at Pali

Week No.	Water deficit (mm) 1994		Water deficit (mm) 1995	
	Observed	Predicted	Observed	Predicted
1	17.20	17.20	14.50	14.50
2	14.40	14.40	13.70	13.70
3	16.90	16.90	10.20	10.20
4	20.30	26.47	18.60	24.83
5	20.50	28.32	20.20	27.10
6	22.70	32.12	22.50	29.85
7	23.60	37.59	21.40	36.70
8	27.60	41.56	21.50	40.82
9	27.70	46.30	23.50	45.45
10	31.40	51.51	24.80	51.07
11	33.00	56.41	31.70	56.04
12	36.50	61.41	30.80	61.07
13	37.60	66.29	36.10	66.08
14	40.30	70.79	38.70	70.61
15	37.60	74.90	37.50	74.76
16	22.00	78.50	36.10	78.40
17	41.10	81.47	36.60	81.39
18	44.70	83.77	40.10	83.70
19	43.60	85.34	41.00	85.30
20	48.70	86.16	47.10	86.13
21	46.70	86.22	37.80	86.19
22	50.60	85.52	45.30	85.50
23	45.80	84.09	43.50	84.08
24	48.80	81.97	41.50	81.95
25	44.20	79.20	40.50	79.19
26	43.20	75.88	44.80	75.87
27	0.00	72.09	46.40	72.08
28	2.90	67.95	26.10	67.95
29	9.40	63.60	9.00	63.60
30	0.00	59.19	0.00	59.19
31	4.90	54.87	0.00	54.87
32	11.60	50.80	4.00	50.80
33	0.00	47.14	12.40	47.14
34	0.00	44.01	10.00	44.01
35	0.00	41.51	1.90	41.51
36	0.20	39.72	15.10	39.72
37	2.90	38.64	12.80	38.64
38	4.80	38.25	16.40	38.25

contd.....

Table 6.62 – contd...

Week No.	Water deficit (mm) 1994 Observed	Water deficit (mm) 1994 Predicted	Water deficit (mm) 1995 Observed	Water deficit (mm) 1995 Predicted
39	7.00	38.48	16.10	38.48
40	8.90	39.22	17.20	39.22
41	11.40	40.31	21.10	40.31
42	11.70	41.58	16.20	41.58
43	13.90	42.87	16.10	42.87
44	13.00	44.00	16.20	44.00
45	14.60	44.85	17.40	44.85
46	14.40	45.30	16.20	45.30
47	13.60	45.30	17.10	45.30
48	13.00	44.84	19.30	44.84
49	14.80	43.99	16.00	43.99
50	14.10	42.83	14.80	42.83
51	15.00	41.49	14.00	41.49
52	13.20	40.16	14.50	40.16

Table 6.63: Comparison between predicted and observed weekly water deficit for the year 1994 and 1995 at Jaipur

Week No.	Water deficit (mm) 1994 Observed	Water deficit (mm) 1994 Predicted	Water deficit (mm) 1995 Observed	Water deficit (mm) 1995 Predicted
1	7.70	7.70	10.80	10.80
2	0.00	0.00	0.00	0.00
3	8.80	8.80	6.90	6.90
4	12.20	18.58	12.40	19.69
5	11.80	18.61	20.20	18.59
6	15.70	24.61	14.20	23.89
7	18.50	30.81	0.80	31.20
8	18.40	33.23	13.90	33.21
9	20.40	37.65	16.80	37.38
10	18.20	42.06	21.90	42.19
11	19.70	45.13	22.40	45.12
12	33.10	49.15	26.10	49.04
13	24.50	53.52	10.00	53.57
14	18.60	57.86	26.00	57.85
15	36.60	63.00	35.20	62.96
16	40.00	68.62	33.50	68.64
17	42.60	74.38	41.60	74.37
18	51.60	80.27	45.30	80.25
19	65.30	85.84	51.60	85.84

contd.....

Table 6.63 – contd...

	Water deficit (mm) 1994		Water deficit (mm) 1995	
Week No.	Observed	Predicted	Observed	Predicted
20	40.80	90.60	58.80	90.59
21	54.10	94.23	35.00	94.22
22	73.20	96.34	63.00	96.34
23	48.00	96.64	72.50	96.63
24	32.00	95.03	54.90	95.03
25	32.10	91.54	37.60	91.54
26	0.00	86.34	46.40	86.34
27	10.30	79.74	46.80	79.74
28	0.00	72.17	0.00	72.17
29	0.00	64.07	0.00	64.07
30	0.00	55.94	0.00	55.94
31	0.00	48.21	0.00	48.21
32	0.00	41.27	0.00	41.27
33	0.00	35.41	2.70	35.41
34	0.00	30.82	0.00	30.82
35	0.00	27.59	0.00	27.59
36	0.00	25.71	0.10	25.71
37	0.00	25.10	2.80	25.10
38	2.70	25.57	9.00	25.57
39	8.70	26.90	13.20	26.90
40	16.00	28.83	18.30	28.83
41	8.40	31.07	20.40	31.07
42	17.40	33.33	20.90	33.33
43	17.50	35.33	14.80	35.33
44	12.40	36.88	16.90	36.88
45	15.30	37.80	17.80	37.80
46	11.10	38.04	16.70	38.04
47	10.20	37.60	12.10	37.60
48	10.80	36.60	10.90	36.60
49	12.50	35.21	12.20	35.21
50	10.30	33.67	12.60	33.67
51	10.10	32.23	12.00	32.23
52	14.20	31.12	12.70	31.12

Table 6.64 : Comparison between predicted and observed weekly water deficit for the year 1994 and 1995 at Udaipur

	Water deficit (mm) 1994		Water deficit (mm) 1995	
Week No.	Observed	Predicted	Observed	Predicted
1	12.90	12.90	11.20	11.20
2	6.60	6.60	0.00	0.00
3	12.30	15.69	9.50	13.95
4	14.50	16.82	10.90	15.07
5	15.00	19.89	10.70	19.16
6	16.20	22.03	12.20	21.51
7	17.40	24.73	13.70	24.47
8	20.50	27.66	14.30	27.49
9	20.30	31.00	16.30	30.91
10	21.10	34.64	17.40	34.59
11	25.70	38.52	20.70	38.49
12	31.00	42.49	21.20	42.47
13	32.40	46.40	25.50	46.39
14	31.70	50.09	26.20	50.09
15	26.90	53.39	29.20	53.39
16	31.60	56.14	28.20	56.14
17	35.90	58.21	34.20	58.21
18	36.60	59.47	36.90	59.47
19	36.40	59.86	44.40	59.86
20	40.00	59.32	32.10	59.32
21	39.70	57.85	41.50	57.85
22	44.60	55.51	39.20	55.51
23	39.50	52.36	37.40	52.36
24	0.00	48.52	43.90	48.52
25	19.80	44.14	0.00	44.14
26	0.00	39.39	29.70	39.39
27	0.00	34.46	0.00	34.46
28	0.00	29.52	31.00	29.52
29	2.60	24.76	0.00	24.76
30	11.90	20.36	0.00	20.36
31	8.50	16.46	0.00	16.46
32	3.00	13.17	18.50	13.17
33	0.00	10.59	23.00	10.59
34	0.00	8.77	9.50	8.77
35	0.00	7.69	19.40	7.69
36	0.00	7.35	10.90	7.35
37	0.10	7.67	20.60	7.67

contd.....

Table 6.64 – contd...

Week No.	Water deficit (mm) 1994		Water deficit (mm) 1995	
	Observed	Predicted	Observed	Predicted
38	0.00	8.55	21.70	8.55
39	1.60	9.88	21.80	9.88
40	4.00	11.53	19.30	11.53
41	6.80	13.36	22.70	13.36
42	8.10	15.22	20.20	15.22
43	9.50	16.99	22.00	16.99
44	11.10	18.57	21.50	18.57
45	11.10	19.85	19.10	19.85
46	10.90	20.80	18.30	20.80
47	10.80	21.39	17.40	21.39
48	10.60	21.63	16.40	21.63
49	10.80	21.56	15.80	21.56
50	10.80	21.25	14.40	21.25
51	10.40	20.81	14.10	20.81
52	10.50	20.33	8.60	20.33

Table 6.65 : Comparison between predicted and observed weekly water deficit for the year 1994 and 1995 at Banswara

Week No.	Water deficit (mm) 1994		Water deficit (mm) 1995	
	Observed	Predicted	Observed	Predicted
1	9.30	9.30	7.40	7.40
2	0.00	0.00	0.00	0.00
3	8.40	18.10	3.40	17.36
4	10.20	18.31	4.20	18.16
5	11.40	25.93	5.10	25.61
6	10.60	27.92	6.00	27.81
7	13.00	31.78	6.10	31.63
8	15.20	33.82	6.80	33.74
9	14.70	36.34	7.10	36.26
10	18.00	38.41	7.70	38.37
11	19.50	40.88	12.50	40.84
12	23.00	43.65	14.10	43.62
13	24.20	47.09	20.20	47.07
14	27.20	51.27	27.60	51.26
15	26.80	56.28	27.10	56.27
16	32.80	62.02	28.90	62.01
17	39.40	68.25	34.30	68.25

contd.....

Table 6.65 – contd...

	Water deficit (mm) 1994		Water deficit (mm) 1995	
Week No.	Observed	Predicted	Observed	Predicted
18	28.20	74.60	39.80	74.60
19	46.40	80.57	17.70	80.57
20	53.00	85.59	48.80	85.59
21	57.70	89.11	45.90	89.11
22	55.00	90.64	52.60	90.63
23	53.50	89.85	52.10	89.85
24	0.00	86.62	56.90	86.62
25	6.40	81.06	56.50	81.06
26	0.00	73.50	55.70	73.50
27	0.00	64.47	59.40	64.47
28	0.00	54.62	0.00	54.62
29	0.00	44.66	0.00	44.66
30	0.00	35.25	0.00	35.25
31	0.00	26.96	0.00	26.96
32	0.00	20.19	0.40	20.19
33	0.00	15.14	0.40	15.14
34	0.00	11.84	0.00	11.84
35	0.00	10.15	0.00	10.15
36	0.00	9.81	0.00	9.81
37	0.00	10.51	0.00	10.51
38	0.00	11.89	0.60	11.89
39	0.60	13.68	1.60	13.68
40	1.80	15.62	2.80	15.62
41	3.10	17.56	1.60	17.56
42	3.80	19.40	3.30	19.40
43	5.20	21.09	4.40	21.09
44	5.60	22.63	4.60	22.63
45	6.70	24.03	7.00	24.03
46	4.40	25.30	5.80	25.30
47	7.30	26.45	0.00	26.45
48	7.30	27.48	5.40	27.48
49	8.00	28.40	6.40	28.40
50	7.80	29.22	7.00	29.22
51	8.80	29.96	7.20	29.96
52	8.40	30.67	0.00	30.67

Table 6.66 : Comparison between predicted and observed weekly water deficit for the year 1994 and 1995 at Kota

	Water deficit (mm) 1994		Water deficit (mm) 1995	
Week No.	Observed	Predicted	Observed	Predicted
1	11.00	11.00	13.60	13.60
2	4.20	4.20	0.00	0.00
3	12.70	12.70	12.90	12.90
4	14.80	17.68	14.40	17.94
5	15.50	17.76	14.10	16.48
6	16.80	22.28	15.60	22.30
7	18.00	25.75	17.10	25.65
8	21.10	28.63	17.60	28.22
9	21.20	33.07	19.10	33.04
10	19.50	37.46	20.10	37.37
11	26.70	42.02	11.70	41.89
12	31.90	47.10	24.20	47.07
13	33.50	52.10	13.30	52.05
14	21.80	56.97	31.40	56.92
15	33.30	61.62	30.50	61.60
16	34.10	65.73	34.30	65.71
17	38.10	69.18	40.50	69.16
18	38.70	71.84	40.80	71.83
19	38.60	73.57	41.10	73.56
20	42.30	74.29	47.10	74.28
21	42.20	73.98	34.90	73.98
22	47.10	72.65	44.30	72.64
23	18.00	70.33	42.10	70.33
24	41.90	67.13	40.20	67.13
25	2.90	63.15	42.10	63.15
26	0.00	58.55	33.10	58.55
27	19.10	53.48	33.20	53.48
28	1.30	48.11	13.40	48.11
29	11.00	42.62	0.00	42.62
30	0.00	37.19	0.00	37.18
31	0.50	31.97	0.00	31.97
32	1.30	27.12	0.80	27.12
33	0.00	22.78	5.40	24.78
34	0.30	19.05	0.00	19.05
35	1.00	16.02	0.00	16.02

contd.....

Table 6.66 – contd...

Week No.	Water deficit (mm) 1994		Water deficit (mm) 1995	
	Observed	Predicted	Observed	Predicted
36	0.00	13.74	0.00	13.74
37	0.50	12.24	1.70	12.24
38	2.90	11.50	4.40	11.50
39	5.70	11.48	7.50	11.48
40	7.30	12.10	9.00	12.10
41	8.80	13.26	12.40	13.26
42	10.60	14.83	13.10	14.83
43	11.90	16.68	15.20	16.68
44	13.30	18.66	0.00	18.66
45	13.30	20.62	10.30	20.62
46	13.10	22.43	11.20	22.43
47	12.90	23.99	11.70	23.99
48	12.80	25.21	11.90	25.21
49	13.00	26.07	7.80	26.07
50	13.10	26.56	11.70	26.56
51	12.80	26.71	12.00	26.71
52	12.80	26.63	0.00	26.63

This may be due to the significant variations in the amount of rainfall received during 1994 and 1995 and also variations caused by other seasonal climatic parameters at different stations. Thus, it may be used for long-term prediction of weekly water deficit.

Table 6.67 : Statistical parameters of observed and generated weekly water deficit for the year 1994 and 1995

Station	Year	Nature of series	Mean (mm)	Standard deviation (mm)	Coefficient of variation	Skewness	Kurtosis
Jodhpur	1994	Observed	18.106	14.241	0.787	0.893	1.026
		Generated	51.379	19.361	0.377	-0.112	-0.144
	1995	Observed	20.625	11.200	0.543	0.110	-0.511
		Generated	51.671	18.852	0.365	0.029	-0.477
Sriganga-nagar	1994	Observed	18.838	15.952	0.847	0.897	-0.420
		Generated	35.416	20.855	0.589	0.394	-0.755
	1995	Observed	17.379	14.736	0.848	1.072	0.356
		Generated	35.327	20.970	0.594	0.377	-0.754

contd.....

Table 6.67 – contd...

Station	Year	Nature of series	Mean (mm)	Standard deviation (mm)	Coefficient of variation	Skewness	Kurtosis
Fatehpur	1994	Observed	20.525	13.706	0.668	0.810	0.916
		Generated	41.279	21.042	0.510	0.650	-0.686
	1995	Observed	23.056	12.196	0.529	0.417	-0.112
		Generated	41.658	20.518	0.493	0.767	-0.772
Pali	1994	Observed	21.192	15.872	0.749	0.435	-1.100
		Generated	53.140	19.915	0.375	0.229	-0.862
	1995	Observed	23.198	13.034	0.562	0.331	-0.915
		Generated	52.758	20.421	0.387	0.131	-0.735
Jaipur	1994	Observed	17.925	17.859	0.996	1.301	1.311
		Generated	47.245	25.764	0.545	0.582	-0.667
	1995	Observed	20.206	18.542	0.918	1.043	0.398
		Generated	47.276	25.715	0.544	0.588	-0.667
Udaipur	1994	Observed	15.033	13.176	0.876	0.695	-0.652
		Generated	28.387	17.156	0.604	0.615	-1.006
	1995	Observed	19.475	11.487	0.590	0.246	-0.185
		Generated	28.124	17.435	0.620	0.572	-0.990
Banswara	1994	Observed	13.129	16.334	1.244	1.510	1.381
		Generated	38.612	25.583	0.663	0.767	-0.600
	1995	Observed	14.662	19.137	1.305	1.317	0.254
		Generated	38.541	25.645	0.665	0.762	-0.604
Kota	1994	Observed	16.254	13.474	0.829	0.721	-0.466
		Generated	35.611	21.696	0.609	0.569	-1.117
	1995	Observed	17.092	14.485	0.847	0.638	-0.810
		Generated	35.545	21.780	0.613	0.550	-1.092

7. Crop Planning on the Basis of Rainfall Analysis and Water Balance

7.1 Introduction

India receives adequate amount of rainfall annually through the four different types of weather phenomena viz., south-west monsoon (74 per cent), north-east monsoon (3 per cent), premonsoon (13 per cent) and post monsoon (10 per cent). The distribution in time and space is erratic thus resulting in a limitation on the length of crop growing periods or the occurrence of droughts. Rajasthan is located in the western part of the country, which often experiences moisture stress conditions or dry spells affecting stable agricultural production. It is essential to follow a profitable cropping system under rainfed areas for proper utilization of available rain water. Rainfall onset, duration and withdrawal are some of the important factors in the preparation of crop production strategies. Suitable cropping systems have been reported from different parts of the country. Probability analysis of rainfall offers a better scope for crop planning. Estimation of water balance is equally important for crop planning in an area.

Sharma *et al.,* (1979) analysed the rainfall data of 17 years (1961–77) at Pantnagar. Weekly data of rainfall were found to be more useful for planning of cropping programme as well as water management practices than monthly, seasonal or annual data. They worked out 80 per cent, 50 per cent and 10 per cent, chances of rainfall in a week, month, season and year. Normal, abnormal and yearly drought estimations were also identified.

Biswas and Maske (1981) analysed rainfall data, which can directly be used for dry land crop planning, land use and water management. Coefficient of variation brought out the variation of rainfall about its mean, Probability analysis quantified the minimum amounts of rainfall expected for growth of crops and also identified time of life saving irrigation. Runs of wet and dry spells gave the duration of stress period within wet periods. Return period indicated the occasions of excess rainfall even in the low rainfall areas, which may damage the crops.

Sastri *et al.,* (1984) studied the frequencies of occurrence of agricultural droughts in respect of pearl millet and Kharif pulses for the three stations in western Rajasthan representing semi-arid, arid and extremely arid types of climates. The study brought out that Kharif pulses are more susceptible to drought than pearl millet. Also the vulnerability of pearl millet to drought under early, normal and late receipt of rains under Jodhpur conditions was presented.

Ramakrishna *et al.,* (1985) computed variability of the average length of growing season in the western Rajasthan based on weekly water balance (1901–1980). The date of commencement of rains was found to influence considerably, the periods of water availability for crop growth. Therefore, to assess the impact of commencement of rains to the growing season on crop performance, length of the growing season and its variability under early, normal and late commencement of the monsoon rains were also studied. Probabilities of occurrence of growing season of varying lengths were investigated.

Possible adjustments in the cropping patterns of different regions in the western Rajasthan were projected along with possible changes in the cropping system.

Bhandarkar *et al.,* (1986) found that rainfall data in the form of simple aritrunetic averages provide a general understanding, whereas statistical analysis is helpful in deciding suitable cropping pattern and proper water management practices. Based on the probability analysis recommendations for various agricultural practices have also been given.

Sharma (1986) worked out the probability analysis of 30 years (1965–85) rainfall data at Barapani, Shillong. He indicated that maximum six days rainfall or five year return period is 31.0 cm with 3 days as the corresponding length of dry spell and this could be taken as the design storm. Transplanting of paddy should be undertaken during May in the pre-monsoon period and adequate drainage should be provided by the middle of June. It was observed that winter rainfall of 3–6 cm with 2–10 years return period can meet the demand partly to the Rabi crops whereas there is no need of providing any irrigation to the Kharif crops.

Subramaniam and Rao (1986) analysed daily meteorological data from 1954 to 1981 of Chirala. Potential evapotranspiration was calculated by Thornthwaite (1948) method. For crop planning they used water balance studies by Thornthwaite and Mather (1955) method and identified the water availability periods.

Prakash and Rao (1986) analysed 25 years (1956–80) rainfall data and indicated that Kota received average annual rainfall of 800 mm and average monthly rainfall in monsoon season was 71, 285, 555 and 117 mm in June, July, August and September respectively. For assessing utilization of rainfall for crop planning weekly, monthly, seasonal and annual values at different probability levels (50%–90%) were computed by Weibull's method. Results indicated that assured rainfall of atleast 10 mm at 70 per cent chance is expected from 27th to 30th meteorological weeks (2nd July to 9th September). Hence, dry sowing of Kharif crops in 20th week (end of June) and choosing crop varieties with growing season of 11–14 weeks duration is most suitable for Kota division under rainfed conditions.

Verma and Sarma (1988) pointed out that distribution of rainfall, the only source of water supply for rainfed crop, is quite erratic in time and space. Historical rainfall data of rainfed Kandi belt of Punjab were analysed to estimate the lowest assured weekly rainfall at different probability levels using incomplete gamma distribution. The results indicated that chance of drought is more at later stages of maize growth and there is a scope for *in situ* moisture conservation measures and run off collection in tanks for supplemental irrigation. They stressed that the Rabi crops has to be grown under moisture stress conditions and a pre sowing irrigation is essential for establishing crops of short duration and drought resistant varieties.

Bhattacharjee and Chandra (1989) found that weekly rainfall analysis have been more useful for crop planning. Probability of the period of onset and termination of water availability has been determined for Purulia district of West Bengal. Recommendations of rice based crop-planning have also been given.

Sengupta and Chanda (1989) worked out that in West Bengal the most important cropping season is Kharif which more or less coincides with monsoon months. During this season the prevalent crop is transplanted rice which performs better under partial submergence. For rice-based cropping pattern a modified water balance equation has been developed which is applicable for fields of different topographic situations and soil types for estimation of water-availability period at different probability levels.

Patil *et al.,* (1989) analysed monthly rainfall data (1955–1984) of Solapur and worked out decrease in coefficient of variation in the month of June, which is ideal for seeding and establishment of rainy season crops. The analysis of weekly rainfall data indicated better chances of getting rainfall during 25th meteorological week for sowing of rainy season crops and during 38th and 39th meteorological weeks for sowing of post rainy season crops.

Sharma and Bohara (1990) worked out the probability analysis of 31 years (1958–88) rainfall data of Bikaner and indicated that maximum two days rainfall of 1.01 to 5 years return period vary from 16 to 82 mm with 2 to 24 days as the corresponding length of dry spell and these could be taken as the design storms for rainwater harvesting. Probability of growing pasture in this area is 99 per cent, whereas for rainfed crop growing it is only 40 per cent and that too by the middle of July only.

Tambile *et al.,* (1991) analysed the rainfall data of 45 years (1945–1989) at Parbhani (Maharashtra). The predication of weekly rainfall at different probability levels provided valuable guidelines for crop planning especially under dry land situation. They used Weibull's method for frequency analysis and predicted rainfall amounts at Parbhani on weekly intervals at nine levels of probability.

Dash and Senapati (1992) found that for agricultural management and planning is necessary to know the sequence of dry and wet periods along with forward and backward accumulation of rainwater. They used the Markov chain model method to know the probability of having different conditions of dry and wet weeks and also forward backward accumulation of rainwater suitable for crop production at Bhubaneswar.

Singh (1992) indicated that the rainfall plays a decisive role in crop planning under dry land farming conditions. Fifteen years data recorded at Jalaun (Uttar Pradesh) have been analysed to the onset, withdrawal and duration of monsoon, rainfall distribution and variation of rainfall. Cropping systems evolved through years of experiences of heavy textured soil have been suggested.

Rana and Thakur (1998) analsyed daily rainfall data for a period of 23 years (1973–95) in Kulu Valley, Himachal Pradesh. They used Weibull's method for computation of weekly rainfall amounts at 20, 50 and 80 per cent probability levels. Based on this analysis they recommended crop plan for Rabi season.

Agnihotri (1999) analysed rainfall data of 39 years (1958–96) at Chandigarh. He worked out precipitation concentration index (PCF) and modified precipitation concentration index (MPCI) and found that Kharif and Rabi season rainfall series are non-random as tested through non-parametric test at 5 per cent level of significance. The MPCI values were used to classify drought and flood years. Mean Kharif season rainfall during, 1958–67, 1968–77, 1978–87 and 1988–96 was found to be significantly different from each other as tested through statistical 't' test at 5 per cent level of significance.

7.2 Selection of Crops and Cropping Pattern Under Different Agro-Climatic Zones of Rajasthan

Selection of crops and cropping pattern depends on the length of growing season if temperature is not a limiting factor. Length of growing season is determined on the basis of period of moisture availability for crop growth. The length of growing season was determined by the Thornthwaite and Mather (1955) method. In this method length of growing season was computed based on the water balance. Length of growing season starts when $I_m > 0.5$ and ends when $I_m < 0.25$, where I_{ma} is the moisture adequacy index, which is the ratio of $\frac{AET}{ET_0}$. The crops and cropping pattern were selected on the basis of water availability period as under:

I_{ma}	Water availability condition
0.00–0.24	Dry
0.25–0.49	Semidry
0.5–0.74	Sub-moist
0.75–0.99	Moist
> 100	Humid

In Rajasthan, droughts occur more frequently in arid and semi-arid regions. Droughts generally occur in 40 to 60 per cent area of the state either due to deficit in seasonal rainfall during the main cropping season or due to inadequate soil moisture availability during the period of prolonged dry spell in between the successive rainfall events. The arid areas are almost subjected to permanent drought and the semi-arid, sub-humid areas are likely to be subjected intermittent droughts even during the years with good seasonal rainfall. Therefore, prediction of drought is essential in terms of dry spell, aridity and water deficit. Based on drought prediction, the drought management strategies for short or long periods can be identified under different agro-climatic zones of Rajasthan. This will further assist sustainable crop production in the rainfed areas of the state.

7.3 Discussion

7.3.1 Crop Planning Under Different Agro-climatic Zones of Rajasthan

The selection of crops and crop varieties for a given region depends on the pattern of the growing season and the quantum and distribution of rainfall. It becomes essential to match the crop peak water need with the most probable periods of good moisture availability to achieve higher yields. Such an analysis not only helps ill identifying the risks involved in the production of a specific crop/varieties in a given region and also enables identification of not only the choice of crops but also sustainability of crop production.

7.3.1.1 Length of growing season

The length of growing season is computed following the Thornthwaite and Mather (1955) method as mentioned in Section 3.13.3.5 for different stations and are given in Table 4.91.

It can be seen from Table 7.1 that the longest length of growing season of 29th week or 203 days is observed at Banswara while the shortest length of growing season of 8 weeks (56 days) is observed at Sriganganagar. It can also be seen that in the eastern Rajasthan length of growing season ranges from 147 days to 203 days, while in the western Rajasthan it ranges from 56 days to 91 days.

It is further inferred from Table 7.1 that the earliest normal date of start of growing season is observed as 24th week at Udaipur and Banswara, while the latest date has been observed as 29th week at Jodhpur. Similarly, the earliest ending of length of growing season is observed as 33rd week at Sriganganagar and latest (52nd week) at Banswara and Kota during the climatic year.

It is interesting to note that while comparing Table 2.4 and Table 7.1 the start of length of growing season coincides with the start of rainy season at Pali, Jaipur, Udaipur, Banswara and Kota, while at Jodhpur, Sriganganagar and Fatehpur, the length of growing season starts later than the start of the rainy season. This might be due to high evaporative demand at these stations.

Table 7.1 Normal length of growing season under different agro-climatic zones of Rajasthan

Zone	Starting week	Ending week	Duration in weeks
Jodhpur (I_a)	27	37	10 (70)
Sriganganagar (I_b)	26	33	8 (56)
Fatehpur (II_a)	27	39	13 (91)
Pali (II_b)	26	37	12 (84)
Jaipur (III_a)	25	45	20 (140)
Udaipur (IV_a)	24	45	21 (147)
Banswara (IV_b)	24	52	29 (203)
Kota (V)	25	52	28 (196)

Figures in parenthesis indicates number of days.

It can be seen from Figrues 7.1 to 7.8 that the length of growing season does not increase as annual rainfall increases at all the stations under study. It can also be inferred that the maximum length of growing season was observed at the maximum annual rain fall at Pali, Udaipur, Banswara and Kota, while at other stations it was observed at rainfall less than the maximum rainfall. This might be due to erratic distribution of rainfall.

It can be seen from Figures 7.1 and Figure 7.2 that at Jodhpur and Sriganganagar the length of growing of 10 weeks and 8 weeks respectively observed at mean annual rainfall with 35.7% probability during the climatic normal year. Similarly, it can also be seen from Figures 7.3 and 7.4 that at Fatehpur and Pali the length of growing season at mean annual rainfall with 48.57% and 46.66% probability respectively during the climatic normal year.

It is interesting to note that at Jaipur the length of growing season of 20 weeks was observed at 950 mm rainfall with 18.57% probability while at mean annual rainfall of 687.00 mm with 42.86% probability it was observed as 20 weeks. This might be due to high variability in rainfall distribution. (Figure 7.5)

It is evident from Fig 7.6 that at Udaipur the length of growing season of 21 weeks was observed at mean annual rainfall with 56-43% probability during the climatic normal year.

Figure 7.7 and Figure 7.8 show that at Banswara and Kota the length of growing season of 29 and 28 weeks was observed at annual rain fall higher than mean annual rainfall of 950 mm with 38.57 per cent probability and 835.3 mm with 8.57 per cent probability respectively.

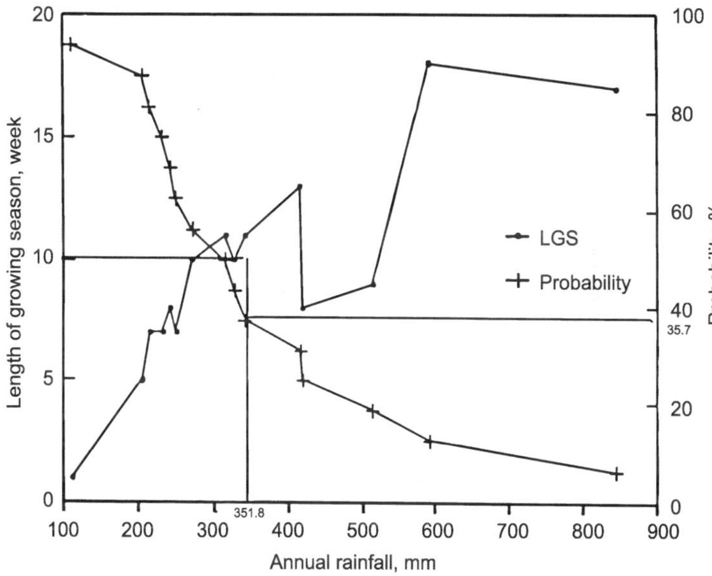

Fig. 7.1 : Appropriate cropping pattern at Jodhpur

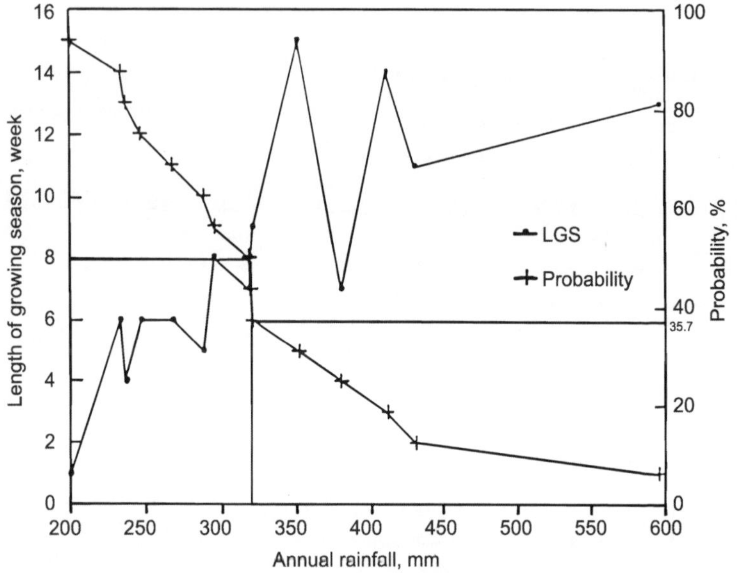

Fig. 7.2 : Appropriate cropping pattern at Sriganganagar

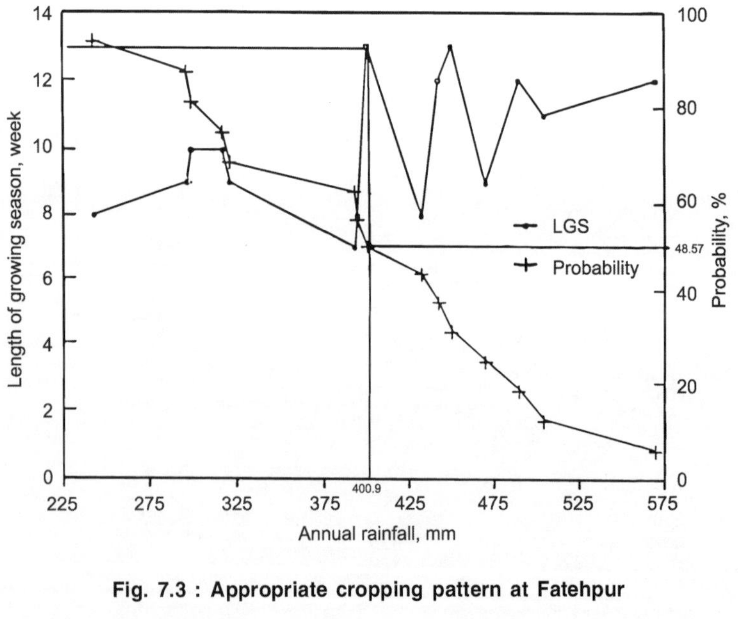

Fig. 7.3 : Appropriate cropping pattern at Fatehpur

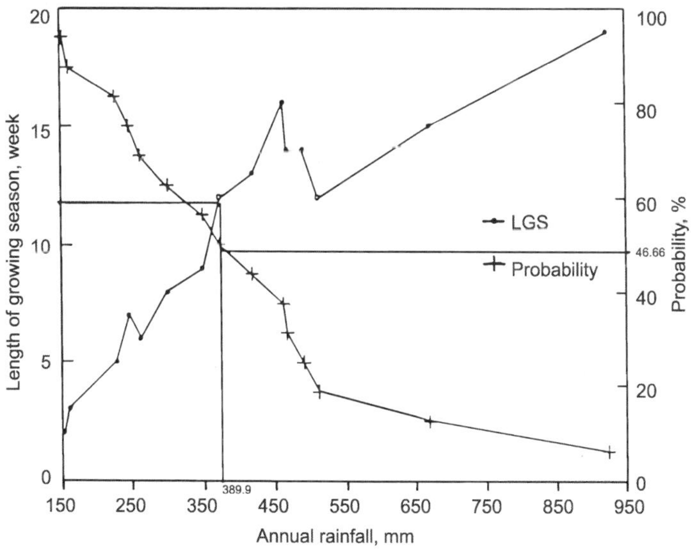

Fig. 7.4 : Appropriate cropping pattern at Pali

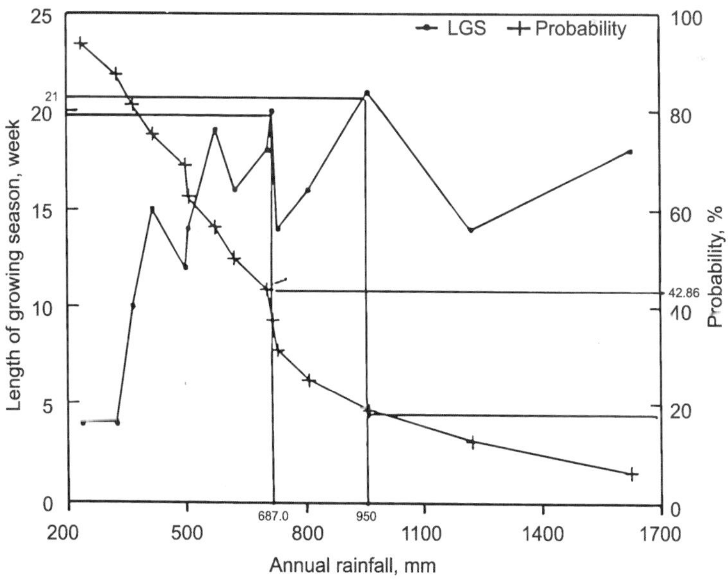

Fig. 7.5 : Appropriate cropping pattern at Jaipur

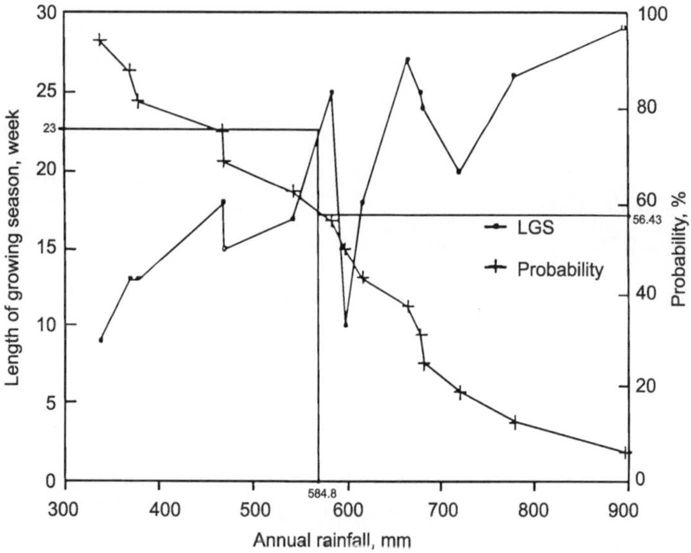

Fig. 7.6 : Appropriate cropping pattern at Udaipur

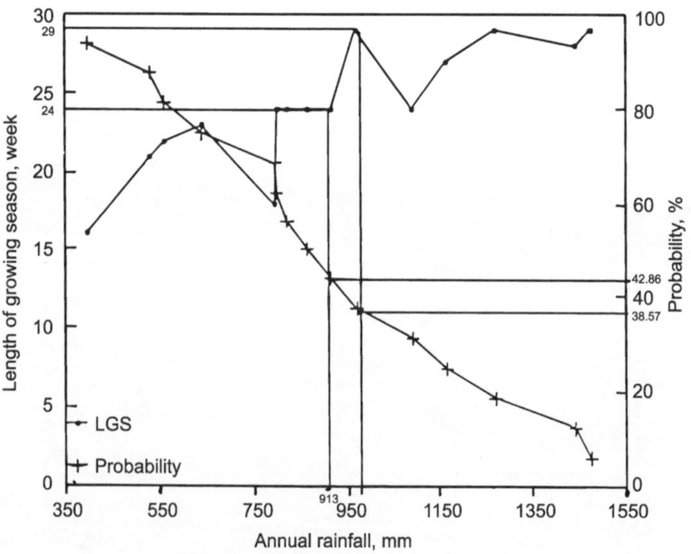

Fig. 7.7 : Appropriate cropping pattern at Banswara

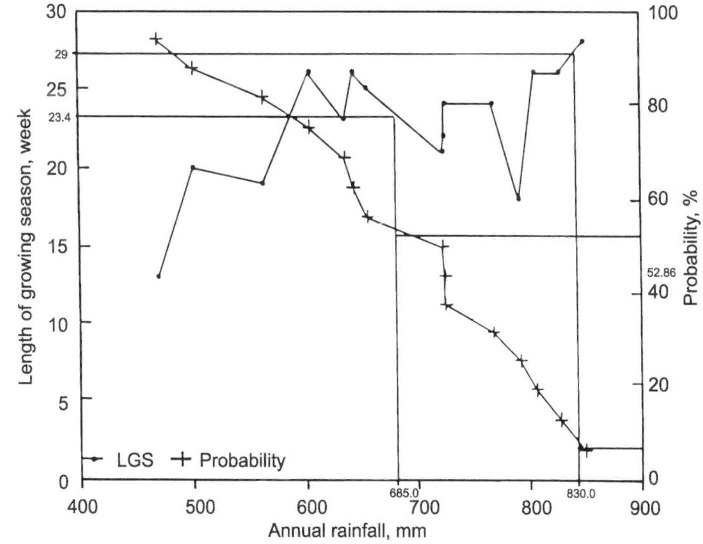

Fig. 7.8 : Appropriate cropping pattern at Kota

7.3.1.2 Choice of crops and cropping systems

The choice of crops and cropping system depends on the length of the growing season of a region. The length of growing season and cropping system for different station of Rajasthan is presented in Table 7.2.

It can be seen from Table 7.2 that double cropping system is suitable only in Banswara and Kota regions, whereas at Sriganganagar grasses, shrubs, trees and short duration crops can be taken with soil mulching as possibility of applying supplemental irrigation is nil and no surplus water is available. Monocropping system is suitable for Jaipur and Udaipur regions. At Jodhpur, Fatehpur and Pali, pearl millet pulse crops are suitable. The surplus water may be harvested at, Jodhpur, Fatehpur, Pali, Jaipur, Udaipur, Banswara and Kota and may be applied during the dry spell periods.

7.3.1.3 Crop planning

The proposed crop plan is presented in Table 7.3 for different stations of Rajasthan. It can be seen from Table 7.2 that the major crops of the Jodhpur region are pearl millet, green gram, mothbean, clusterbean. The preparatory tillage can be started in 25-26 week for sowing of crops. The Kharif crops can be sown in 27th week and intercultural operations may be performed in 28th and 31st week after sowing of crops. Table 7.3 shows that there may be drought-spell after 36 weeks onwards so supplemental irrigation could be applied with the harvested run off water.

Table 7.3 shows that at Sriganganagar, the major crops are grasses, trees and bajra. The preparatory tillage operation can be done in 24th to 25th week and sowing can be done in 26th week. The intercultural operations can be done in 28th and 31st week. As no surplus water is available (Table 3.6 in chapter 3) from runoff so supplemental irrigation can not be applied.

Table 7.2 : Length of growing season and suitability of cropping system for different stations under rainfed conditions

Station	Length of growing season (days)	Cropping system	Major crops	
			Kharif	Rabi
Sriganga-nagar	Less than 75 days	Grasses, shurbs, trees and short duration crops	Grasses trees, and bajra, moth bean	
Jodhpur, Fatehpur and Pali	75 to 130 days	Short duration crops, drought resistant crops pulse crop, intercropping	Pearl millet, green gram, cluster, bean maize, moth bean sesamum	Barley, taramira, mustard, gram
Jaipur and Udaipur	130-180 days	Monocropping, drought resistant crops and inter cropping	Maize, Bajra, sorghum clusterbean cowpea, gram gram and blackgram	Barley, mustard gram and taramira
Banswara and Kota	More than 180 days	Double cropping and intercropping	Maize, paddy, Kharif, pulses, small mollets, cotton, soyabeen (Kota) and Sorghum (Kota)	Wheat, barley, gram, linseed, Rabi maize (Banswara)

It is also inferred from Table 7.2 that the major crops of Fatehpur and Pali are pearlmillet, green gram, moth bean, cluster bean and maize and sesamum. For sowing of crops in these regions the preparatory tillage operation can be started From 24th to 25th week (Table 7.1). The crops can be sown in 27th week at Fatehpur and in 26th week at Pali region.

Table 7.3 : Cultural operations schedule under different agro-climatic zones of Rajasthan during Kharif season

Station	Operation (week)			
	Preparatory tillage (week)	Sowing (week)	Inter culture (week)	Supplemental irrigation (week)
Jodhpur	25-26	27	28, 31	36 onwards
Sriganganagar	24-25	26	28, 31	–
Fatehpur	24-25	27	29, 31	35 onwards
Pali	24-25	26	28, 31	37 onwards
Jaipur	23-24	25	27, 30	37 onwards
Udaipur	22-23	24	27-28	36 onwards
Banswara	23	24	27-28	36 onwards
Kota	23-24	25	27-28	37 onwards

The interculture operations can be done in 29th and 31st week at Fatehpur and 28th and 31st week at Pali. The supplemental irrigation can be applied during 35th week onwards and 37th week onwards at Fatehpur and Pali respectively, during the dry spell.

At Jaipur and Udaipur the major crops are maize, sorghum, clusterbean, green gram, black gram, and Bajra (Table 7.2). It can be seen from Table 7.3 that the preparatory tillage can be started from 23rd week at Udaipur. The Kharif crops can be sown in 24th week at Udaipur and in 25th week at Jaipur. After sowing of crops the intercultural operations can be done in 27th-28th week at Udaipur and in 27th and 30th week at Jaipur. The supplemental irrigation may be applied in 36th week at Udaipur and 37th week at Jaipur from harvested run off water.

It can be seen form Table 7.2 that at Banswara the major crops are maize, paddy, Kharif pulses, small millets and at Kota paddy and sorghum. Table 7.3 shows that the preparatory tillage for sowing of crops can be started in 23rd week at Banswara and 23rd-24th week at Kota. The crops can be sown in 24th week at Banswara and in 25th week at Kota. After sowing of crops the intercultural operation can be done during 27th-28th week. There may a terminal drought during 36th and 37th week at Banswara and Kota respectively. Therefore, the supplemental irrigation can be applied with the harvested runoff water.

Thus, it is interesting to note that in western Rajasthan the preparatory tillage operation can be started form 24th to 26th week, while in eastern Rajasthan it can be started from 22nd to 24th week. Similarly, sowing of crops can be started from 26th to 27th week in western Rajasthan and from 24th to 25th week in eastern Rajasthan.

8. Summary and Conclusions

The main objective of the study is to predict drought under different agro-climatic zones of Rajasthan state. An attempt has been made to predict the drought in terms of dry spell, aridity and water deficit. The weekly climatic data of rainfall, air temperature, relative humidity, wind speed and bright sunshine hours were collected for eight stations *viz.,* Jodhpur, Sriganganagar, Fatehpur, Pali, Jaipur, Udaipur, Banswara and Kota for the period of 15 years (1981–95).

Soils data were also obtained from eight Agricultural Research Stations located in different agro-climatic zones of the state. The rainfall data were analysed for onset and withdrawal of monsoon and initial and conditional probabilities of dry spells were determined by Markov Chain process. Weekly and yearly reference evapotranspiration values were estimated by the Penman-Monteith equation for determination of water balance under different agro-climatic zones of the state. Water deficit data were analysed to develop and evaluate appropriate stochastic models. The results of the study are summarised as follows:

1. The highest average annual rainfall of 913.9 with coefficient of variation of 34.3 per cent was recorded at Banswara and lowest average annual rainfall of 325.2 mm with coefficient of variation of 29.6 per cent was recorded at Sriganganagar. The coefficient of variation of annual rainfall ranged from 16.5 per cent at Kota, to 51.1 per .cent at Jaipur, which indcates that the annual rainfall pattern is very much erratic from zone to zone in the state.

2. Banswara received the highest amount of seasonal rainfall having an average of 833.2 mm with coefficient of variation of 39.14 per cent and contributed 91.5 per cent to annual rainfall. Sriganganagar received the lowest seasonal rainfall having an average of 248.9 mm with

coefficient of variation of 34.99 per cent and contributed 76.26 per cent to annual rainfall. The coefficient of variation of seasonal rainfall rainged from 19.12 per cent at Kota to 58.58 per cent at Jodhpur.

3. The maximum mean weekly rainfall ranged from 29.1 mm at Sriganganagar to 139.1 mm at Banswara. The value of minimum weekly rainfall was found to be zero at all the stations.
4. In eastern Rajasthan the rainy season may start from 27th to 28th week. In western Rajasthan the rainy season may start from 28th to 30th week. The minimum length of rainy season of 3 weeks was recorded at Jodhpur and the maximum length of rainy season of 24 weeks was recorded at Udaipur and Banswara. The mean length of rainy season was observed as 8.4 to 12.4 weeks in western Rajasthan and 11.1 to 17.0 weeks in eastern Rajasthan.
5. The probability of occurrence of dry week decreased from 26th to 28th week at all the stations except at Jodhpur, Sriganganagar and Banswara.
6. Mid season drought may likely to occur at Jaipur, Udaipur, Banswara and Kota during the period 31–32, 31–32, 31–33 and 31–33 weeks respectively with a probability of less than 10 per cent and also during the period 33-35, 30, 30 and 30–35 weeks respectively with a probability of 10-25 per cent due to break in monsoon conditions during rainy season.
7. The annual reference evapotranspiration varied from 1176.8 mm at Banswara to 1792.0 mm at Jodhpur. The mean annual reference evapotranspiration varied from 1363.8 mm with coefficient of variation of 6.80 per cent at Banswara to 1628.9 mm with a coefficient of variation of 7.83 per cent at Jodhpur. The least coefficient of variation (2.97 per cent) in annual reference evapotranspiration was observed at Udaipur.
8. The mean weekly reference evapotranspiration in Rajasthan varied from 8.8 mm in 2nd week at Sriganganagar to 54.6 mm in 23rd week at Jaipur. The water need during the normal year in Rajasthan varied

from 1337.8 mm at Banswara to 1625.6 mm at Jodhpur, while during wettest and driest years it ranged from 1176.8 mm (Banswara 1994) to 1572.0 mm (Jodhpur, 1987) respectively. The value of actual evapotranspiration during the normal year ranged from 326.4 mm at Sriganganagar to 643.5 mm at Kota, while during wettest and driest years it ranged from 354.4 mm at Fatehpur (1987) to 695.7 mm at Kota (1982) and 111.1 mm at Jodhpur (1987) to 449.5 mm at Kota (1987) respectively.

9. The annual water deficit during the normal year ranged from 708.6 mm at Ballswara to 1273.4 at Jodhpur while in wettest and driest years it ranged from 634.1 mm at Banswara (1994) to 1109.0 mm at Jodhpur (1994) and 1046.7 mm at Udaipur (1987) to 1680.9 mm at Jodhpur (1987) respectively. However, the annual surplus water during the normal year appeared only at Jaipur and Banswara.

10. The highest mean aridity index of 79.96 per cent was observed at Jodhpur while minimum was observed at Banswara (57.78%). The mean aridity index decreased from west to east and from north to south in the state.

11. The droughtiness factor increased with increase in the coefficient of variation of aridity index. The low values of droughtiness factor did not coincide with low values of aridity index. High values of droughtiness factor did coincide with low values of aridity index at different longitudes.

12. The year 1987 was found to be severe or disastrous drought year at all the stations except at Fatehpur, where it appeared as a moderate drought year. The frequency of occurrence of drought of different intensities was found to be 7 out of 15 years at all the stations except at Udaipur, where it was 6 out of 15 years. The percentage departure of aridity index from the median was found to be higher at Banswara and Kota, which might have caused 1987 year as a disastrous drought year. Jaipur experienced highest moderate drought year (4) while no moderate drought year was experienced at Kota. The frequency of occurrence of severe drought years varied from one at

Jodhpur, Fatehpur and Jaipur to three at Udaipur and Kota.

13. The surplus water was found likely to exceed 20 mm in 27th and 36th week with probability of 6.67 per cent at Jodhpur, from 29th to 31st weeks at Fatehpur, in 27th, 30th, 32nd, 33rd and 35th week at Pali, from 29th to 36th weeks at Jaipur, from 30th to 32nd weeks at Udaipur, from 28th to 36th week, 39th and 41st week at Banswara and from 30th to 36th weeks at Kota. No surplus water was available in any week during the rainy season with 20 mm or more at Sriganganagar. The surplus water may exceed 100 mm in 29th week at Fatehpur, in 31st and 33rd week at Banswara and in 32nd and 34th week at Kota.

14. The minimum duration of large drought spell of 4 weeks occurred at Fatehpur, whereas maximum duration of large drought spell of 15 weeks occurred at Sriganganagar during the year 1987. The severe drought spell of 2 weeks with 21.99 drought severity index occurred at Pali only in 1987. Jaipur, Banswara and Kota experienced three drought spells, during 1987. All the stations had at least one drought spell of 2 weeks or more during the year 1987. The aridtty was found to be stable at Jodhpur, Sriganganagar, Fatehpur and Pali.

15. The swing from arid to semi-arid conditions was observed as 13.33 per cent at Jodhpur, Fatehpur and Pali and 6.6 per cent at Sriganganagar. The swing from semi-arid to dry sub-humid and from semi-arid to moist sub-humid conditions was observed as 6.6 per cent at Jaipur however, no swing was observed at Udaipur. The swing from semi-arid to dry-sub humid was observed as 26.6% and from semi-arid to moist-sub humid was observed as 13.33% at Banswara, while Kota was found to be unaffected by any swing.

16. The occurrence and spread of drought was not found as a sporadic event in Rajasthan state but had a tendency to originate in the southern region at Banswara from 26th week to spread in south-west direction in 27-28th weeks and to dissipate with a north-eastern movement in 39th week at Fatehpur.

17. The mean weekly water deficit was observed between 2–5 years return period for a maximum duration of 24 weeks at Banswara and minimum duration of 7 weeks at Sriganganagar. The mean weekly water deficit during the rainy season in Rajasthan was observed between 2–5 years return period while during the rest of the period it was observed either at less than 2 years return period or at 2 years return period.
18. Log Pearson Type III distribution gave the best fit for the predicted weekly water deficit at Jodhpur, Sriganganagar, Pali, Jaipur, Udaipur and Banswara, while for Kota log normal distribution gave the best fit for the predicted values of weekly water deficit. At Fatehpur both log normal and Log Pearson Type III distributions gave the best fit for the predicted weekly water deficit
19. Stochastic modelling was made for water deficit by using 13 years (1981–93) data for all the stations under study. The performed statistical test indicated that the time series of the water deficit was trend free. The stochastic models for water deficit (DFF) for different stations can be expressed as under:

Jodhpur,

$$DEF_{(w)} = 25.085 - 4.98 \text{ Cos } (2\pi t/p) + 8.514 \text{ Sin } (2\pi t/p) + 0.330 \text{ Cos } (4\pi t/p) - 5.345 \text{ Sin } (4\pi t/p) - 1.868 \text{ Cos } (6\pi t/p) + 0.563 \text{ Sin } (6\pi t/p) + 0.275\, S_{t-1} + 0.176\, S_{t-2} + 0.81\, S_{t-3} + a_t$$

Sriganganagar,

$$DEF_{(w)} = 19.381 - 13.420 \text{ Cos } (2\pi t/p) + 4.049 \text{ Sin } (2\pi t/p) - 0.490 \text{ Cos}(4\pi t/p) - 7.495 \text{ Sin } (4\pi t/p) + 0.255 \text{ Cos } (6\pi t/p) + 4.954 \text{ Sin } (6\pi t/p) + 0.374\, S_{t-1} + 0.008\, S_{t-2} + 0.083\, S_{t-3} + a_t$$

Fatehpur,

$$DEF_{(w)} = 23.787 - 9.397 \text{ Cos } (2\pi t/p) + 8.513 \text{ Sin } (2\pi t/p) - 0.987 \text{ Cos } (4\pi t/p) - 9.926 \text{ Sin } (4\pi t/p) - 1.076 \text{ Cos } (6\pi t/p) + 4.071 \text{ Sin } (6\pi t/p) + 0.335\, S_{t-1} + 0.098\, S_{t-2} + a_t$$

Pali

$$DEF_{(w)} = 25.68.1 - 6.694 \text{ Cos } (2\pi t/p) + 9.106 \text{ Sin } (2\pi t/p) - 1.352 \text{ Cos } (4\pi t/p) - 4.357 \text{ Sin } (4\pi t/p) + 0.228\, S_{t-1} + 0.124\, S_{t-2} + 0.116\, S_{t-3} + a_t$$

Jaipur,

$$DEF_{(w)} = 20.432 - 7.201 \text{ Cos } (2\pi t/p) + 11.209 \text{ Sin } (2\pi t/p) + 0.295 \text{ Cos } (4\pi t/p) - 9.594 \text{ Sin } (4\pi t/p) - 2.387 \text{ Cos } (6\pi t/p) + 3.219 \text{ Sin } (6\pi t/p) + 0.418\, S_{t-1} + 0.071\, S_{t-2} + 0.102\, S_{t-3} + a_t$$

Udaipur,

$$DEF_{(w)} = -16.912 - 4.249 \text{ Cos } (2\pi t/p) + 11.484 \text{ Sin } (2\pi t/p) - 1.090 \text{ Cos } (4\pi t/p) - 6.870 \text{ Sin } (4\pi t/p) + 0.211\, S_{t-1} + 0.208\, S_{t-2} + a_t$$

Banswara,

$$DEF_{(w)} = 16.607 - 5.03G \text{ Cos } (2\pi t/p) + 11.616 \text{ Sin } (2\pi t/p) + 1.886 \text{ Cos}(4\pi t/p) - 8.775 \text{ Sin } (4\pi t/p) - 0.257 \text{ Cos } (6\pi t/p) + 3.906 \text{ Sin } (6\pi t/p) + 0.394\, S_{t-1} + 0.190\, S_{t-2} + a_t$$

and Kota,

$$DEF_{(w)} = 18.040 - 5.398 \text{ Cos } (2\pi t/p) + 12.987 \text{ Sin } (2\pi t/p) + 0.040 \text{ Cos } (4\pi t/p) - 7.575 \text{ Sin } (4\pi t/p) + 0.314\, S_{t-1} + 0.137\, S_{t-2} + 0.062\, S_{t-3} + a_t$$

However, the prediction of water deficit by stochastic models was not appeared to be adequate for yearly data of 1994 and 1995.

20. The longest length of growing season of 29 weeks (203 days) was observed at Banswara, while the shortest length of growing season of 8 weeks (56 days) was observed at Sriganganagar. The length of growing season did not increase with increase in annual rainfall at all the stations. But maximum length of growing season was observed at Pali, Udaipur, Banswara and Kota under conditions of maximum annual rainfall while at other stations it was observed even at less than maximum annual rainfall conditions. The growing season starts from 24 to 25th week in eastern Rajasthan, while in western, Rajasthan it starts from 26th to 27th week.

21. Double cropping system was found to be suitable in Banswara and Kota regions, whereas in Sriganganagar, grasses, shrubs, trees and short duration crops may be taken up. Monocropping system was found to be suitable in Jaipur and Udaipur regions, while in Jodhpur, Fatehpur and Pali, pearl millet, pulses were found to be suitable.

22. The preparatory tillage in western Rajasthan may be started from 24th to 26th weeks and in eastern Rajasthan it may be started from 22nd to 24th weeks. The interculture operations may be started from 28th to 29th weeks, in western Rajasthan, while in eastern Rajasthan. It may start in 27th week. The supplemental irrigations from harvested runoff water may be given from 35 to 37th weeks in western Rajasthan and from 36th to 37th weeks in eastern Rajasthan.

References

Abul, F., M. Salah and I.B. Sadiqual. 1996. Analysis of agricultural drought in a rainfed lowland rice system: A case study in north in north-west Bangladesh, Journal of Indian Water Resources Society. 2(1): 51-56.

Adhikari, R.N., P.B. Rao, M.S. Rao, and R.M. Rao. 1988. Autoregressive models for drought management. Proceedings of National Symposium held at Hyderabad from September 13-16. pp. 1-12.

Agnew, C. T. and A. Chappell, 1999. A drought in the Sahel. Geography Journal. 48(4): 299-311.

Agnihotri, Y. 1999. Trend analysis of short-term seasonal rainfall and crop planning in Shivalik foothill region. Indian Journal of Soil Conservation. 27(1): 64-69.

Agnihotri. Y. and R.C. Bansal, 1983. Distribution and frequency of rainy days and wet speels in Chandigarh. Indian Journal of Soil Conservation. 11(1): 30-34.

Air Ministry Meteorological Office. 1936. British Rainfall. London Meteorological Office. London.

Allen, R.G., C.E. Brockway, and J.L. Wright. 1983. Weather station siting and consumpative use estimates. J. Water Resources Planning and Management. ASCE. 109(2): 134-146.

Allen, R.G., I.S., Pereira, D. Daes and M. Smith., 1998. Crop Evapotranspiration, Guideline for Computing Crop Water Requirements. FAO Irrigation and Drainage Paper 56, Rome' Italy. p.300.

Alley, W.M. 1984. The Palmer drought severity index: Limitaitons and assumption. Journal of Climate and Applied Meteorology, 3: 1100-1109.

Awade, S.T., M.Y. Totagi, S.M. Bawikar and D.R. Sikka, 1984. Dynamics of the large scale tropospheric circulation during summer monsoon and tropical droughts. Mausam (35): 315-322.

Banerji, S. and B.M. Chhabra. 1963. Drought condition in the Telengana Division (A.P.) during the southwest monsoon season. Indian Journal of Meteorology and Geophysics. .14(4): 403-415.

Barakat, F. and A. Handoufe, 1998. Agroclimatic assessment of agricultural drought in Morocco. Approche agroclimatique de la secheresse agricole au moroc Secheresse 9(3): 201-208.

Barger, G.L. and H.C.S. Thorn 1949. Evaluation of drought hazard. Agronomy Journal. 41(11): 519-526. .

Bhttau,D. 2000. Drought on the Emili Romagana Plains, areas of low rainfall and series of drought events. Situazioni disiccita nella emiliono-romagnola. Le agree-menoplovose egli eventi plu gravosi. Irrigazionee Drenaggio 47(2): 52-56.

Bhakar, S.R. 2000. Modelling of Evaporation and Evapotranspiration under Climatic Conditions of Udaipur, Unpublished Ph.D. Thesis, Maharana Pratap University of Agriculture and Technology, Udaipur.

Bhakar, S.R. and R.V. Singh. 2001. Computer application in estimation of evapotranspiration. World Congress in Agriculture and Natural Resources, Paper accepted for presentation in the International Conference on World Congress of Computers in Agriculture and Natural Resources to be held at Iguassu Falls, Brazil from September 19–21, 2001, pp.6.

Bhalme, H.N. and D.A. Mooley, 1979. On the performance of modified Palmer index. Proceedings of International Symposium on Hydrological Aspects of Droughts held at New Delhi Vol. 1 & 2: 373-382.

Bhalme, H.N. D.A. Molley, and S.K. Jadhav, 1984. Large-scale April pressure index connected with the southern oscillation and its potential for prediction of large-scale drought over India. Mausam. 35(3): 355-360.

Bhandarkar, D.M., L.P. Gite, and S.S. Mandvikar, 1986. Rainfall analysis for crop planning in Bhopal region. International Journal of Tropical Agriculture. 5 (3-4): 182-189.

Bhatt, V.K., A.K. Tiwari and A.K. Sharma. 1996. Probability models for predictions of annual maximum daily rainfall for Datia. Indian J. of Soil Conservation 24(1): 25-27.

Bhattacharjee, B.K. and B.N. Chandra. 1989. Systematic study of rainfall of Pumiia district with a view to proper crop planning and irrigation. Journal of Institution of Engineers (Agriculture). 69: 45-55.

Bishnoi, O.P. 1984. Crop moisture index in characterizing the agricultural drought at Hisar. Mausam. 35(3): 266.

Biswas, B.C. and P.S. Nayar. 1984. Quantification of drought and crop potential. Mausam. 35(3): 281-286.

Biswas, B.C. and S.J. Maske. 1981. Rainfall analysis for use in dryland agriculture. Indian J. Soil conservation.

Blaney, H.F. and W.D. Criddle. 1962. Evaporation from free water surface at high altitudes. Trans. ASCE. 123:235-265.

Box, O.E.P. and O.M. Jenkins. 1976. Time Series Analysis, Forecasting and Control. Holden Day San Franciso., California. P. 535.

Brignall, A.P. T.E., Douning, M.D. Favis, P.A. Harrison, and J.L. Orn. 1999. Agricultural drought in Europe: Site, regional and national effects of climate change. In Climate Change and Risk. 79-124.

Brutsaeert, W. 1982. Evaporation into the Atmoshpere. R. Defiled Publishing Company, Dordreecht, Holland, p. 299.

Chakraborty, P.B, 1998, Rainfall characteristics of Kakdwip, West Bengal. Indian J. Soil Conservation. 26(2): 166-169.

Choudhary, A and V.P. Abhyankar, 1984a. On some climatological aspects of drought in India. Mausam, 35 (3): 375-378.

Choudhary, A. and V.P. Abhyankar, 1984b. A Markov Chain model for the probability of drought incidence in India. Mausam. 35(3): 403-405.

Choudhary, A., S.S. Ookhale, and O.S. Rentala, 1979. Spells of dry day related to agricululral drought in India. Mausam. 80(4): 501-510.

Chow, V. T. 1951. A general formula for hydrological frequency analysis. Trans. American Geographical Union. 32(2):231-237.

Chowdhary, A., M.M. Dandekar and P.S. Raut. 1989. Variability in drought incidence over India. A statistical approach. Mausam. 40(2): 207-214.

Chowdhury, A and G.S. Sarwade, 1984. Drought prediction in India from radiation parameters. Mausam, 35(3): 345-348.

Dabral, P.P. 1996, Meteorological drought analysis based on rainfall data. Indian J. Soil Conservation. 24(1): 37-40.

Das, P.K. 1983. Droughts and famines in India-Historical prospective. Mausam 34(2): 123-130.

Doorenbos, J. and W.O. Priutt. 1977. "Guidelines for Predicting Crop Water Requirement: FAD Irrigation and Drainage Paper No. 24, FAD, Rome, Italy. p. 156.

Foley, J.C. 1957, Drought in Australia. Book No. 45, Melbourne, Bureau of Meteorology.

Gahlot, J.S., 1937. The History of Rajputana. Vol. II, Pub, RUCL. Jodhpur., p.122.

George, C.J., V.P. Abhyankar, and O.S. Rentala, 1975. Areal extent of south-west monsoon droughts in India. Journal of Meteorology and Geo-Physics. 26(2): 199-202.

George, C.J. and K.S. Ramasastri. 1975. Agricultural drought of 1972 Kharif season. Indian Journal of Meteorology. Hydrology and Geo-Physics. 26(1): 89-96.

Ghambete, N.N. and B.C. Biswas. 1984. Application of Markov chain model in determining drought proneness. Mausam 35(3): 407-410.

Ghosh, T .K. and G.K. Tripathy, 1996. Analysis of precipitation and satellite data over drought. Journal of Applied Hydrology. 9 (1 and 2): 113-120.

Gibbs, W.J. and J. V. Maher, 1967. Rainfall decile as drought indicators. Bull. 48, Bur, Met., Melbourne.

Gowarikar, V.,V. Thapliyal, S.M. Kulshrestra, G.S. MandaI, N. Sen Roy and D.R. Sikka. 1991. A power regression model for long range forecast of south west monsoon rainfall over India. Masuam. 42(2): 125-130.

Grismer, M.E. 2000. Long-term evapotranspiration from coastal avocado/citrus archard. J. Irri. and Drain. Engg. ASCE. 126(1): 1-7.

Hargreaves, G.H. 1968. Consumptive use derived from evaporation pan data. J. Irri. and Drain. Div. ASCE, 97(1):97-105.

Handa, D.P. and P.R. Mishra. 1968. Studies on rainfall characteristics of Kota. Indian Forester. October; 745-752.

Hatfield, J.L. and R.G. Allen. 1997. Evapotranspiration estimates under deficient water supplies. J. Irri. Drain. Engg. ASCE. 122(5): 301-308.

Herbst, P.H., D.B. Bredenkamp, and H.M.G. Barker. 1966. A technique for the evaluation of drought from reainfall data. Journal of Hydrology. 4: 264-272.

Hipel, K.W. and A.l. McLeed. 1994. Time Series Modelling of Water Resources and Environmental Systems Elsevier Science, B.V. Amsterdam. P. 1013.

Hosking, J.R.M. 1985. Fractional differencing modelling in hydrology. Water Resources Bulletin. 21: 677-682.

Hounam, L.E., J.J. Burgos, M.S. Kalik, W.C. Palmer, and J. Rodd, 1975. Drought and Agriculture U.M.O. Technical Note No. 138.

Jeevarathnam, K. and M. Jayakumar. 1979. Probability analysis of annual maximum daily rainfall for Ootacamund, Indian J. Soil Conservation. 7(1): 10-16.

Jensen, D. T., G.H. Hargreaves, R.G. Temesgen and R.G. Allen. 1999. Computation of ET_0 under nonideal conditions. J Irri, and Drain. Engg., ASCE. 123(5): 394-400.

Jensen. M.E. and H.R. Haise. 1963. Estimating evapotranspiration from solar radiation. J. Irri. and Drain Div., ASCE. 89(4): 15-14.

Jensen, M.E., R.D. Burman and R.G. Allen (ed). 1990. Evapotranspiration and Irrigation Water Requirements.ASCE Manuals and Reports on Engineering Practices No. 70, ASCE. New York. p. 332.

Keledhonkar. M.J., S.K. Kamra, R.K. Yadav and K.N. Tiwari, 1996. Probabilities of historic droughts for south western Orissa in India. Journal of Indian Water Resources Society. 2(4): 19-24.

Kamte, P.P. and S.D. Dahale, 1984. A stochastic model of drought. Mausam. 35(3): 387-390.

Karl, T., I. Quinlan and D.S. Ezell. 1987. Drought termination and amelioration: Its climatological probability. Journal of Climate and Applied Meteorology, 26: 1198-1209.

Khambete, N.N. and B.C. Biswas, 1984. Application of Markov chain model in determining drought proneness. Mausam. 35(3): 407-410.

Klugm, M.R. 1978. Drought in upper midwest 1931-1969. Journal of Applied Meteorology. 17 (10): 1425-1431.

Kottegoda, N. T. 1980. Stochastic Water Resources Technology. The MacMillan Press Ltd., London. p. 384.

Krishnan, A. and K.P. Thanvi. 1982. Study on the variability of rainfall and evaluation of major rainfall types in western Rajasthan. Geological Survey of India. Misc. Pub. No. 49. p. 156-160.

Krishnan, A. 1984. An analysis of trends in the rainfall and droughts occurring in the southwest monsoon and north-east monsoon systems in the Southern Peninsular India. Mausam. 35(3): 379-386.

Krishnan, A. 1988. Delineation of soil-climatic zones of India and its practical application in agriculture. Fertilizer News. 33 (4): 18-33.

Krishnan, A. and J .S., Rao. 1979. Classification of major rainfall types in western Rajasthan. Forage Research. 5: 19-30.

Krishnan, A. and K.P. Thanvi. 1969. Water budget in arid zone of Rajasthan during 1941-60. Annals of Arid Zone. 8(2): 291-299.

Krishnan, A. and P. Rakhecha. 1965. Potential evapotranspiration by Thornthwaite and other methods. Annals of Arid Zone 4 (1): 32-35.

Krishnan, A. and K.P. Thanvi, 1977. Quantification of rainfall in different regions of western Rajasthan. Annals of Arid Zone. 16(2): 185-194.

Kumar, A. 2000. Prediction of annual maximum daily rainfall of Ranichauri (Tehri Gharhwa1) based on probability analsysis. Indian J. Soil Conservation 28(2): 180-200.

Kumar, D. 1995. Variation in rainfall pattern in Tarai region of Uttar Pradesh. Indian Journal of Soil Conservations. 23(2): 103-106.

Kumar, D., R. Kumar, and S. Kumar. 1998. Monthly rainfall probability analysis of Aonla (Bareilly) Rohilkhand Region. Indian Journal of Soil Conservation. 26(2): 156-161.

Kumar, D., Y. Chandola and A. Rastogi. 2000. Comparison of probability distributions for analysis of data. Indian Journal of Soil Conservation. 28(2):110-1.16.

Kumar, K.K. and D.Y., Chandola, and A. Rastogi. 2000. Comparison of probability distributions for analysis of data. Indian Journal of Soil Conservation. 28(2): 110-116.

Kumar, K.K. and V.U. Rao. 1983. Rainfall in Andhra Pradesh - A preliminary study. Indian J. Soil Conservation. 11(1): 36-45.

Landberg, H.E, 1974. Drought - Summary of Lectures and Discussion, Grneva, W.M.O. Bulletin XXIII (4) pp. 226.

Lewis, P.A.W. 1985. Some simple models for continuous variate time series. Water Resources Bulletin; 21 : 635-644.

Lohani, V.K., G. V. Loganathan, and S. Mastaghimi. 1998. Long-term analsysis and short term foreasting of dry spells by Palmer Drought severity index. Nordic Hydrology. 29(1): 21-40. .

Malani, B .H. 1981. Patterns of aridity spread and severity of drought in Andhra Pradesh. The Indian Geographical Journal. 56(1): 30-36.

Mandelbet, B.B. 1971. A fast fractional Gaussion noise generator. Water Resour. Res. 7: 543-553.

Mishra, K.K. and T. Satyanarayan. 1987. Effect of extreme events on statistical parameters of rainfall at Kharagpur, West Bengal. Journal of Agricultural Engineering. ISAE. 24(1): 74-86.

Mockus, V. 1960. "Selecting a Flood-Frequency method". Am Soc. Agr. Eng. Trans. 3: 48-51,54.

Mooley, D.A., B.Parthasarathy and N.K. Sontakha. 1984. Droughts over peninsular India during 1861-1980 and associated circulation features. Mausam, 35 (3): 337-342. Geophysics. 25 (3 and 4): 411-416. .

Morris, R.A. and H.G. Zandstra. 1979. Land and Climate in relation to cropping patterns. In rainfed low land rice, selected papers from 1970 Int. Rice Res. Conf. IRRI, Losbanos, P. 255-274.

Mutua, F.M. 1989. Transfer function hydrologic modelling. A case study. Jr. of Applied Hydrology. 11(2): 11-15.

Narain, P., P.N. Bhargava, and A. Sakeena, 1984. A statistical study on incidence of drought in relation to agricultural production. Mausam. 35(3): 391-396.

Pandarinath, N. 1991. Markov chain model probability of dry and wet weeks during monsoon period over Andhra Pradesh. Mausam; 42(4): 393-400.

Panu, U.S. and T.E. Unny 1980. Stochastic synthesis of hydrologic data based on concepts of pattern recognition. II Application of natural watershed. Jr. of Hydrology of 46: 197-217.

Parthasarathy, B. and P. Rakhecha. 1972. Fluctuation in aridity over Rajasthan and neighbourhood. Annals of Arid Zone. 11 (3 and 4): 161:168.

Parthasarthy, B. N.A., Sonatakke, A.A., Monot, and D.R., Kothwala. 1987. Droughts and floods in the summer monsoon season over different meteorological sub-divisions of India for the period 1871-1984. Jr. of Climatology. 7 (11): 75-70.

Patil, C.B., D.K. Kathmale and S.P. Kale, 1989. Monsoon management in drought prone areas of Maharashtra. A case study of Solapur Indian Journal of Dryland Agricultural Research and Development, 4(2): 133-139.

Patil, P.P. and M.M. Patil. 1991. Relationship between the extreme events and the statistical parameter of rainfall at Dhule, Maharashtra State. Indian J. Soil Conservation. 19 (1 and 2): 75-82.

Paul, J.C., A.P. Sahu and S.D. Sharma 1999. Studies on water balance and incidence of drought at Bhubaneswar. Journal of Applied Hydrology. 12(2 and 3): 8-14.

Penman, H.L., 1948. Natural evaporation from open water bare soil and grass Proc. R. Soc. London. (Ser. A) 193: 120-145.

Prakash. C. and D.H. Rao, 1986. Frequency analysis of rainfall data for crop planning - Kota Indian Journal of Soil Conservation. 14(2): 23-26.

Rakhecha, P. 1974. The study of droughts by water budget method over Andhra Pradesh. Indian Journal of Meteorology, Hydrology and Geophysics. 25 (3 and 4): 411-416. .

Ram Mohan H.S., S.M. Juliet, and P. Valsala, 1984. An agroclimatic approach to the assessment of tropical drought. Mausam. 35(3): 291-294.

Ram Mohan, H.S., P. Vaisala, and B.V. Appa Rao, 1983. Drought spells over north-central India during the 1979 south west monsoon. Mausam. 34(3): 299-302.

Rama Krishna, Y.S. and A.S.R.A.S. Sastri, 1980. Studies on the incidence of droughts over western Rajasthan. The National Geographical Journal of India. 26 (1-2): 44-47.

Rama Krishna. Y.S., A.S.R.A.S. Sastri, G.G.S.N. Rao and B.V. Ramana Rao, 1984. On prediction of drought in the Indian Geographical Journal of India. 26(1-2): 44-47.

Rama Krishna. Y.S., A.S.R.A.S. Sastri, G.G.S.N. Rao 1985. Water balance and crop planning. A case study of western Rajasthan. Annals of Arid Zone. 24(2): 144-119.

Ramana Rao, B. V., A.S.R.A.S. Sastri, and Y.S. Ramakrishna. 1984. Variation in the moisture regime of the Indian arid zone. The National Geographical Journal of India. 30(2): 73-80.

Ramdas, L.A. 1960. Crops and Weather in India. Indian Council of Agricultural Research, New Delhi, pp. 127.

Ramdas, L.A. and A.K. Malik, 1948. The objective assessment of the crop outlook on the basis of the progress of water week by week during the growing season. Agricultural Situation in India. 3: 1- 5.7

Rana, R.S. and D.R. Thakur, 1998. Rainfall analysis for crop planning in Kullu Valley, Himachal Pradesh. Indian J. Soil Cons., 26. (2): 144-146.

Rao, A.S. and A.R. Subramaniam. 1986. An analysis of droughts in Maharashtra by a modified Palmer's approach. Mausam. 37(3): 377-384.

Rao, D., N. Bhaskar, C. V. Naidu, B.R.S. Rao and Bhagyalakshmi; R. 1998. Variability of summar monsoon rainfall over India. Journal of Applied Hydrology 11(1): 49-57.

Rao, G.N. 1999. Monsoon rainfall and its variability in Godavari river basin. Proceedings of the Indian Academy of Sciences, Earth and Planetary Sciences. Atmospheric Science. 108(4): 327-332.

Rao, G.R. and D.G. Phulari., 1984. Extreme probability analysis of drought periods in monsoon in relation to water need of crop stages. Mausam. 35(3): 397-402.

Rao, K.N., C.J. Gearge, P.E; Moray, and P.C. Mehta. 1973. Spectral analysis of drought index (Palmer) for India. Indian Journal of Meteorology and Geo-Physics. 24(3): 257-270.

Rao, N.H. 1987. Estimation of probable weekly rainfall. Journal of Indian Resources Society 7(1): 15-20.

Rao, P.S., K. Jayarathanam and M.R. Sethuraj. 1990. Water balance studies of the rubber growing regions of South India. Journal of Applied Hydrology. 3(2): 23-30.

Ratnam, B.P. and S.N., Joshi 1977. Drought frequency at Hagari and Bijapur. Annals of Arid Zone 16(2): 171-175.

Ratnam. B.P., M.M. Hosmani and S.N. Joshi. 1973. Drought climatology at Dharwar. Indian Journal of Agricultural Sciences. 43(8): 799-801.

Ray. C.R., P .C. Senapati and R. Lal, 1980. Rainfall analysis for crop planning at Gopalpur (Orissa). Journal of Agricultural Engineering, ISAE. 17(3 & 4): 1-8.

Ray, C.R., P .C. Senapati and R.Lal, 1987. Investigation of drought from rainfall data at Goplapur (Orissa). Indian Journal of Soil Conservation. 15(1): 15-19.

Rosenberg, N.J., H.E. Hart, and K.W. Brown. 1968. Evapotranspiration-Review of Research Nebr. Agr.Exp. Station. Misc. Bull. No. 20. p. 80.

Salas, J.D., J.W. Delleur, V. Yevjevich, and W.L. Lane. 1985. Applied Modelling of Hydrologic Time Series. Water Resources Publication, Littleton, Colorado. P. 482.

Samani, N.,E. Raeisu and A.R. Soltani. 1995. Stochastic synthesis of droughts for reservoir storage design. Journal of Applied Hydrology; 8 (1 to 4): 9-18.

Samani, Z. 2000. Estimating solar radiation and evapotranspiration using minimum climatological data J. Irri. and Drain. Engg. ASCE. 126(4) 265-267.

Sankaranm Vasumati, 1993. Rainfall differentials and their impact on agricultural practices in two dryland farming talukas of Gujarat A micro-climatic zone approach. Annals of Arid Zone 32(2): 121-123.

Sarma, A.A.I.N. and M. Ravindarnath, 1984. A study of aridity and droughts in the climatic specturm of the south Indian region. Mausam. 35 93): 287-290.

Sastri, A.S.R.A.S. and A.R. Malakar, 1981. Climatological analysis of drought over northwest India during the year 1979. Mausam. 35(3): 259-262.

Sastri, A.S.R.A.S, Y.S. Rama Krishna. G.G.S.N. Rao and B.V. Ramana Rao, 1984. Agricultural Drought and crop planning. A case study for western Rajasthan. Mausam 35(3): 261-264.

Sastri, A.S.R.A.S. 1991. Drought climatology in sub-humid climates and its impact on agricultural- A case study for Raipur district. Bhagirath. 38: 113-123.

Sastri, A.S.RA.S. and Y.S. Rama Krishna. 1980. A modified scheme of drought classification applicable to the arid zone of western Rajasthan. Annals of Arid Zone 19 (1 & 2) : 65-72.

Sastri, C. V.S. 1984. The drought year of 1979. A critical appraisal with reference to the semi-arid region of Delhi. Mausam. 35(3): 273-276.

Sastry, P.S.N. and N.V.K. Charavarty. 1984. Assessment of atmospheric drought during monsoon cropping season. Mausam, 35(3): 267-272.

Sen, S.R. 1970. Droughts in India. Certain dimensional conditions in growth and instability. Indian Agriculture. p. 29-35.

Sen, Z. 1978. Autorum analysis of hydrologic time series Jr. of Hydrol. 36: 75-85.

Sen, Z., 1980a. Regional drought and flood frequency analysis: Theoretical consideration. Journal of Hydrology. 46: 265-279.

Sen, Z. 1980b. Adaptive Frourier analysis of periodic-stochastic hydrologic squences. Jr. of Hydrol. 46:239-249.

Sen, Z. 1980c. Adaptive Frourier analysis of periodic-stochastic processes. Jr. of Hydrol. 46: 125-263.

Senapati, P .C., S.D. Sharma and R. Lal. 1989. Probability analysis of annual rainfall data at Bhubneswar. Orissa. Journal of Agricultural Engg. ISAE. 16(4): 131-139.

Sengupta, A.K. and B.N. Chanda. 1989. Water balance Studies for rice based cropping pattern and its application of Purulia district. Institution of Engineers Agril Division 69:97.

Sharma, H.C. and N.D. Dastane, 1969. Assessing evapo-transpiration with pan evaporimeters. Annuals of Arid Zone. 8 (1): 23-26.

Sharma, H.C., H.S. Chauhan, and Sews Rama, 1979. Probability analysis of rainfall for crop planning. Journal of agricultural Engineering. ISAE. 16(3): 87-94.

Sharma, K.D. 1986. Rainfall analysis for rainwater management in the N.E.H. Region of Indian. India Journal of Soil Conservation. (42): 51-55.

Sharma, K.D. and D.N. Bohara 1990. Rainfall analysis for rainwater management in a drought prone area. Annals of Arid Zone. 29(3): 165-169.

Sharma, T.C. 1997. A drought frequency formula. Hydrological Sciences Journal. 42 (6): 803-814.

Singh, R.A. 1992. Rainfall characterisation for crop planning of Rendhar watershed in Bundelkhand region (I.P.) Journal at Indian Water Resources Society 12 (3&4): 233-235.

Singh, R.S., Y.S. Ramakrishna and R.S. Purohit. 1991. Rainfall characteristics analysis of Kutch (Gujarat) region of India. Annals of Arid Zone 30:93-100.

Singh, R.V. 2000. Watershed Planning and Management. Yash Publishing House, Bikaner. p. 470.

Singh. S.V. R.H. Kripalani, P. Saha and K.D. Prasad, 1984. Analysis and prediction of short period droughts during summer monsoon over India. Mausam. 35(3): 361-366.

Sostaric, J. and S. Madar, 1997. Application of the Palfai aridity index for drought periods determination in agriculture. Poljoprivreda. 1(3): 35-42.

Srinivasan, K. and S.K. Ramesh. 1993. Stochastic stream flow models for over year drought studies J. of Indian Water Resources Society 13 (3 & 4): 186-197.

Subrahamanyam, V.P. and C.V.S. Sastri, 1969a. Some aspects of drought limatology of the dry sub humid zones of south India. Journal of Meteorological Society of Japan 47(4): 239-244.

Subrahmanyam. V.P. and C.V.S. Sastri. 1969b. A study of aridity and droughts at Visakhapatnam Annals of Arid Zone 8(1): 18-22.

Subrahmanyam, V.P. and C.V.S. Sastri, 1971. New techniques in drought analysis. Annals of Arid Zone 10(2 & 3): 120-135.

Subrahmanyam, V.P. 1982. Water balance and its application (with special reference to India). Andhra University Press, Visakhapatnam. pp. 79.

Subrahmanyam. V.P. and A.R. Subarahmaniam. 1964. Application of Water balance concepts tor the climatic study of drought in South India. Annals of Arid Zone. 4(1): 46-55.

Subrahmanyam. A.R. and A.V. Rao. 1983. Agricultural Droughts at Peddapuram, East Godavari district, Andhra Pradesh. Mausam. 34(4): 439-442.

Subrahmanyam. A.R. and G.S.L.H. V. Prasada Rao. 1980a. Climatic study of water balance, aridity and droughts in Rajasthan state Annals of Arid Zone. 19: 371-377.

Subrahmanyam. A.R. and G.S.L.H. V. Prasada Rao. 1980b. Meriditional variation of water balance in Rajasthan state. Annals of Arid Zone. 19:51-57.

Subramaniam. A.R. and K. Uma Devi 1983. Water availability and crop management in Orissa. Indian J. Soil Conservation. 11(1): 54-59.

Subramaniam, A.R., G.S.L.H. V. Prasad Rao, and K.B. Rao. 1984. Floods drought in Rajasthan. Agroclimatic study. Indian Journal of Soil Conservation. i2(2&3): 1-16.

Subramaniam, A.R. and P.A. Narasimha Raju. 1986. Water availability periods for crop management in North Coastal Andhra. Indian. J. Soil Conservation. 14(1): 7-11.

Subramaniam, A.R. and P .S. Rao. 1986. A note on water balance studies for agricultural planning at Chirala. Indian Journal of Soil Conservation. 12 (2 and 3): 1-16.

Tambile, R.H. U.G. Deshmukh, H.S. Acharaya, A.R. Mantri and B.P. Sawant, 1991. Probability analysis of annual rainfall data at Prabhani (Maharashtra) for crop Planning. Journal of Indian Water Resources Society. 11(2): 33-35.

Thapliyal, V. 1981 ARIMA Model for long range prediction of monsoon rain fall in Peninsular India. Met. Monogr Climatology No, 12, India Met. Dep.

Thapliyal, V. 1984. Prediction of Indian droughts with lower strato - spheric winds, Mausam. 35(3): 367-374.

Thapliyal, V. 1982. Stochastic dynamic model for long range prediction of monsoon rainfall in penisulate India. Masuam. 33(4): 399-404.

Thomas, R.K. 1983. Some spatial characteristics of drought duration in the United States. Journal at Climate and Applied Meteorology. 22: 1356-1365.

Thornthwaite. C. W. 1948. An approach towards a rational classification of climate. Geographical Review, 38: 55-94.

Thornthwaite, C.W. and J.R. Mather 1955. The Water Balance Climate. 8(1). Dreseal Inst. of Tech., New Jersey, U.S.A. p.p. 104.

Thornthwaite, C.W . and J.R. Mather. 1957. Instructions and Tables for computing potential evapotranspiration and the water balance. Pub. in Clio Lab. of Cli. Elmar., New Jersey. Vol. 10. No.3.

Tiwari, A.K. A.K. Sharma, V.K. Bhatt, and M.M. Srivastava, 1998. Drought estimation through rainfall analysis for Bundelkhand region. Indian Journal of Soil Conservation. 26 (3): 280-283.

Tripathi, R.P. 1992. Probability of sequences of wet and dry days at Pantnagar. Journal of Indian Water Resources Society. 12 (1 and 2): 85-88.

Tripathi, R.P. 1996. Probability of occurrence of rainfall at Pantnagar, Northern India. Journal of Indian Water Resources Society. 2(2): 76-80.

Varshney, K.G. 1989. Frequency of weekly rainfall, totals using analytical techniques. Proceeding of the Silver Jubilee Convention, ISAE pp. 140-146.

Verma, H.N. and P.B.S. Sarma, 1988. Analysis of short duration rainfall for planning rainfed crops. Journal of Agricultural Engineering. ISAE. 25(3): 14-19.

Victor, U.S. and P.S.N. Sastry, 1984. Evaluation of agricultural drought using probability distribution of soil moisture index. Mausam. 35(3): 259-260.

Victor, U.S., N.N. Srivastava and B.V. Ramana Rao, 1991. Moisture regime, aridity and drought in the arid region of Andhra Pradesh. Annals of Arid Zone. 30(2): 81-91.

Walsh, J.S. 1987. Comparison of NOAA AVHRR data to meteorological drought indices. Photogrammetric Engineering and Remote Sensing. 53(8): 1069-1074.

WMO, 1975. Drought Special Environment-Report No. 5 WMO No. 403, Geneva, Switzerland.

Yadav, Y.P., H.B. Vasisth, and N. Ram. 1987. Trend of rainfall in Balason catchement of Darjeeling hills. Indian Journal of Soil Conservation. 15(1): 25-29.

Yevjevich, V. 1967. Objective approach to definitions and investigation of continental hydrological droughts, Hydrology Papers, Colorade State University, Fort Collins, Colorado, p. 1-18.

Appendices

Appendix A : Normal weekly rainfall (RF) and coefficient of variation (VC) for different stations of Rajashan

Std.	Jodhpur		Sriganganagar		Fatehpur		Pali		Jaipur		Udaipur		Banswara		Kota	
Week	RF mm	CV %	RF mm	CV %	RF mm	CV %	RF mm	CV %	RF mm	CV %	RF mm	CV %	RF mm	CV %	RF mm	CV %
1	0.3	290	3.5	308.6	0.6	298.3	1.3	329.2	1.6	318.8	0.1	372.6	0.1	350.0	0.4	210.0
2	2.9	234.5	2.2	192.7	4.8	262.5	1.7	204.7	4.5	184.4	4.0	238.5	8.6	310.5	4.2	190.5
3	0.7	214.3	1.4	254.3	1.3	276.9	0.6	200.0	0.4	175.0	0.2	315.0	0.3	300.0	0.4	270.0
4	0.1	160.0	2.6	300.0	0.8	350.0	0.9	348.9	0.8	250.0	0.4	280.0	0.6	400.0	2.5	258.4
5	0.4	325.0	1.9	204.1	0.8	337.5	0.4	350.0	1.9	210.5	1.1	194.5	0.3	266.7	1.0	312.0
6	0.9	367.8	4.0	170.0	4.1	334.1	0.3	360.7	0.3	333.3	0.0	0.0	0.0	0.0	4.2	370.0
7	0.2	345.0	3.2	161.9	4.7	22.7.7	0.1	420.0	2.3	256.5	0.1	272.6	0.0	0.0	1.0	357.0
8	1.2	375.0	4.5	155.8	2.7	174.1	1.5	260.0	3.1	206.5	0.5	328.0	0.0	0.0	0.1	320.0
9	1.5	276.7	2.8	188.9	0.8	387.5	1.5	282.7	3.9	266.7	0.7	382.9	0.1	300.0	1.3	208.5
10	0.3	326.7	1.1	201.8	3.7	335.1	0.2	275.0	0.1	300.0	0.1	170.0	0.0	0.0	0.5	180.0
11	0.6	416.7	4.3	179.3	1.6	287.5	0.0	0.0	0.1	200.0	0.2	295.0	0.1	200.0	1.0	339.0
12	0.3	396.7	3.4	107.4	1.0	258.0	0.3	300.0	2.9	217.2	0.4	240.0	0.4	350.0	0.4	275.0
13	0.6	260.0	0.8	150.0	3.8	326.3	0.2	425.0	1.4	257.1	0.0	0.0	0.2	350.0	1.4	293.6
14	4.6	188.7	3.5	202.3	1.6	243.7	1.0	290.0	2.2	240.9	1.3	279.2	0.0	0.0	0.7	272.9
15	2.1	355.2	6.6	189.5	4.4	275.0	2.4	374.0	6.3	360.3	0.3	396.7	0.0	0.0	1.6	275.6

contd....

Appendix A – *contd....*

Std.	Jodhpur		Sriganganagar		Fatehpur		Pali		Jaipur		Udaipur		Banswara		Kota	
Week	RF mm	CV %	RF mm	CV %	RF mm	CV %	RF mm	CV %	RF mm	CV %	RF mm	CV %	RF mm	CV %	RF mm	CV %
16	1.8	291.1	3.8	227.6	1.6	306.3	1.5	294.7	2.2	368.2	1.1	214.5	0.7	328.6	0.8	202.5
17	8.5	384.1	1.8	258.9	1.4	321.4	1.3	228.5	1.8	377.8	1.7	250.6	2.1	261.9	4.6	367.6
18	0.9	348.9	6.0	250.3	1.7	264.7	0.2	310.0	0.5	360.0	0.6	233.3	0.5	400.0	0.5	222.0
19	6.7	215.9	1.8	276.1	9.8	215.3	2.5	315.6	2.8	189.3	6.4	332.7	5.0	264.0	1.0	285.0
20	0.9	205.6	1.5	210.7	6.7	225.3	0.9	206.7	1.9	373.7	0.3	273.3	0	0	0.1	52.00
21	2.1	253.8	5.4	239.4	7.9	234.2	1.3	366.2	7.5	278.7	1.1	260.9	0.0	0.0	3.2	102.9
22	7.8	211.5	2.5	169.6	2.3	213.0	0.1	222.9	6.7	214.9	5.4	233.9	1.6	331.3	1.7	183.5
23	0.9	245.6	6.8	176.5	8.4	191.7	1.9	180.5	3.9	148.7	5.5	325.5	1.4	364.3	4.5	172.2
24	6.0	177.7	12.8	123.5	7.7	135.1	12.0	148.7	7.3	127.4	13.5	110.0	25.2	126.6	7.8	106.7
25	4.1	200.9	7.2	236.5	11.4	164.9	6.9	156.5	17.3	254.3	14.1	132.6	26.4	128.8	13.1	149.2
26	7.5	250.9	13.9	169.8	3.9	143.6	12.5	171.8	22.3	116.8	23.2	106.5	37.3	108.3	27.7	179.0
27	43.4	284.0	23.4	164.1	23.7	128.7	50.2	210.2	15.3	218.3	33.5	161.6	49.4	128.9	28.9	126.9
28	8.0	94.6	20.1	112.4	26.7	131.1	22.7	125.4	39.2	88.3	40.3	45.8	55.7	116.9	36.6	93.9
29	31.8	100.2	23.5	122.9	45.7	137.9	27.6	74.0	95.2	164.4	49.4	83.8	53.9	66.9	50.2	69.5
30	48.4	130.9	29.1	111.4	46.2	101.7	55.5	134.5	86.7	119.0	57.0	97.2	106.1	143.3	82.1	97.9
31	20.5	134.1	26.6	119.7	22.9	152.8	33.2	152.0	38.6	80.6	41.9	107.0	139.1	106.5	58.1	82.2
32	20.3	118.2	12.2	158.4	21.6	81.9	26.8	172.9	55.2	75.0	57.8	77.3	50.4	92.1	54.1	87.4
33	29.4	164.3	21.9	145.8	18.1	122.7	29.3	156.6	38.7	85.3	48.3	78.3	88.7	122.7	65.5	72.3

contd....

Appendix A – *contd....*

Std.	Jodhpur		Sriganganagar		Fatehpur		Pali		Jaipur		Udaipur		Banswara		Kota	
Week	RF mm	CV %	RF mm	CV %	RF mm	CV %	RF mm	CV %	RF mm	CV %	RF mm	CV %	RF mm	CV %	RF mm	CV %
34	19.6	122.8	7.3	227.4	24.7	115.4	21.3	100.9	102.2	221.5	48.8	109.0	56.7	92.2	56.9	95.3
35	12.4	133.5	9.5	151.1	12.5	147.2	16.3	138.1	42.5	100.9	28.4	94.4	39.6	121.5	44.6	98.9
36	24.3	215.8	15.9	211.5	9.9	135.4	13.8	157.1	29.6	147.3	26.6	147.7	42.8	157.0	46.9	123.1
37	7.6	180.9	74	202.7	5.4	188.9	6.4	192.0	6.2	188.7	7.4	125.7	17.8	199.1	8.1	165.3
38	4.3	298.1	8.2	184.1	5.9	162.7	1.2	295.0	3.3	224.2	14.2	139.4	19.3	127.5	12.5	131.8
39	9.7	255.5	2.2	310.0	9.7	213.4	16.8	198.8	3.3	263.6	16.1	113.0	26.7	179.4	8.8	189.3
40	1.2	247.5	0.1	380.0	1.1	281.8	1.9	252.6	0.9	377.8	9.4	254.3	5.1	213.7	5.6	254.5
41	2.6	309.6	0.9	225.5	1.5	286.7	1.0	377.0	5.6	355.4	3.0	280.0	20.9	245.9	13.9	206.5
42	0.6	266.7	2.2	284.1	1.1	481.8	0.2	300.0	0	0	0.3	280.0	6.5	263.1	3.5	369.7
43	1.1	266.4	0	0	0	0	0.7	245.7	0.3	266.7	1.4	370.0	0.5	400.0	0.4	387.5
44	2.6	379.2	0	0	3.7	378.4	0	0	6.9	257.9	3.6	376.9	6.3	374.6	6.8	263.7
45	0	0	1.0	316.0	0.3	333.3	0.8	376.3	1.1	336.4	8.1	344.4	7.3	375.3	7.5	361.6
46	0.1	41.0	0	0	0.3	333.3	0	0	0.1	400.0	0.1	0.1	1.1	263.6	0.1	450.0
47	0	0	0.5	288.0	0.9	155.6	0	0	0.2	350.0	0.5	396.0	7.2	222.2	0	0
48	0	0	1.5	353.0	0	0	0	0	0	0	0	0	0	0	0	0
49	0	0	0.2	285.0	0.3	366.7	0	0	0	0	0	0	0	0	0.5	350.0
50	0	0	0.8	206.3	0.5	300.0	0.5	400.0	0.02	500.0	0.3	363.3	0	0	0.4	312.5
51	0.1	160.0	1.5	253.3	0.8	225.0	0.04	275.0	.8	266.7	0.4	287.5	2.5	376.0	2.2	288.2

Appendix B1 : Weekly rainfall distribution at different probability (%) levels by using Gamma distribution at Jodhpur

Week	Rainfall (mm) for the probabilities					Mean (mm)
	90	75	50	25	10	
1	0.6	0.9	1.2	1.7	2.1	0.3
2	0.2	0.8	2.4	5.4	9.5	2.9
3	0.5	0.9	1.4	2.2	3.1	0.7
4	0.0	0.0	0.0	0.0	0.0	0.1
5	0.5	0.8	1.2	1.8	2.4	0.4
6	0.0	0.0	0.9	4.2	6.3	0.9
7	0.0	0.3	0.2	2.0	2.8	0.2
8	0.0	0.0	1.2	5.4	8.2	1.2
9	0.3	0.7	1.8	3.5	5.7	1.5
10	0.6	0.8	1.2	1.6	2.1	0.3
11	0.4	0.7	1.3	2.2	3.4	0.6
12	0.5	0.8	1.2	1.7	2.3	0.3
13	0.5	0.8	1.4	2.2	3.1	0.6
14	0.3	1.0	3.2	7.6	14.0	4.6
15	0.2	0.6	1.9	4.2	7.5	2.0
16	0.3	0.8	1.9	3.9	6.5	1.8
17	0.0	0.5	3.2	11.8	27.0	8.5
18	0.3	0.7	1.4	2.5	3.9	0.8
19	0.2	0.9	3.7	10.3	20.5	6.7
20	0.5	0.9	1.6	2.6	3.7	0.9
21	0.3	0.8	2.1	4.3	7.3	2.1
22	0.2	1.0	4.1	11.7	23.7	7.8
23	0.4	0.8	1.5	2.5	3.8	0.9
24	0.3	1.3	4.0	9.6	17.7	6.0
25	0.3	1.0	3.0	7.0	12.6	4.1
26	0.1	0.8	3.1	11.1	23.1	7.5
27	0.0	1.4	12.3	52.4	129.4	43.3
28	1.0	2.7	6.4	12.5	20.5	8.0
29	2.4	7.6	20.9	45.4	79.1	31.8
30	0.7	4.9	22.2	65.2	134.0	48.4
31	0.8	3.5	11.8	29.3	55.0	20.5
32	0.9	3.7	12.0	29.1	54.0	20.3
33	0.3	2.6	12.8	39.6	83.7	29.4
34	0.8	3.5	11.6	28.2	52.4	19.6
35	0.4	2.0	7.2	18.2	34.5	12.4
36	0.2	1.6	9.5	32.2	71.0	24.3
37	0.2	1.0	4.1	11.5	22.9	7.6

contd....

Appendix B1 – *contd...*

Week	Rainfall (mm) for the probabilities					Mean (mm)
	90	75	50	25	10	
38	0.1	0.6	2.6	7.0	14.0	4.3
39	0.1	1.0	4.6	14.0	29.3	9.7
40	0.4	0.8	1.7	3.0	4.7	1.2
41	0.2	0.7	2.1	5.0	9.2	2.7
42	0.5	0.8	1.4	2.1	3.1	0.6
43	0.4	0.8	1.6	2.8	4.4	1.1
44	0.1	0.6	1.9	4.9	9.2	2.6
45	0.5	0.7	0.9	1.3	1.6	0.0
46	0.7	0.9	1.1	1.3	1.5	0.1
47	0.5	0.7	0.9	1.3	1.6	0.0
48	0.5	0.7	0.9	1.3	1.6	0.0
49	0.5	0.7	0.9	1.3	1.6	0.0
50	0.4	0.7	1.3	2.0	2.9	0.5
51	0.5	0.7	0.9	1.3	1.6	0.0
52	0.0	0.0	0.0	0.0	0.0	0.0
ANN	163.5	231.3	327.0	446.4	575.4	351.8

Appendix B2 : Weekly rainfall distribution at different probability (%) levels by using Gamma distribution at Sriganganagar

Week	Rainfall (mm) for the probabilities					Mean (mm)
	90	75	50	25	10	
1	0.2	0.7	2.4	6.1	11.6	3.5
2	0.4	1.1	2.7	5.4	9.0	2.9
3	0.4	0.9	1.8	3.3	5.2	1.4
4	0.2	0.7	2.1	4.9	8.9	2.6
5	0.3	0.9	2.0	4.1	6.7	1.9
6	0.0	0.2	4.0	9.8	14.1	4.0
7	0.0	0.6	3.2	7.9	11.2	3.2
8	0.0	0.5	4.5	10.4	14.9	4.5
9	0.4	1.0	2.6	5.3	8.8	2.8
10	0.5	0.9	1.7	2.9	4.4	1.1
11	0.3	1.1	3.2	7.3	13.2	4.3
12	0.7	1.6	3.3	6.0	9.5	3.4
13	0.7	1.0	1.6	2.4	3.2	0.8
14	0.3	1.1	2.9	6.3	10.9	3.5
15	0.2	1.1	3.9	10.3	19.8	6.6

contd....

Appendix B2 – *contd...*

Week	Rainfall (mm) for the probabilities					Mean (mm)
	90	75	50	25	10	
16	0.4	0.9	2.9	6.7	12.1	3.8
17	0.3	0.8	1.9	3.9	6.6	1.8
18	0.2	0.9	3.5	9.4	18.5	6.0
19	0.3	0.8	2.0	3.9	6.5	1.8
20	0.3	0.8	1.8	3.4	5.5	1.5
21	0.2	0.9	3.3	8.6	16.7	5.4
22	0.5	1.1	2.5	4.8	7.8	2.5
23	0.2	1.1	4.0	10.5	20.3	6.8
24	0.7	2.7	8.1	18.9	34.3	12.8
25	0.2	0.9	3.9	10.9	21.9	7.2
26	0.2	1.5	6.7	19.7	40.3	13.9
27	0.5	3.0	11.9	32.7	64.8	23.4
28	1.2	4.4	12.8	29.0	51.9	20.0
29	1.0	4.1	13.7	33.5	62.5	23.5
30	2.0	6.7	18.8	41.7	73.4	29.1
31	1.3	5.0	15.4	36.1	65.6	25.2
32	0.3	1.7	6.6	17.7	34.8	12.2
33	1.0	4.0	13.0	31.3	57.9	21.9
34	0.2	0.9	3.9	11.0	22.1	7.3
35	0.4	1.8	5.9	14.4	26.6	9.5
36	0.2	1.3	6.9	21.9	46.7	15.9
37	0.2	1.1	4.2	11.3	22.2	7.4
38	0.2	1.1	4.5	12.4	24.6	8.2
39	0.2	0.7	2.0	4.5	7.8	2.2
40	0.8	0.9	1.1	1.3	1.6	0.1
41	0.3	0.7	1.5	2.7	4.1	0.9
42	0.2	0.7	2.0	4.4	7.9	2.2
43	0.5	0.7	0.9	1.3	1.6	0.0
44	0.5	0.7	0.9	1.3	1.6	0.0
45	0.3	0.7	1.5	2.8	4.5	1.1
46	0.5	0.7	0.9	1.3	1.6	0.0
47	0.5	0.8	1.3	2.0	2.9	0.5
48	0.2	0.7	1.7	3.4	5.8	1.5
49	0.7	0.9	1.2	1.5	1.8	0.2
50	0.5	0.9	1.7	2.8	4.1	1.0
51	0.5	0.9	1.5	2.4	3.5	0.8
52	0.3	0.8	1.8	3.5	5.7	1.5
ANN	218.2	262.6	318.6	382.1	445.9	325.8

Appendix B3 : Weekly rainfall distribution at different probability (%) levels by using Gamma distribution at Fatepur

Week	Rainfall (mm) for the probabilities					Mean (mm)
	90	75	50	25	10	
1	0.4	0.8	1.4	2.2	3.1	0.6
2	0.1	0.6	2.7	7.7	15.6	4.8
3	0.3	0.7	1.7	3.2	5.2	1.3
4	0.3	0.7	1.4	2.5	3.8	0.8
5	0.3	0.7	1.4	2.4	3.7	0.8
6	0.0	0.0	4.1	14.7	23.3	4.1
7	0.0	0.0	4.7	13.2	20.0	4.7
8	0.0	0.3	2.7	7.1	10.1	2.7
9	0.3	0.7	1.4	2.5	3.9	0.8
10	0.1	0.6	2.4	6.4	12.5	3.8
11	0.3	0.7	1.8	3.7	6.2	1.7
12	0.4	0.9	1.6	2.8	4.2	1.0
13	0.1	0.7	2.5	6.5	12.5	3.8
14	0.3	0.8	1.8	3.6	6.0	1.6
15	0.2	0.8	2.9	7.3	13.9	4.4
16	0.3	0.7	1.8	3.7	6.2	1.7
17	0.3	0.7	1.7	3.3	5.4	1.4
18	0.3	0.8	1.9	3.7	6.2	1.7
19	0.2	1.3	5.3	14.6	28.8	9.9
20	0.2	1.1	4.1	10.7	20.6	6.9
21	0.1	0.9	4.1	11.8	24.1	7.9
22	0.3	0.8	2.2	4.7	8.1	2.4
23	0.1	1.0	4.3	12.5	25.6	8.5
24	0.5	1.8	5.3	12.0	21.5	7.7
25	0.3	1.7	6.4	16.8	32.5	11.4
26	0.4	1.3	3.3	6.8	11.6	3.9
27	0.5	2.9	12.0	33.1	65.9	23.7
28	2.1	7.7	2.8	52.0	93.2	36.7
29	4.0	12.0	31.0	64.7	110.3	45.7
30	3.5	11.9	33.2	73.1	128.4	51.9
31	0.5	3.0	12.2	33.3	65.	23.8
32	3.0	7.5	16.6	31.4	50.2	1.7
34	0.6	3.3	12.9	34.7	67.9	24.7
35	0.4	1.9	7.0	18.3	35.3	2.5
36	0.5	1.9	6.2	15.0	27.8	9.9
37	0.2	1.0	3.5	8.7	16.4	5.4
38	0.3	1.1	3.8	9.5	17.9	6.0

contd....

Appendix B3 – *contd...*

Week	Rainfall (mm) for the probabilities					Mean (mm)
	90	75	50	25	10	
39	0.2	1.2	5.2	14.7	29.6	10.0
40	0.3	0.7	1.6	3.0	4.7	1.1
41	0.3	0.8	1.8	3.6	5.9	1.6
42	0.4	0.8	1.7	3.0	4.6	1.2
43	0.5	0.7	0.9	1.3	1.6	0.0
44	0.1	0.5	2.2	6.3	12.8	3.7
45	0.6	0.8	1.2	1.6	2.1	0.3
46	0.7	0.9	1.1	1.4	1.6	0.1
47	0.6	1.0	1.7	2.5	3.6	0.9
48	0.5	0.7	0.9	1.3	1.6	0.0
49	0.5	0.8	1.2	1.7	2.2	0.3
50	0.5	0.7	0.9	1.3	1.6	0.0
51	0.5	0.8	1.3	2.0	2.7	0.5
52	0.5	0.8	1.5	2.4	3.5	0.8
ANN	291.1	337.7	395.1	458.7	521.4	400.9

Appendix B4 : Weekly rainfall distribution at different probability (%) levels by using Gamma distribution at Pali

Week	Rainfall (mm) for the probabilities					Mean (mm)
	90	75	50	25	10	
1	0.3	0.7	1.6	3.2	5.2	1.3
2	0.4	0.9	2.0	3.7	6.0	1.7
3	0.6	0.9	1.4	2.1	2.8	0.6
4	0.3	0.7	1.4	2.6	4.0	0.9
5	0.5	0.8	1.2	1.8	2.5	0.4
6	0.0	0.3	0.3	2.3	3.3	0.3
7	0.0	0.4	0.1	1.9	2.6	0.1
8	0.0	0.0	1.5	5.3	7.8	1.5
9	0.3	0.7	1.7	3.5	5.8	1.5
10	0.7	0.9	1.1	1.4	1.7	0.2
11	0.8	0.9	1.1	1.2	1.3	0.1
12	0.6	0.8	1.2	1.7	2.2	0.3
13	0.6	0.8	1.1	1.5	2.0	0.2
14	0.3	0.7	1.5	2.8	4.3	1.0
15	0.1	0.5	1.9	4.6	8.7	2.4
16	0.3	0.7	1.7	3.5	5.7	1.5
17	0.4	0.8	1.7	3.1	4.9	1.3

contd....

Appendix B4 – *contd...*

Week	Rainfall (mm) for the probabilities					Mean (mm)
	90	75	50	25	10	
18	0.7	0.9	1.1	1.4	1.7	0.2
19	0.2	0.7	2.1	4.8	8.6	2.5
20	0.5	0.9	1.6	2.6	3.8	0.9
21	0.2	0.6	1.6	3.2	5.3	1.3
22	0.1	0.9	4.1	12.0	24.8	8.1
23	0.4	0.9	2.1	4.0	6.5	1.9
24	0.4	1.8	6.9	18.4	35.8	12.6
25	0.3	1.3	4.4	10.9	20.5	7.0
26	0.1	1.1	5.6	17.5	37.2	12.5
27	0.2	2.4	16.9	63.3	146.7	50.2
28	0.6	3.1	12.0	31.9	62.3	22.7
29	3.6	9.1	20.6	39.6	64.3	27.6
30	1.0	6.2	26.6	75.3	151.6	55.5
31	0.6	3.7	16.0	45.5	92.1	33.2
32	1.2	5.0	15.9	38.1	70.2	26.8
33	0.4	2.6	12.9	39.6	83.2	29.3
34	1.1	4.2	13.0	30.6	55.9	21.3
35	0.3	1.8	8.0	22.9	46.5	16.3
36	0.3	1.8	7.3	19.9	39.3	13.8
37	0.2	1.0	3.8	10.0	19.4	6.4
38	0.3	0.8	1.6	3.1	4.9	1.2
39	0.1	1.2	6.8	22.7	49.9	16.8
40	0.3	0.8	1.9	4.0	6.8	1.9
41	0.3	0.6	1.5	2.8	4.5	1.0
42	0.7	0.9	1.1	1.4	1.7	0.2
43	0.4	0.8	1.4	2.3	3.3	0.7
44	0.5	0.7	0.9	1.3	1.6	0.0
45	0.3	0.7	1.4	2.5	3.9	0.8
46	0.5	0.7	0.9	1.3	1.6	0.0
47	0.5	0.7	0.9	1.3	1.6	0.0
48	0.5	0.7	0.9	1.3	1.6	0.0
49	0.5	0.7	0.9	1.3	1.6	0.0
50	0.3	0.7	1.7	3.3	5.4	1.4
51	0.4	0.7	1.3	2.1	3.0	0.5
52	0.0	0.3	0.0	1.7	2.4	0.0
ANN	180.8	256.0	362.3	494.8	638.1	389.9

Appendix B5 : Weekly rainfall distribution at different probability (%) levels by using Gamma distribution at Jaipur

Week	Rainfall (mm) for the probabilities					Mean (mm)
	90	75	50	25	10	
1	0.3	0.7	1.8	3.6	6.0	1.6
2	0.2	0.9	3.1	7.5	13.9	4.5
3	0.7	0.9	1.3	1.7	2.1	0.4
4	0.4	0.8	1.5	2.4	3.5	0.8
5	0.3	0.8	2.0	4.1	6.8	1.9
6	0.0	0.3	0.3	2.3	3.1	0.3
7	0.0	0.0	2.3	7.5	11.2	2.3
8	0.0	0.0	3.1	8.7	12.8	3.1
9	0.2	0.8	2.8	6.8	12.7	4.0
10	0.8	0.9	1.1	1.2	1.4	0.1
11	0.8	0.9	1.1	1.2	1.3	0.1
12	0.3	0.9	2.5	5.4	9.6	2.9
13	0.3	0.8	1.8	3.4	5.5	1.5
14	0.2	0.7	2.0	4.4	7.6	2.2
15	0.1	0.6	3.0	9.5	20.4	6.3
16	0.2	0.6	1.9	4.4	8.1	2.2
17	0.2	0.6	1.7	3.9	7.0	1.8
18	0.4	0.7	1.3	2.0	2.9	0.5
19	0.3	1.0	2.5	5.4	9.2	2.9
20	0.2	0.6	1.7	4.0	7.2	1.9
21	0.1	0.8	3.8	11.2	23.1	7.5
22	0.1	0.8	3.5	10.2	20.8	6.7
23	0.3	1.1	3.1	6.8	11.9	3.9
24	0.5	1.7	5.0	11.4	20.3	7.3
25	0.1	0.9	6.2	22.7	52.1	17.3
26	0.4	2.4	10.6	30.8	63.0	22.3
27	0.3	1.7	7.5	21.6	43.9	15.3
28	2.9	9.4	25.6	55.6	96.9	39.2
29	8.7	25.4	64.5	133.3	225.8	95.2
30	1.8	10.4	42.6	117.4	233.4	86.7
31	4.0	11.1	27.2	55.0	92.0	38.7
32	12.4	24.3	45.4	76.5	114.2	55.2
33	2.5	8.5	24.5	54.9	97.3	38.7
34	1.1	8.9	45.5	142.7	303.3	108.9
35	2.8	9.6	27.2	60.1	105.9	42.5
36	0.4	2.9	13.6	40.4	83.4	29.6
37	0.2	1.0	3.7	9.7	18.9	6.2

contd....

Appendix B5 – *contd...*

Week	Rainfall (mm) for the probabilities					Mean (mm)
	90	75	50	25	10	
38	0.2	0.8	2.5	5.9	10.7	3.3
39	0.2	0.7	2.4	5.9	11.0	3.3
40	0.3	0.7	1.4	2.7	4.2	0.9
41	0.1	0.6	2.8	8.6	18.1	5.6
42	0.5	0.7	0.9	1.3	1.6	0.0
43	0.6	0.9	1.2	1.6	2.0	0.3
44	0.1	0.6	3.2	10.3	22.0	7.0
45	0.3	0.7	1.6	3.0	4.8	1.1
46	0.8	0.9	1.1	1.3	1.6	0.1
47	0.7	0.9	1.2	1.6	1.9	0.3
48	0.5	0.7	0.9	1.3	1.6	0.0
49	0.5	0.7	0.9	1.3	1.6	0.0
50	0.3	0.6	1.5	2.8	4.6	1.1
51	0.0	0.3	0.0	1.7	2.4	0.0
52	0.2	0.7	1.8	3.8	6.6	1.8
ANN	313.5	446.8	636.1	873.0	1130.0	687.0

Appendix B6 : Weekly rainfall distribution at different probability (%) levels by using Gamma distribution at Udaipur

Week	Rainfall (mm) for the probabilities					Mean (mm)
	90	75	50	25	10	
1	0.8	0.9	1.1	1.3	1.5	0.1
2	0.2	0.8	2.7	6.8	12.8	4.0
3	0.7	0.9	1.2	1.5	1.9	0.2
4	0.5	0.8	1.2	1.8	2.4	0.4
5	0.4	0.9	1.6	2.8	4.2	1.0
6	0.0	0.3	0.0	1.7	2.3	0.0
7	0.0	0.4	0.1	1.8	2.5	0.1
8	0.0	0.1	0.5	2.8	4.0	0.5
9	0.4	0.7	1.4	2.4	3.7	0.8
10	0.0	0.0	0.0	0.0	0.0	0.1
11	0.7	0.9	1.1	1.4	1.7	0.2
12	0.6	0.9	1.2	1.8	2.3	0.4
13	0.5	0.7	0.9	1.3	1.6	0.0
14	0.3	0.8	1.7	3.1	4.9	1.2
15	0.5	0.8	1.2	1.7	2.3	0.3

contd....

Appendix B6 – *contd...*

Week	Rainfall (mm) for the probabilities					Mean (mm)
	90	75	50	25	10	
16	0.4	0.9	1.6	2.8	4.2	1.1
17	0.2	0.7	1.8	3.7	6.3	1.7
18	0.5	0.8	1.3	2.1	2.9	0.6
19	0.1	0.7	3.2	9.7	20.3	6.4
20	0.6	0.9	1.2	1.6	2.1	0.3
21	0.4	0.8	1.6	2.8	4.4	1.1
22	0.2	0.8	3.2	8.7	17.0	5.4
23	0.1	0.7	3.0	8.6	17.5	5.5
24	1.7	4.4	10.3	20.1	32.9	13.5
25	0.5	2.2	8.0	20.5	39.1	14.1
26	1.4	4.9	14.6	33.4	59.8	23.2
27	0.7	3.9	16.4	46.0	92.2	33.5
28	1.5	6.6	22.7	56.4	16.1	40.3
29	6.1	15.7	36.1	69.8	113.8	49.4
30	2.7	10.6	33.5	79.6	146.0	57.0
31	1.9	7.9	26.0	63.3	117.5	45.2
32	9.6	21.5	44.7	81.0	126.6	57.8
33	7.3	17.1	36.8	68.1	18.0	48.3
34	1.8	7.9	27.2	67.9	127.9	48.8
35	2.2	7.0	18.9	40.7	70.6	28.4
36	0.3	2.5	11.9	36.2	75.7	26.6
37	0.5	1.8	5.2	11.5	20.5	7.4
38	0.4	1.9	7.6	20.5	40.3	14.2
39	0.6	2.8	9.5	23.4	43.8	16.1
40	0.2	1.1	4.9	13.9	28.1	9.4
41	0.2	0.7	2.3	5.5	10.2	3.0
42	0.6	0.9	1.2	1.7	2.2	0.3
43	0.2	0.6	1.6	3.3	5.7	1.4
44	0.1	0.5	2.2	6.2	12.4	3.6
45	0.1	0.6	3.3	11.5	25.5	8.1
46	0.5	0.7	0.9	1.3	1.6	0.0
47	0.4	0.7	1.3	2.1	3.0	0.5
48	0.5	0.7	0.9	1.3	1.6	0.0
49	0.5	0.7	0.9	1.3	1.6	0.0
50	0.1	0.6	2.0	4.9	9.1	2.6
51	0.6	0.8	1.2	1.7	2.3	0.3
52	0.5	0.8	1.2	1.8	2.3	0.4
ANN	393.2	472.0	571.4	683.9	796.8	584.8

Appendix B7 : Weekly rainfall distribution at different probability (%) levels by using Gamma distribution at Banswara

Week	Rainfall (mm) for the probabilities					Mean (mm)
	90	75	50	25	10	
1	0.8	0.9	1.1	1.3	1.5	0.1
2	0.1	0.7	3.7	12.3	26.8	8.6
3	0.6	0.8	1.2	1.6	2.1	0.3
4	0.4	0.7	1.3	2.2	3.4	0.6
5	0.7	0.9	1.1	1.4	1.7	0.2
6	0.0	0.3	0.0	1.7	2.3	0.0
7	0.0	0.3	0.0	1.7	2.3	0.0
8	0.0	0.3	0.0	1.7	2.3	0.0
9	0.8	0.9	1.1	1.2	1.4	0.1
10	0.5	0.7	0.9	1.3	1.6	0.0
11	0.8	0.9	1.1	1.2	1.3	0.1
12	0.5	0.8	1.3	1.9	2.6	0.4
13	0.6	0.8	1.1	1.5	1.9	0.2
14	0.5	0.7	0.9	1.3	1.6	0.0
15	0.5	0.7	0.9	1.3	1.6	0.0
16	0.4	0.8	1.4	2.3	3.4	0.7
17	0.2	0.7	1.9	4.3	7.5	2.1
18	0.4	0.7	1.3	2.1	3.0	0.5
19	0.1	0.7	2.8	8.0	16.1	5.0
20	0.5	0.7	0.9	1.3	1.6	0.0
21	0.5	0.7	0.9	1.3	1.6	0.0
22	0.2	0.7	1.7	3.7	6.2	1.6
23	0.2	0.6	1.6	3.3	5.6	1.4
24	0.4	2.4	10.8	31.5	64.4	22.8
25	0.6	3.6	14.6	40.0	79.2	28.8
26	1.0	5.1	19.1	50.0	96.8	36.0
27	0.6	4.4	21.5	65.9	138.4	49.4
28	3.1	1.3	34.0	78.2	140.7	55.7
29	13.8	5.6	45.6	74.2	108.2	53.9
30	7.6	4.8	68.2	148.2	258.5	106.1
31	4.6	1.3	75.2	190.5	361.8	139.1
32	8.7	9.3	39.4	70.7	109.9	50.4
33	4.9	7.9	53.8	123.7	222.6	88.8
34	5.5	5.7	39.1	79.9	134.6	56.7
35	1.6	6.7	22.6	55.4	103.6	39.6
36	0.2	2.4	15.4	54.9	124.3	42.8
37	0.7	3.1	10.5	25.8	48.2	17.9

contd....

Appendix B7 – *contd...*

Week	Rainfall (mm) for the probabilities					Mean (mm)
	90	75	50	25	10	
38	0.6	2.8	10.5	27.4	53.0	19.3
39	0.2	1.8	10.6	35.5	77.9	26.8
40	0.2	0.8	3.1	8.2	15.9	5.1
41	0.0	0.9	6.8	26.6	63.3	20.9
42	0.1	0.7	3.2	9.8	20.4	6.5
43	0.4	0.7	1.3	2.1	3.0	0.5
44	0.1	0.5	2.8	9.3	20.5	6.3
45	0.0	0.5	3.0	10.5	23.6	7.3
46	0.3	0.7	1.6	3.0	4.7	1.1
47	0.1	0.8	3.6	10.7	22.3	7.2
48	0.5	0.7	0.9	1.3	1.6	0.0
49	0.5	0.7	0.9	1.3	1.6	0.0
50	0.3	0.7	1.3	2.3	3.4	0.7
51	0.5	0.7	0.9	1.3	1.6	0.0
52	0.1	0.5	1.9	4.8	9.1	2.5
ANN	534.4	682.5	877.5	1106.7	1343.8	913.9

Appendix B8 : Weekly rainfall distribution at different probability (%) levels by using Gamma distribution at Kota

Week	Rainfall (mm) for the probabilities					Mean (mm)
	90	75	50	25	10	
1	0.6	0.9	1.3	1.8	2.3	0.4
2	0.3	1.0	3.1	7.1	12.8	4.1
3	0.6	0.8	1.3	1.8	2.4	0.4
4	0.1	0.6	2.0	4.8	8.9	2.5
5	0.3	0.7	1.4	2.7	4.4	1.0
6	0.0	0.0	4.2	16.0	25.8	4.2
7	0.0	0.0	1.0	4.5	6.9	1.0
8	0.0	0.4	0.1	1.8	2.5	0.1
9	0.4	0.9	1.8	3.1	4.9	1.3
10	0.6	0.9	1.4	1.9	2.5	0.5
11	0.3	0.7	1.4	2.7	4.2	0.9
12	0.5	0.8	1.3	1.9	2.5	0.4
13	0.3	0.8	1.7	3.3	5.4	1.4
14	0.4	0.8	1.4	2.3	3.3	0.7
15	0.2	0.7	1.8	3.6	6.1	1.6

contd....

Appendix B8 – *contd...*

Week	Rainfall (mm) for the probabilities					Mean (mm)
	90	75	50	25	10	
16	0.5	0.9	1.5	2.4	3.4	0.8
17	0.1	0.5	2.4	7.3	15.1	4.5
18	0.6	0.9	1.4	2.0	2.7	0.5
19	0.4	0.8	1.6	2.8	4.2	1.0
20	0.7	0.9	1.1	1.4	1.6	0.1
21	0.1	0.7	2.5	6.5	12.4	3.8
22	0.4	0.9	2.0	3.7	5.9	1.7
23	0.3	1.0	3.2	7.6	13.9	4.5
24	0.7	2.1	5.7	12.2	21.0	7.8
25	0.3	1.8	7.0	18.9	37.2	13.1
26	0.3	2.3	11.8	37.3	79.5	27.7
27	0.5	3.3	14.0	39.8	80.3	28.9
28	4.8	12.0	27.1	52.1	84.4	36.6
29	12.7	23.8	42.8	70.2	102.9	50.8
30	5.1	17.7	51.1	114.8	203.8	82.1
31	5.1	15.3	39.3	82.0	139.6	58.1
32	7.5	18.3	40.4	76.2	122.2	54.1
33	10.7	24.1	50.4	91.6	143.6	65.5
34	2.0	8.9	31.3	78.8	149.3	56.9
35	2.1	8.3	26.3	62.6	115.0	44.6
36	1.1	5.9	23.6	64.3	127.0	46.9
37	0.3	1.4	4.9	12.4	23.6	8.1
38	0.5	2.1	7.3	18.3	34.7	12.5
39	0.2	1.2	4.8	13.1	25.9	8.8
40	0.1	0.6	2.9	8.7	17.9	5.6
41	0.1	0.8	5.2	18.7	42.6	14.0
42	0.1	0.5	2.1	6.0	11.9	3.5
43	0.5	0.8	1.2	1.9	2.6	0.4
44	0.1	0.6	3.2	10.1	21.7	6.8
45	0.1	0.5	3.2	10.8	23.9	7.5
46	0.7	0.9	1.1	1.3	1.6	0.1
47	0.5	0.7	0.9	1.3	1.6	0.0
48	0.5	0.7	0.9	1.3	1.6	0.0
49	0.4	0.7	1.3	2.0	2.8	0.5
50	0.1	0.6	2.5	6.9	13.6	4.1
51	0.5	0.8	1.2	1.8	2.4	0.4
52	0.2	0.7	1.9	4.4	7.8	2.2
ANN	21.3	21.9	22.3	22.6	22.7	685.1

Appendix C1 : Computer programme for calculating initial and conditional probability of dry and wet week by Markov Chain Process

```
C       MARKOV CHAIN -WEEKLY
        DIMENSION PPT(100, 52) ,ALIM(10)
        CHARACTER FILE1*12,NAME*13
        , WRITE (*,' (A \) ') , PPT FILENAME:'
        READ (*,306)FILE1
  306   FORMAT (A12)
C            WRITE (*,'(A\)') , CENTRE NAME:' .
C            READ (*,13)NAME
C 13         FORMAT (A13)
        WRITE (*,' (A \) ') , NY, NL, NSW, NEW, LBY, LEY: ' .-
        READ (*, *),NY, NL, NSW, NEW, LBY, LEY
        WRITE (*, ) (A \) ') , LIMIT:'
        READ (*,*) (ALIM(JJ) ,JJ=1,NL)
        OPEN(5,FILE=FILE1,STATUS='OLD')
        OPEN(6,FILE= OUT2 :MKV',STATUS='NEW')
        READ (5, , (A13) ')NAME
        DO 61 NYS=1,NY
        READ (5,*)LYR
        DO 64 NWK=1,52
        READ (5,*)APPT
        PPT(NYS,NWK)=APPT
64      CONTINUE
61      CONTINUE
        CLOSE (5)
        DO 60 M=1,NL
        WRITE (6,1'4)NAME
14      FORMAT (5X, 'INITIAL AND CONDITIONAL PROBABILITIES'
        1     ' OF RAINFALL AT ',IX,A13/)
        WRITE (6,62)ALIM(M) , LBY, LEY
62      1 FORMAT (1OX, 'LIMIT=',F7.1, 1X, 'mm',22, 'DATABASE: "
        1 ' I5, '-', I5)
        WRITE (6,38)
        WRITE (6,35)
35      FORMAT (15X,'INITIAL',24X, 'CONDITIONAL'/12X, 'PROBABILITIES',
        1     20X, 'PROBABILITIES'/11X,14 (1H-) ,6X,40(1H-))
        WRITE (6,36)
36      FORMAT (1X, 'WEEK',6X, 'P (W) I ,6X, 'P (D)' ,7X, 'p (w/w), 5X, 'P (D/w)',
        1     5X, 'P (D/D) , , 5X, 'P (W/D) ,)
        WRITE (6,38)
        DO 81 I=NSW,NEW
        J=I-1 . ..
        I.F {J.EQ.0)J=52 .
        NDD=O
        NWD=O
```

```
          NWW=O
      NDW=O
      IF (J.EQ.52)GOTO 53
          DO 101 K=1,NY
      IF (PPT(K,I)-ALIM(M)) 21,22,22
21        IF (PPT(K,J)-ALIM(M)) 23,24,24
23    NDD=NDD+1
      GOTO 101
24    NDW=NDW+1
      GOTO 101
22    IF (PPT(K,J) -ALIM(M)) 25,26,26
25    NWD=NWD+1
      GOTO 101
26    NWW-NWW+1
101   CONTINUE .
          GOTO 54
53        DO 102 KL=1,NY
      KK=KL+1
      IF (KK.GT.NY)GOTO 54
      IF (PPT(KK,I)-ALIM(M) 41,42,42
41    IF (PPT(KL,J)-ALIM(M) 43,44,44
43    NDD=NDD+1
      GOTO 102
44    NDW=NDW+1
      GOTO 102
42    IF (PPT(KL,J) -ALIM (M) ) 45,46,46
45    NWD=NWD+1
      GOTO 102
46    NWW=NWW+11
102       CONTINUE
54    CONTINUE
      WD=NWD
          DW=NDW
          WW=NWW
          DD=NDD
      TW=WW+DW
      TD=WD+DD
      PW=(WW+WD)/(TW+TD)
      PD=1.0-PW
      IF (TW.EQ.O.O)GOTO 300
      IF (TD.EQ.O.O)GOTO 301
      PWD=WD/TD
      PWW=WW/TW
      PDD=1.0-PWD
      PDW=l.0-PWW
      GOTO 302
301   PDD=O.O
```

```
      PWD=O.O
      PWW=WW/TW
    PDW=1.0-PWW
    GOTO 302
300 PWW=O.O
      PDW=O.O
      PWD=WD/TD
    PDD=1.0-PWD
302 WRITE (6,37)I,PW,PD,PWW,PDW,PDD,PWD
37  FORMAT (1X,I4,6F11.4)
81  CONTINUE
    WRITE (6,38)
38  FORMAT (IX,70(IH-)
    WRITE (6,40)
40  FORMAT (1H1)
60  CONTINUE
    CLOSE (6)
    STOP
    END
```

Appendix C2 : Computer programme for computing consecutive dry and wet weeks by Markov-chain process

```
C           MARKOV CHAIN -WEEKLY (SPELLS)
      DIMENSION PPT(100,S2) ,ALIM(10) ,PW(S2) ,PD(S2) ,PWW(S2)
            DIMENSION PDD(S2) ,PDW(S2) ,PWD(S2) ,P2D(S2) ,P3D(S2)
,P4D(S2)
            DIMENSION P2W(S2) ,P3W(S2) ,P4W(S2)
      CHARACTER FILE1*12,NAME*13
      WRITE (*,' (A\) ') ' PPT FILENAME: '
      READ (*,306)FILE1
306   FORMAT (A12) .
C     WRITE (*,' (A \) ,) ' CENTRE NAME: '
C     READ (*,13)NAME
C13   FORMAT (A13)
      WRITE (*, *(A\) ,) ' NY,NL,NSW,NEW, LBY, LEY: '
      READ (*,*)NY,NL,NSW,NEW,LBY,LEY
      WRITE (*, I (A \) ') ' LIMIT : .
      READ (*,*) (ALIM(JJ) ,JJ=1,NL)
      OPEN(5,FILE=FILE1,STATUS='OLD')
      OPEN(6,FILE='OUT4.MSL,STATUS='NEW')
            READ(5, ' (a13) ')Name
      DO 61 NYS=1,NY
      READ (5,*)LYR
      DO 64 NWK=1,S2
      READ (5,*)APPT
      PPT(NYS,NWK)=APPT
64    CONTINUE
61    CONTINUE
      CLOSE(5)
      DO 60 M=1,NL
      WRITE (6,14)NAME
14    FORMAT (1X, 'PROBABILITIES OF DRY AND WET SPELLS OF'
      1     ' * CONSECUTIVE WEEKS ,AT', 1X,A13/)
      WRITE (6,62)ALIM(M) ,LBY,LEY
62    FORMAT (//10X, 'LIMIT=',F7.1,1X, 'mm',22X, 'DATABASE: ',
      1     I5, '-', I5/)
      WRITE (6,38)
      WRITE (6,35)
35    FORMAT (17X, 'CONSECUTIVE DRY' ,18X, 'CONSECUTIVE WET'/18X,
      1     'PROBABILITIES' ,20X, 'PROBABILITIES'/11X,27 (1H-), 6X,27(1H-
))
      WRITE (6,36)
36    FORMAT (1X, 'WEEK' ,6X, 'P(2D)' ,6X, 'P(3D)' ,6X, 'P(4D) 1 ,6X,
      'P(2W) ".
      1     6X,,'P(3W)',6X,'P(4W)')
      WRITE (6,38)
            DO 81 I=NSW,NEW
      J=I-1
```

```
      IF "(J.EQ.0)J=52
      NDD=O
      NWD=O
           NWW=O
      NDW=O
      IF (J.EQ.52)GOTO S3
           DO IOI K=I,NY
      IF (PPT(K,I)-ALIM(M)21,22,22)
21         IF (PPT(K,J) ALIM (M))23,24,24
23    NDD=NDD+1
      GOTO 101
24    NDW=NDW+1
      GOTO 101
22    IF (PPT(K,J)-ALIM(M)25,26,26
25    NWD=NWD+1
      GOTO 101
26    NWW=NWW+1 '
101   CONTINUE
      GOTO 54
53         DO 102 KL=1,NY
      KK=KL+1
      IF (KK.GT.NY)GOTO 54
      IF (PPT(KK,I)-ALIM(M)41,42,42
41    IF (PPT(KL,J)-ALIM(M)43,44,44
43    NDD=NDD+1
      GOTO 102
44    NDW=NDW+1
      GOTO 102
42    IF (PPT(KL,J).-ALIM(M)45,46,46
45    NWD=NWD+1
      GOTO 102
46    NWW=NWW+1
102   CONTINUE
54    CONTINUE
      WD=NWD
           DW=NDW
           WW=NWW
           DD=NDD
      TW=WW+DW
      TD=WD+DD
      PW(I)=(WW+WD)/(TW+TD)
      PD ( I) = 1 . 0 -PW ( I)
      IF (TW.EQ.O.O)GOTO 300
      IF (TD.EQ.O.O)GOTO 301
      PWD(I)=WD/TD
      PWW(.I)=WW/TW .
      PDD(I)=1.0-PWD(I)
      PDW(I)=1.0-PWW(I)
```

```
      GOTO 302
301   PDD(I)=O.O
            PWD(I)=O.O
            PWW(I)=WW/TW
      PDW(I)=1.0-PWW(I)
      GOTO 302
300   PWW(I)=O.O
            PDW(I)=O.O
            PWD(I)=WD/TD
      PDD(I)=1.0-PWD(I)
302   CONTINUE
C     WRITE (6,37)I,PW,PD,PWW,PDW,PDD,PWD
C     FORMAT (1X,I4,6F11.4)
81    CONTINUE
      L=NEW-1
      DO 111 IJ=NSW,L
      P2D (IJ) =0 . 0
      P3D (IJ) =0 . 0
      P4D (IJ) =0 . 0
      P2W (IJ) = 0 . 0
      P3W(IJ)=0.0
      P4W(IJ)=0.0
111   CONTINUE .
      DO 502 IN=NSW,L
      P2D(IN)=PD(IN)*PDD(IN+1)
      P2W(IN)=PW(IN.) *PWW(IN+1)
502   CONTINUE
      LL=NEW-2
      DO 103 IK=NSW,LL '
      P3D(IK)=PD(IK) *PDD(IK+1) *PDD(IK+2)
      P3W(IK)=PW(IK) *PWW(IK+1) *PWW(IK+2)
103   CONTINUE
      N=NEW-3
      DO 104 IL=NSW,N
      P4D(IL) =PD(IL) *PDD(IL+1) *PDD(IL+2) *PDD(IL+3)
      P4W(IL) =PW(IL) *PWW(IL+1) *PWW(IL+2) *PWW(IL+3)
104         CONTINUE
      WRITE (6,71) (IM,P2D(IM) ,P3D(IM) ,P4D(IM) ,P2W(IM),
      1     P3W(IM) ,P4W(IM) ,IM=NSW,L)
71    FORMAT (1X,I4',6F11.4)
      WRITE (6,38)
38    FORMAT (IX, 70'(1H=) /)
      WRITE (6,40)
40    FORMAT (1H1)'
60    CONTINUE .-
      CLOSE (6) .
      STOP
      END
```

Appendix D : Computer programme for computing reference evapotranspiration by Penman-Monteith FAO-56 Method

```
C       FORTRAN PROGRAMME FOR DETERMINATION OF ETO
C       FAO 56 PENMEN-MONTEITH EQUATION
C       E10 REFERENCE EVAPOTRANSPIRATION, MM/DAY
C       TMAX, TMIN ARE MAXIMUM AND MINIMUM TEMPERATURE °C
C       TAV AVERAGE TEMPERATUREoC
C       RH RELATIVE HUMIDITY, %
C       U2 WIND VELOCITY AT2MHEIGHT, M/SEC .
C       AN, NMAX ARE ACTUAL AND MAXIMUM SUNSHINE HOURS
C       MSL MEAN SEA LEVEL, M
C       EMAX, EMIN ARE MAX. AND MIN. VAPOUR PRESSURE, KPA
C       EMIN MINIMUM SATURATION VAPOUR PRESSURE, KPA
C       EA AVERAGE AIR VAPOUR PRESSURE, KPA
C       RS IS INCOMING SOLAR RADIATION, MJ/MI/2/DAY
C       RNS, RNL ARE SHORT & LONG WAVE RADIATION, MJ/MI\2/DA Y
C       RN IS NET RADIATION, MJ/MI\2/DA Y
        REAL TMAX, TMIN,RH,U2KMPH,AN,ANMAX,MSL,RS,A 1,B 1
        REAL RAIN,EVA,LAT ,J,A,B,LAMBDA
        WRITE(*,*)' STATION=UDAIPUR'
        WRITE(*,*) , WEEK NO. ETO'
        DO 10I=1,52
        READ(*,*)J,A 1,B 1 ,A,B,LAT ,MSL
        READ(*,*)TMAX, TMIN,RH 1 ,RH2,U2KMPH,AN,RAIN,EV A
        U2=U2KMPH*0.255*0.93
        RH=(RHI+RH2)/2
        TAV=(TMAX+TMIN)/2.0
        Z=(17.27*TAV)/(TAV+237.3)
        ZI=EXP(Z) .
        DELTA=(2504.0*Z 1 )/TA V+237 .3)**2)
        P = 101.3*290.4-0.0065*MSL)/290.4)**5.26
        LAMBDA = 2.501-(2.361 E-03*TAV)
        GAMMA=0.00163*P/LAMBDA
        EMAX=0.61 1 *EXP(17 .27*TMAX)/(TMAX+237.3)
        EMIN = 0.61 l*EXP(17.27*TMIN)/(TMIN+237.3)
        EA=(EMAX +EMIN)/2.0
        ED=RH/( (50. 0/EMAX)+( 50/EM IN+
        DEL=0.409*SIN(0.0 172* J-1.39)
        PHI=LAT*PI/180.00
        AWS=-TAN(PHI)*TAN(DEL)
        ANMAX= ACOS(AWS)*7.64
        DR=I+0.33*COS(0.0172* J)
        RA=37.6*DR *(WS * SIN(PHI)* SIN(DEL )+COS(DEL )*COS(PHI)*
SIN(WS))
```

```
      RS=(A+B* AN/ANMAX)*RA
      RNS=(1-0.05)*RS
      TKAL=TAV+273.0
      RNL=4.9E:-09*(TKAL **4.0)*(A 1-B 1*SQRT(ED+*(O.1 +(0.9*AN/AN
      MAX)
      RN = RNS-RNL.
      AX=0.40S*DELTA *(RN-0.0)
      BX=GAMMA *(900.0/(TAV+273.0))
      CX=U2*(EA-ED)
      DX=DELTA+GAMMA *(1.0+0.34*U2)
      ET0=(AX +BX*CX)/DX
      WRITE(*,*) I,ETO,EV A
10    CONTINUE
      STOP
      END
```

Appendix E1 : Normal weekly water balance at Jodhpur (1981-1995)

(A WC=100 mm)

Week (mm)	PPT (mm)	ET_0 (mm)	AET (mm)	SPL (mm)	DEF (mm)	I_{ma}	I_a
1	0.3	19.6	0.3	0.0	19.3	0.0	1.0
2	2.9	20.1	2.9	0.0	17.2	0.1	0.9
3	0.7	20.9	0.7	0.0	20.2	0.0	1.0
4	0.1	23.5	0.1	0.0	23.4	0.0	1.0
5	0.4	24.0	0:4	0.0	23.6	0.0	1.0
6	0.9	24.2	0.9	0.0	23.3	0.0	1.0
7	0.2	24.5	0.2	0.0	24.3	0.0	1.0
8	1.2	26.7	1.2	0.0	25.5	0.0	1.0
9	1.5	27.8	1.5	0.0	26.3	0.1	1.0
10	0.3	30.6	0.3	0.0	30.3	0.0	1.0
11	0.7	33.2	0.7	0.0	32.5	0.0	1.0
12	0.3	34.2	0.3	0.0	33.9	0.0	1.0
13	0.6	35.1	0.6	0.0	34.5	0.0	1.0
14	4.6	37.8	4.6	0.0	33.2	0.1	0.9
15	2.1	38.1	2.1	0.0	36.0	0.1	0.9
16	1.8	39.1	1.8	0.0	37.3	0.1	1.0
17	8.5	41.2	8.5	0.0	32.7	0.2	0.8
18	0.9	41.6	0.9	0.0	40.7	0.0	1.0
19	6.7	41.2	6.7	0.0	34.5	0.2	0.8
20	0.9	43.8	0.9	0.0	42.9	0.0	1.0
21	2.1	42.4	2.1	0.0	40.3	0.1	1.0
22	7.8	42.5	7.8	0.0	34.7	0.2	0.8
23	0.9	40.4	0.9	0.0	39.5	0.0	1.0
24	6.0	39.5	6.0	0.0	33.5	0.2	0.9
25	4.0	42.3	4.0	0.0	38.3	0.1	0.9
26	7.5	44.8	7.5	0.0	0.37	0.2	0.8
27	43.4	40.9	40.0	0.0	0.0	10.0	0.0
28	8.0	40.6	8.7	0.0	31.9	0.2	0.8
29	31.8	38.3	31.9	0.0	6.4	0.8	0.2
30	48.4	40.4	40.4	0.0	0.0	10.0	0.0
31	20.5	36.2	21.9	0.0	14.3	0.6	0.4
32	20.3	36.1	21.5	0.0	14.6	0.6	0.4
33	29.4	33.4	29.7	0.0	3.7	0.9	0.1
34	19.6	31.8	20.4	0.0	11.4	0.6	0.4
35	12.4	30.4	13.4	0.0	17.0	0.4	0.6
36	24.3	28.1	24.5	0.0	3.6	0.9	0.1
37	7.6	27.1	8.5	0.0	18.6	0.3	0.7

contd...

Appendix E1 – *contd...*

Week (mm)	PPT (mm)	ET_0 (mm)	AET (mm)	SPL (mm)	DEF (mm)	I_{ma}	I_a
38	4.3	25.7	5.1	0.0	20.6	0.2	0.8
39	9.7	26.4	10.2	0.0	16.2	0.4	0.6
40	1.2	28.1	1.8	0.0	26.3	0.1	0.9
41	2.6	27.7	3.1	0.0	24.6	0.1	0.9
42	0.6	27.5	1.0	0.0	26.5	0.0	1.0
43	1.1	25.9	1.4	0.0	24.5	0.1	1.0
44	2.6	26.5	2.8	0.0	23.7	0.1	0.9
45	0.0	24.3	0.2	0.0	24.1	0.0	1.0
46	0.1	23.0	0.2	0.0	22.8	0.0	1.0
47	0.0	22.4	0.1	0.0	22.3	0.0	1.0
48	0.0	22.9	0.1	0.0	22.8	8.0	1.0
49	0.0	20.8	0.1	0.0	20.7	0.0	1.0
50	0.5	20.2	0.5	0.0	19.7	0.0	1.0
51	0.0	20.8	0.0	0.0	20.8	0.0	1.0
52	0.0	21.0	0.0	0.0	21.0	0.0	1.0
ANN	352.3	1625.6	352.2	0.0	1273.4		

Appendix E2 : Normal weekly water balance at Sriganganagar (1981–1995)
(AWC = 100 mm)

Week (mm)	PPT (mm)	ET_0 (mm)	AET (mm)	SPL (mm)	DEF (mm)	I_{ma}	I_a
1	3.5	9.0	3.5	0.0	5.5	0.4	0.6
2	2.2	8.8	2.2	0.0	6.6	0.3	0.8
3	1.4	9.8	1.4	0.0	8.4	0.1	0.9
4	2.6	11.8	2.6	0.0	9.2	0.2	0.8
5	1.9	12.3	1.9	0.0	10.4	0.2	0.9
6	4.0	13.5	4.0	0.0	9.5	0.3	0.7
7	3.2	14.2	3.2	0.0	11.0	0.2	0.8
8	4.5	13.8	4.5	0.0	9.3	0.3	0.7
9	2.8	16.6	2.8	0.0	13.8	0.2	0.8
10	1.1	20.9	1.1	0.0	19.8	0.1	1.0
11	4.3	22.1	4.3	0.0	17.8	0.2	0.8
12	3.4	22.8	3.4	0.0	19.4	0.2	0.9
13	0.8	26.4	0.8	0.0	25.6	0.0	1.0
14	3.5	30.3	3.5	0.0	26.8	0.1	0.9
15	6.6	31.5	6.6	0.0	24.9	0.2	0.8
16	3.8	36.9	3.8	0.0	33.1	0.1	0.9

contd...

Appendix E2 – *contd...*

Week (mm)	PPT (mm)	ET_0 (mm)	AET (mm)	SPL (mm)	DEF (mm)	I_{ma}	I_a
17	1.8	41.8	1.8	0.0	40.0	0.0	1.0
18	6.0	46.7	6.0	0.0	40.7	0.1	0.9
19	1.8	46.8	1.8	0.0	45.0	0.0	1.0
20	1.5	49.4	1.5	0.0	47.9	0.0	1.0
21	5.4	51.5	5.4	0.0	46.1	0.1	0.9
22	2.5	53.2	2.5	0.0	50.7	0.1	1.0
23	6.8	53.8	6.8	0.0	47.0	0.1	0.9
24	12.8	50.1	12.8	0.0	37.3	0.3	0.7
25	7.2	49.7	7.2	0.0	42.5	0.1	0.9
26	13.9	48.9	13.9	0.0	35.0	0.3	0.7
27	23.4	40.5	23.4	0.0	17.1	0.6	0.4
28	20.1	36.8	20.1	0.0	16.7	0.6	0.5
29	23.5	34.8	23.5	0.0	11.3	0.7	0.3
30	29.1	31.8	29.1	0.0	2.7	0.9	0.1
31	26.6	33.4	26.6	0.0	6.8	0.8	0.2
32	12.2	32.6	12.2	0.0	20.4	0.4	0.6
33	21.9	32.4	21.9	0.0	10.5	0.7	0.3
34	7.3	33.0	7.3	0.0	25.7	0.2	0.8
35	9.5	32.5	9.5	0.0	23.0	0.3	0.7
36	15.9	30.7	15.9	0.0	14.8	0.5	0.5
37	7.4	32.3	7.4	0.0	24.9	0.2	0.8
38	8.2	31.0	8.2	0.0	22.8	0.3	0.7
39	2.2	28.5	2.2	0.0	26.3	0.1	0.9
40	0.1	27.7	0.1	0.0	27.6	0.0	1.0
41	0.9	24.7	0.9	0.0	23.8	0.0	1.0
42	2.2	23.1	2.2	0.0	20.9	0.1	0.9
43	0.0	20.5	0.0	0.0	20.5	0.0	1.0
44	0.0	17.8	0.0	0.0	17.8	0.0	1.0
45	1.1	16.0	1.1	0.0	14.9	0.1	0.9
46	0.0	13.9	0.0	0.0	13.9	0.0	1.0
47	0.5	12.6	0.5	0.0	12.1	0.0	1.0
48	1.5	11.5	1.5	0.0	10.0	0.1	0.9
49	0.2	10.9	0.2	0.0	10.7	0.0	1.0
50	1.0	10.3	1.0	0.0	9.3	0.1	0.9
51	0.8	9.7	0.8	0.0	8.9	0.1	0.9
52	1.5	9.8	1.5	0.0	8.3	0.2	0.9
ANN	326.4	1431.4	326.4	0.0	1105.0		

Appendix E3 : Normal weekly water balance at Fatehpur (1981-1995)

(AWC=100 mm)

Week (mm)	PPT (mm)	ET_0 (mm)	AET (mm)	SPL (mm)	DEF (mm)	I_{ma}	I_a
1	0.6	14.0	0.6	0.0	13.4	0.0	1.0
2	4.8	13.2	4.8	0.0	8.4	0.4	0.6
3	1.3	14.2	1.3	0.0	12.9	0.1	0.9
4	0.8	16.8	0.8	0.0	16.0	0.1	1.0
5	0.8	17.2	0.8	0.0	16.4	0.1	1.0
6	4.1	19.3	4.1	0.0	15.2	0.2	0.8
7	4.7	19.6	4.7	0.0	14.9	0.2	0.8
8	2.7	20.0	2.7	0.0	17.3	0.1	0.9
9	0.8	21.9	0.8	0.0	21.1	0.0	1.0
10	3.7	25.6	3.7	0.0	21.9	0.1	0.9
11	1.6	26.7	1.6	0.0	25.1	0.1	0.9
12	1.0	31.3	1.0	0.0	30.3	0.0	1.0
13	3.8	32.8	3.8	0.0	29.0	0.1	0.9
14	1.6	35.5	1.6	0.0	33.9	0.1	1.0
15	4.4	38.1	4.4	0.0	33.7	0.1	0.9
16	1.6	39.9	1.6	0.0	38.3	0.0	1.0
17	1.4	42.5	1.4	0.0	41.1	0.0	1.0
18	1.7	43.6	1.7	0.0	41.9	0.0	1.0
19	9.8	44.8	9.8	0.0	35.0	0.2	0.8
20	6.7	47.6	6.7	0.0	40.9	0.1	0.9
21	7.9	47.4	7.9	0.0	39.5	0.2	0.8
22	2.3	48.0	2.3	0.0	45.7	0.1	1.0
23	8.4	47.8	8.4	0.0	39.4	0.2	0.8
24	7.7	46.3	7.7	0.0	38.6	0.2	0.8
25	11.4	47.9	11.4	0.0	36.5	0.2	0.8
26	3.9	47.1	3.9	0.0	43.2	0.1	0.9
27	23.7	43.1	23.7	0.0	19.4	0.6	0.5
28	36.7	36.5	36.5	0.0	0.0	10.0	0.0
29	45.7	35.1	35.1	0.0	0.0	10.0	0.0
30	46.2	32.0	32.0	0.0	0.0	10.0	0.0
31	22.9	36.0	26.0	0.0	10.0	0.7	0.3
32	21.6	34.3	24.2	0.0	10.1	0.7	0.3
33	18.1	34.3	21.0	0.0	13.3	0.6	0.4
34	24.7	29.7	25.5	0.0	4.2	0.9	0.1
35	12.5	27.5	14.7	0.0	12.8	0.5	0.5
36	9.9	30.5	12.4	0.0	18.1	0.4	0.6
37	5.4	32.4	8.0	0.0	24.4	0.3	0.8

contd...

Appendix E3 – *contd...*

Week (mm)	PPT (mm)	ET_0 (mm)	AET (mm)	SPL (mm)	DEF (mm)	I_{ma}	I_a
38	5.9	29.7	7.7	0.0	22.0	0.3	0.7
39	9.7	27.8	10.8	0.0	17.0	0.4	0.6
40	1.1	25.3	2.3	0.0	23.0	0.1	0.9
41	1.5	24.8	2.4	0.0	22.4	0.1	0.9
42	1.1	26.8	1.9	0.0	24.9	0.1	0.9
43	0.0	25.0	0.6	0.0	24.4	0.0	1.0
44	3.7	21.7	4.0	0.0	17.7	0.2	0.8
45	0.3	21.9	0.6	0.0	21.3	0.0	1.0
46	0.3	18.6	0.5	0.0	18.1	0.0	1.0
47	0.9	17.9	1.1	0.0	16.8	0.1	0.9
48	0.0	17.9	0.2	0.0	17.7	0.0	1.0
49	0.3	14.7	0.4	0.0	14.3	0.0	1.0
50	0.0	13.6	0.1	0.0	13.5	0.0	1.0
51	0.5	14.9	0.6	0.0	14.3	0.0	1.0
52	0.8	13.4	0.9	0.0	12.5	0.1	0.9
ANN	393.0	1534.5	392.5	0.0	1142.0		

Appendix E4 : Normal weekly water balance at Pali (1981-1995)

(A WC=90 mm)

Week (mm)	PPT (mm)	ET_0 (mm)	AET (mm)	SPL (mm)	DEF (mm)	I_{ma}	I_a
1	1.3	17.9	1.3	0.0	16.6	0.1	0.9
2	1.7	18.2	1.7	0.0	16.5	0.1	0.9
3	0.6	19.0	0.6	0.0	18.4	0.0	1.0
4	0.9	21.2	0.9	0.0	20.3	0.0	1.0
5	0.4	21.8	0.4	0.0	21.4	0.0	1.0
6	0.8	22.3	0.3	0.0	22.0	0.0	1.0
7	0.1	23.2	0.1	0.0	23.1	0.0	1.0
8	1.5	25.5	1.5	0.0	24.0	0.1	0.9
9	1.5	26.7	1.5	0.0	25.2	0.1	0.9
10	0.2	29.2	0.2	0.0	29.0	0.0	1.0
11	0.0	32.5	0.0	0.0	32.5	0.0	1.0
12	0.3	35.1	0.3	0.0	34.8	0.0	1.0
13	0.2	35.6	0.2	0.0	35.4	0.0	1.0
14	1.0	39.1	1.0	0.0	38.1	0.0	1.0
15	2.4	38.3	2.4	0.0	35.9	0.1	0.9
16	1.5	39.8	1.5	0.0	38.3	0.0	1.0

contd...

Appendix E4 – *contd...*

Week (mm)	PPT (mm)	ET_0 (mm)	AET (mm)	SPL (mm)	DEF (mm)	I_{ma}	I_a
17	1.3	41.2	1.3	0.0	39.9	0.0	1.0
18	0.2	41.5	0.2	0.0	41.3	0.0	1.0
19	2.5	42.7	2.5	0.0	40.2	0.1	0.9
20	0.9	44.1	0.9	0.0	43.2	0.0	1.0
21	1.3	44.0	1.3	0.0	42.7	0.0	1.0
22	8.1	43.2	8.1	0.0	35.1	0.2	0.8
23	1.9	43.6	1.9	0.0	41.7	0.0	1.0
24	12.6	41.8	12.6	0.0	29.2	0.3	0.7
25	6.9	44.7	6.9	0.0	37.8	0.2	0.9
26	12.5	45.3	12.5	0.0	32.8	0.3	0.7
27	50.2	40.5	40.5	0.0	0.0	10.0	0.0
28	22.7	41.7	24.6	0.0	17.1	0.6	0.4
29	27.6	39.7	28.6	0.0	11.1	0.7	0.3
30	55.5	41.4	41.4	0.0	0.0	10.0	0.0
31	33.2	39.2	34.6	0.0	4.6	0.9	0.1
32	26.8	38.2	29.1	0.0	9.1	0.8	0.2
33	29.3	35.3	30.4	0.0	4.9	0.9	0.1
34	21.3	33.9	23.4	0.0	10.5	0.7	0.3
35	16.3	29.6	18.2	0.0	11.4	0.6	0.4
36	13.8	28.6	15.6	0.0	13.0	0.6	0.5
37	6.4	26.8	8.5	0.0	18.3	0.3	0.7
38	1.2	26.7	3.2	0.0	23.5	0.1	0.9
39	16.8	27.0	17.5	0.0	9.5	0.7	0.4
40	1.9	26.3	3.2	0.0	23.1	0.1	0.9
41	1.0	28.1	2.1	0.0	26.0	0.1	0.9
42	0.2	27.3	1.0	0.0	26.3	0.0	1.0
43	0.7	25.5	1.3	0.0	24.2	0.1	1.0
44	0.0	24.5	0.4	0.0	24.1	0.0	1.0
45	0.8	22.6	1.1	0.0	21.5	0.1	1.0
46	0.0	21.7	0.2	0.0	21.5	0.0	1.0
47	0.0	20.9	0.2	0.0	20.7	0.0	1.0
48	0.0	20.8	0.1	0.0	20.7	0.0	1.0
49	0.0	20.2	0.1	0.0	20.1	0.0	1.0
50	1.4	18.9	1.5	0.0	17.4	0.1	0.9
51	0.5	18.5	0.6	0.0	17.9	0.0	1.0
52	0.1	18.3	0.2	0.0	18.1	0.0	1.0
ANN	389.0	8 1619.7	389.6	0.0	1230.1		

Appendix E5 : Normal weekly water balance at Jodhpur (1981-1995)

(AWC=130mm)

Week (mm)	PPT (mm)	ET_0 (mm)	AET (mm)	SPL (mm)	DEF (mm)	I_{ma}	I_a
1	1.6	13.7	1.6	0.0	12.1	0.1	0.9
2	4.5	13.8	4.5	0.0	9.3	0.3	0.7
3	0.4	13.8	0.4	0.0	13.4	0.0	1.0
4	0.8	17.1	0.8	0.0	16.3	0.1	1.0
5	1.9	18.1	1.9	0.0	16.2	0.1	0.9
6	0.3	18.8	0.3	0.0	18.5	0.0	1.0
7	2.3	19.9	2.3	0.0	17.6	0.1	0.9
8	3.1	22.7	3.1	0.0	19.6	0.1	0.9
9	3.9	24.6	3.9	0.0	20.7	0.2	0.8
10	0.1	26.5	0.1	0.0	26.4	0.0	1.0
11	0.1	27.7	0.1	0.0	27.6	0.0	1.0
12	2.9	30.0	2.9	0.0	27.1	0.1	0.9
13	1.4	30.6	1.4	0.0	29.2	0.1	1.0
14	2.2	33.4	2.2	0.0	31.2	0.1	0.9
15	6.3	36.7	6.3	0.0	30.4	0.2	0.8
16	2.2	38.4	2.2	0.0	36.2	0.1	0.9
17	1.8	41.5	1.8	0.0	39.7	0.0	1.0
18	0.5	44.7	0.5	0.0	44.2	0.0	1.0
19	2.8	47.4	2.8	0.0	44.6	0.1	0.9
20	1.9	51.8	1.9	0.0	49.9	0.0	1.0
21	7.5	49.3	7.5	0.0	41.8	0.2	0.9
22	6.7	54.6	6.7	0.0	47.9	0.1	0.9
23	3.9	52.8	3.9	0.0	48.9	0.1	0.9
24	7.3	47.3	7.3	0.0	40.0	0.2	0.9
25	17.3	43.5	17.3	0.0	26.2	0.4	0.6
26	22.3	44.2	22.3	0.0	21.9	0.5	0.5
27	15.3	43.9	15.3	0.0	28.6	0.4	0.7
28	39.2	40.2	39.2	0.0	1.0	1.0	0.0
29	95.2	35.4	35.4	0.0	0.0	10.0	0.0
30	86.7	35.2	35.2	0.0	0.0	10.0	0.0
31	38.7	34.2	34.2	0.0	0.0	10.0	0.0
32	55.2	32.9	32.9	8.1	0.0	10.0	0.0
33	38.7	38.1	38.1	0.6	0.0	10.0	0.0
34	102.2	30.1	30.1	72.1	0.0	10.0	0.0
35	42.5	31.1	31.1	11.4	0.0	10.0	0.0
36	29.6	28.9	28.9	0.7	0.0	10.0	0.0
37	6.2	28.7	26.9	0.0	1.8	0.9	0.1

contd...

Appendix E5 – *contd...*

Week (mm)	PPT (mm)	ET_0 (mm)	AET (mm)	SPL (mm)	DEF (mm)	I_{ma}	I_a
38	3.3	29.1	23.0	0.0	6.1	0.8	0.2
39	3.3	26.0	17.7	0.0	8.3	0.7	0.3
40	0.9	27.5	14.9	0.0	12.6	0.5	0.5
41	5.6	26.5	14.7	0.0	11.8	0.6	0.4
42	0.0	25.7	9.4	0.0	16.3	0.4	0.6
43	0.3	25.5	7.9	0.0	17.6	0.3	0.7
44	6.9	22.7	10.9	0.0	11.8	0.5	0.5
45	1.1	21.8	5.7	0.0	16.1	0.3	0.7
46	0.1	20.3	3.9	0.0	16.4	0.2	0.8
47	0.2	17.4	3.0	0.0	14.4	0.2	0.8
48	0.0	17.2	2.5	0.0	14.7	0.1	0.9
49	0.0	18.4	2.3	0.0	16.1	0.1	0.9
50	1.1	15.8	2.7	0.0	13.1	0.2	0.8
51	0.0	14.7	1.5	0.0	13.2	0.1	0.9
52	1.8	15.7	3.0	0.0	12.7	0.2	0.8
ANN	680.1	1565.9	576.4	92.9	989.5		

Appendix E6 : Normal water balance at Udaipur (1981-1995)

(A WC = 150 mm)

Week (mm)	PPT (mm)	ET_0 (mm)	AET (mm)	SPL (mm)	DEF (mm)	I_{ma}	I_a
1	0.1	13.9	0.1	0.0	13.8	0.0	1.0
2	4.0	14.1	4.0	0.0	10.1	0.3	0.7
3	0.2	14.4	0.2	0.0	14.3	0.0	1.0
4	0.4	16.1	0.4	0.0	15.7	0.0	1.0
5	1.1	16.5	1.1	0.0	15.4	0.1	0.9
6	0.0	17.3	0.0	0.0	17.3	0.0	1.0
7	0.1	18.8	0.1	0.0	18.7	0.0	1.0
8	0.5	20.1	0.5	0.0	19.6	0.0	1.0
9	0.8	21.3	0.8	0.0	20.5	0.0	1.0
10	0.1	23.5	0.1	0.0	23.4	0.0	1.0
11	0.2	25.9	0.2	0.0	25.7	0.0	1.0
12	0.4	28.9	0.4	0.0	28.5	0.0	1.0
13	0.0	29.9	0.0	0.0	29.9	0.0	1.0
14	1.2	31.9	1.2	0.0	30.7	0.0	1.0
15	0.3	33.0	0.3	0.0	32.7	0.0	1.0
16	1.1	35.5	1.1	0.0	34.4	0.0	1.0

contd...

Appendix E6 – *contd...*

Week (mm)	PPT (mm)	ET_0 (mm)	AET (mm)	SPL (mm)	DEF (mm)	I_{ma}	I_a
17	1.7	36.4	1.7	0.0	34.7	0.1	1.0
18	0.5	36.0	0.5	0.0	35.5	0.0	1.0
19	6.4	37.5	6.4	0.0	31.1	0.2	0.8
20	0.3	38.3	0.3	0.0	38.0	0.0	1.0
21	1.1	37.1	1.1	0.0	36.0	0.0	1.0
22	5.4	37.4	5.4	0.0	32.0	0.1	0.9
23	5.5	37.3	5.5	0.0	31.8	0.2	0.9
24	13.5	36.9	13.5	0.0	23.4	0.4	0.6
25	14.1	36.2	14.1	0.0	22.1	0.4	0.6
26	23.2	37.5	23.2	0.0	14.3	0.6	0.4
27	33.5	35.5	33.5	0.0	2.0	0.9	0.1
28	40.3	37.6	37.6	0.0	0.0	10.0	0.0
29	49.4	35.5	35.5	0.0	0.0	10.0	0.0
30	57.0	35.7	35.7	0.0	0.0	10.0	0.0
31	41.9	36.8	36.8	0.0	0.0	10.0	0.0
32	57.8	35.7	35.7	0.0	0.0	10.0	0.0
33	48.3	34.1	34.1	0.0	0.0	10.0	0.0
34	48.8	32.6	32.6	0.0	0.0	10.0	0.0
35	28.4	29.5	29.1	0.0	0.4	1.0	0.0
36	26.6	27.5	27.2	0.0	0.3	1.0	0.0
37	7.4	24.6	17.6	0.0	7.0	0.7	0.3
38	14.2	24.0	19.5	0.0	4.5	0.8	0.2
39	16.1	23.7	20.0	0.0	3.7	0.8	0.2
40	9.4	21.6	15.3	0.0	6.4	0.7	0.3
41	3.0	22.7	11.5	0.0	11.2	0.5	0.5
42	0.3	21.5	8.3	0.0	13.2	0.4	0.6
43	1.4	20.1	7.6	0.0	12.6	0.4	0.6
44	3.6	20.0	8.4	0.0	11.6	0.4	0.6
45	8.1	19.3	11.1	0.0	8.2	0.6	0.4
46	0.0	17.8	4.3	0.0	13.5	0.2	0.8
47	0.5	17.0	4.1	0.0	12.9	0.2	0.8
48	0.0	16.3	3.2	0.0	13.2	0.2	0.8
49	0.0	15.4	2.7	0.0	12.7	0.2	0.8
50	2.6	14.5	4.5	0.0	10.0	0.3	0.7
51	0.3	14.2	2.3	0.0	11.9	0.2	0.8
52	0.4	13.9	2.2	0.0	11.7	0.2	0.8
ANN	581.5	1379.0	562.4	0.0	816.6		

Appendix E7 : Normal weekly water balance at Banswara (1981–1995)

(AWC = 270 mm)

Week (mm)	PPT (mm)	ET_0 (mm)	AET (mm)	SPL (mm)	DEF (mm)	I_{ma}	I_a
1	0.1	17.2	0.1	0.0	17.1	0.0	1.0
2	8.6	17.1	8.6	0.0	8.5	0.5	0.5
3	0.3	17.1	0.3	0.0	16.8	0.0	1.0
4	0.6	18.8	0.6	0.0	18.2	0.0	1.0
5	0.3	18.1	0.3	0.0	17.8	0.0	1.0
6	0.0	18.7	0.0	0.0	18.7	0.0	1.0
7	0.0	19.3	0.0	0.0	19.3	0.0	1.0
8	0.0	19.8	0.0	0.0	19.8	0.0	1.0
9	0.1	20.3	0.1	0.0	20.2	0.0	1.0
10	0.0	21.0	0.0	0.0	21.0	0.0	1.0
11	0.1	22.8	0.1	0.0	22.7	0.0	1.0
12	0.4	25.2	0.4	0.0	24.8	0.0	1.0
13	0.2	26.1	0.2	0.0	25.9	0.0	1.0
14	0.0	28.9	0.0	0.0	28.9	0.0	1.0
15	0.0	28.8	0.0	0.0	28.8	0.0	1.0
16	0.7	30.6	0.7	0.0	29.9	0.0	1.0
17	2.1	34.2	2.1	0.0	32.1	0.1	0.9
18	0.5	35.1	0.5	0.0	34.6	0.0	1.0
19	5.0	37.3	5.0	0.0	32.3	0.1	0.9
20	0.0	40.6	0.0	0.0	40.6	0.0	1.0
21	0.0	41.6	0.0	0.0	41.6	0.0	1.0
22	1.6	44.2	1.6	0.0	42.6	0.0	1.0
23	1.4	47.1	1.4	0.0	45.7	0.0	1.0
24	25.2	39.9	25.2	0.0	14.7	0.6	0.4
25	26.4	36.9	26.4	0.0	10.5	0.7	0.3
26	37.3	34.4	34.4	0.0	1.0	0.0	0.0
27	49.4	34.1	34.1	0.0	1.0	0.0	0.0
28	55.7	30.1	30.1	0.0	1.0	0.0	0.0
29	53.9	26.9	26.9	0.0	1.0	0.0	0.0
30	106.1	24.4	24.4	0.0	1.0	0.0	0.0
31	139.1	21.1	21.1	0.5	1.0	0.0	0.0
32	50.4	19.1	19.1	31.3	1.0	0.0	0.0
33	88.7	19.1	19.1	69.6	1.0	0.0	0.0
34	56.7	19.9	19.9	36.8	1.0	0.0	0.0
35	39.6	22.4	22.4	17.2	1.0	0.0	0.0
36	42.8	27.3	27.3	15.5	1.0	0.0	0.0
37	17.8	25.9	25.8	0.0	1.0	0.0	0.0

contd...

Appendix E7 – *contd...*

Week (mm)	PPT (mm)	ET_0 (mm)	AET (mm)	SPL (mm)	DEF (mm)	I_{ma}	I_a
38	19.3	25.5	25.2	0.0	0.3	1.0	0.0
39	26.7	22.4	22.4	0.0	1.0	0.0	0.0
40	5.1	21.8	20.7	0.0	1.1	1.0	0.1
41	20.9	22.8	22.6	0.0	0.2	1.0	0.0
42	6.5	22.0	20.1	0.0	1.9	0.9	0.1
43	0.5	22.6	18.5	0.0	4.1	0.8	0.2
44	6.3	22.1	18.3	0.0	3.8	0.8	0.2
45	7.3	22.7	18.4	0.0	4.3	0.8	0.2
46	1.1	22.4	15.4	0.0	7.0	0.7	0.3
47	7.2	22.1	16.5	0.0	5.6	0.8	0.3
48	0.0	22.0	12.9	0.0	9.1	0.6	0.4
49	0.0	21.8	11.8	0.0	10.0	0.5	0.5
50	0.6	20.1	10.4	0.0	9.7	0.5	0.5
51	0.0	18.5	8.6	0.0	9.9	0.5	0.5
52	2.5	17.6	9.1	0.0	8.5	0.5	0.5
ANN	915.1	1337.8	629.2	170.9	708.6		

Appendix E8 : Normal weekly water balance at Kota (1981–1995)

(AWC = 200 mm)

Week (mm)	PPT (mm)	ET_0 (mm)	AET (mm)	SPL (mm)	DEF (mm)	I_{ma}	I_a
1	0.4	17.0	0.4	0.0	16.6	0.0	1.0
2	4.2	17.3	4.2	0.0	13.1	0.2	0.8
3	0.4	17.3	0.4	0.0	16.9	0.0	1.0
4	2.5	18.9	2.5	0.0	16.4	0.1	0.9
5	1.0	19.5	1.0	0.0	18.5	0.1	1.0
6	4.2	20.3	4.2	0.0	16.1	0.2	0.8
7	1.0.	21.8	1.0	0.0	20.8	0.1	1.0
8	0.1	23.1	0.1	0.0	23.0	0.0	1.0
9	1.3	24.3	1.3	0.0	23.0	0.1	1.0
10	0.5	26.3	0.5	0.0	25.8	0.0	1.0
11	1.0	29.0	1.0	0.0	28.0	0.0	1.0
12	0.4	31.6	0.4	0.0	31.3	0.0	1.0
13	1.4	33.3	1.4	0.0	31.9	0.0	1.0
14	0.7	35.0	0.7	0.0	34.3	0.0	1.0
15	1.6	37.6	1.6	0.0	36.0	0.0	1.0
16	0.8	39.3	0.8	0.0	38.5	0.0	1.0

contd...

Appendix E8 – *contd...*

Week (mm)	PPT (mm)	ET_0 (mm)	AET (mm)	SPL (mm)	DEF (mm)	I_{ma}	I_a
17	4.6	40.6	4.6	0.0	36.0	0.1	0.9
18	0.5	39.1	0.5	0.0	38.6	0.0	1.0
19	1.0	40.8	1.0	0.0	39.8	0.0	1.0
20	0.1	42.1	0.1	0.0	42.0	0.0	1.0
21	3.8	40.2	3.8	0.0	36.4	0.1	0.9
22	1.7	40.2	1.7	0.0	38.5	0.0	1.0
23	4.5	40.5	4.5	0.0	36.0	0.1	0.9
24	7.8	41.5	7.8	0.0	33.7	0.2	0.8
25	13.1	40.1	13.1	0.0	27.0	0.3	0.7
26	27.7	40.2	27.7	0.0	12.5	0.7	0.3
27	28.9	40.0	28.9	0.0	11.1	0.7	0.3
28	36.6	40.2	36.6	0.0	3.6	0.9	0.1
29	50.8	40.0	40.0	0.0	1.0	0.0	0.0
30	82.1	38.4	38.4	0.0	1.0	0.0	0.0
31	58.1	39.8	39.8	0.0	1.0	0.0	0.0
32	54.1	39.1	39.1	0.0	1.0	0.0	0.0
33	65.5	36.9	36.9	0.0	1.0	0.0	0.0
34	56.9	35.3	35.3	0.0	1.0	0.0	0.0
35	44.6	32.5	32.5	0.0	1.0	0.0	0.0
36	46.9	31.1	31.1	0.0	1.0	0.0	0.0
37	8.1	26.9	23.0	0.0	3.9	0.9	0.2
38	12.5	26.9	23.0	0.0	3.9	0.9	0.2
39	8.8	26.6	20.8	0.0	5.8	0.8	0.2
40	5.6	24.6	17.3	0.0	7.4	0.7	0.3
41	13.1	25.5	20.1	0.0	5.4	0.8	0.2
42	3.5	24.5	14.5	0.0	10.1	0.6	0.4
43	0.4	23.1	11.0	0.0	12.1	0.5	0.5
44	6.8	23.0	13.7	0.0	9.3	0.6	0.4
45	7.5	22.3	13.3	0.0	9.0	0.6	0.4
46	0.1	20.8	7.5	0.0	13.2	0.4	0.6
47	0.0	19.9	6.5	0.0	13.4	0.3	0.7
48	0.0	19.3	5.7	0.0	13.6	0.3	0.7
49	0.5	18.4	5.3	0.0	13.1	0.3	0.7
50	4.1	17.5	7.4	0.0	10.1	0.4	0.6
51	0.4	17.2	4.3	0.0	13.0	0.3	0.8
52	2.2	16.9	5.3	0.0	11.6	0.3	0.7
ANN	684.4	1543.7	643.5	0.0	900.2		

Appendix F : Adequacy of length of record tor frequency analysis of water deticit

Standard week	Length of record used	Recommended length of record (Mocus, 1960) Jodhpur	Sriganganagar	Udaipur	Fatehpur	Pali	Jaipur	Banswara	Kota
1	15	9	11	9	-	11	24	10	7
2	15	12	15	9	37	17	8	9	20
3	15	9	18	9	12	12	11	27	19
4	15	8	16	17	12	27	14	20	11
5	15	10	10	14	28	16	15	16	9
6	15	8	13	13	11	20	15	8	12
7	15	10	15	9	-	7	9	8	8
8	15	8	9	8	13	8	10	9	8
9	15	12	14	8	14	12	12	6	14
10	15	8	12	9	-	9	14	11	7
11	15	10	10	11	7	8	2	8	12
12	15	8	15	8	20	8	7	6	7
13	15	9	14	11	25	8	18	26	11
14	15	8	13	10	8	9	9	24	9
15	15	7	12	9	11	8	8	6	13
16	15	9	10	8	8	17	14	37	10
17	15	10	9	7	20	7	8	31	9
18	15	20	12	8	10	18	14	31	10
19	15	14	9	9	-	31	10	9	16
20	15	10	11	11	-	21	8	15	8
21	15	8	12	10	-	14	18	8	17
22	15	15	13	9	-	7	26	8	10
23	15	13	9	8	13	20	29	8	8
24	15	13	9	14	-	36	8	38	10
25	15	11	9	12	20	15	29	25	7
26	15	8	9	11	13	-16	14	41	8
27	15	10	11	41	18	7	16	17	55
28	15	8	15	90	-	35	-	-	50
29	15	56	41	-	-	-	-	-	-
30	15	36	14	-	36	-	-	-	-
31	15	60	71	-	40	-	-	-	-
32	15	14	15	-	26	-	-	-	-
33	15	44	-	-	38	-	-	-	-
34	15	41	10	-	-	-	-	-	-
35	15	65	39	15	47	-	-	-	-
36	15	15	12	11	-	-	-	-	-

contd...

Appendix F – *contd...*

Standard week	Length of record used	Recommended length of record (Mocus, 1960) Jodhpur	Sriganganagar	Udaipur	Fatehpur	Pali	Jaipur	Banswara	Kota
37	15	15	14	26	42	-	-	-	64
38	15	36	14	15	-	-	57	-	47
39	15	51	12	-	-	19	-	-	56
40	15	15	14	13	21	32	45	83	15
41	15	21	11	15	7	21	52	75	15
42	15	13	12	12	7	12	43	73	31
43	15	13	14	13	-	32	26	65	19
44	15	15	8	13	-	25	38	55	31
45	15	11	10	13	-	16	13	46	22
46	15	12	10	14	-	23	16	31	21
47	15	8	13	13	23	13	10	35	12
48	15	11	74	14	13	28	19	23	31
49	15	17	9	10	13	7	11	15	12
50	15	10	12	14	11	13	31	16	21
51	15	12	13	9	47	13	19	16	14
52	15	8	13	11	-	21	11	15	9

Appendix G : Computer programme for stochastic modelling of water deficit

```
# include <stdio.h>
# include <math.h>
# include <conio.h>
# include <stdlib.h>
# include <ctype.h>
# include <alloc.h>

FILE            *f1 ,*f2,*f4,*f5;
char      fin[20],fout[20],fout1[20],blank[60],
title[ 50],fout2[20],t'out3 [20],yy;
          int    nset,is=O;
void main(void);
void get(void);
void get2(void);
void get3(int, double, double*,double*,double
          [5],float[10]); .
void line(FILE*, int);
void main()
{
          clrscr();
          printf("\nTo calculate\n");
          flushall();
          printf(" Input File? ");
scanf("%s",&fin);
          f1 = fopen(fin,"r");
          printf(" Input File for Phi? ");
scanf("%s",&fin);
          f5 = fopen(fin,"r");
          printf(" Output File of ressults ? ");
scanf("%s",&fout);
          f2 = fopen(fout,"w");
          printf(" Output File for Residual series? ");
scanf("%s" ,&fout1);
          f4 = fopen(foutl,"w");
          printf("\nDo you want Validation of AR
          model y/n ? ");
          yy = toupper(getch());
          printf("\nNo of sets? ");
          scanf("%d",&nset);
          while ( ++is <= nset) {
               fscanf(f1;"\n");
               fgets(title,50,f1 );
               puts(title);
               fprintf(f2,"%d. %s",is,title);
```

```
            fprintf(f6,"%d. %s",is,title);
            get2();
        }
        fscanf(f1 ,"\n");
        fgets(title,50,f1 );
        puts(title );
        fcloseall;
}
void get2()

{
        int i,j,k,yrs,weeks,number,legs;
        float sx,**x,*week_m;
        double cv,mean,sd;
        fscanf(f1,"%d%d",&yrs;&weeks);
        printf("Number of legs? ");
        scanf("%d",&legs);
        number = weeks*yrs;
        x = alloc_f2(weeks+2, yrs+2);
        week_m = new float[weeks+2];
        fprintf(t2,"\nTable 4. Turnin point test\n")
        for(i=1; i<=yrs; i++)
            fscanf(f1 ,"%f',&sx);
        for(i=0; i<weeks; i++){
            for(sx=j=0; j<yrs; j++){
                float sx 1 ;
                fscanf(f1 ,"%f',&sx l ):
                x[i][j] = sx1; .
                sx+=x[i][j];
                }
            x[i][j] = sx;
            week_m[i] = sx/yrs;
            }
        for(i=O; i<=yrs; i++){
            for(sx=j=O; j<weeks; j++)
                sx+=x[j][i];
            x[j][i] = sx;
            }
// Turning point test

        mean = 2*(yrs-2)/3.0;
        sd = (16*yrs-29)/90.0;
        int point=O;
        for(i=1; i<yrs-1; i++)
            if(x[weeks][i-1]>=x[weeks][il &, &
        x[weeks][i]>=x[weeks][i+1] II x[weeks][i-
        1 ]<=x[ weeks][i] &&
```

```
x[weeks][i]<=x[weeks][i+ 1])
          ;
     else
          point++;
fprintf(f2,"\nNumber of points = %d",point)
if(sd>0) {
     mean = (point-mean)/sqrt(sd);
     fprintf(t2,"Z of turning point test
%.3f",mean);
     if(mean>2.576)
          fprintf(f2,"** ");
     else if(mean> 1.96)
          fprintf(f2,"*");
     fprintf(f2,"\n");
     }
else
     fprintf(f2,"Variance zero, Z not
possible for turning point test\n");

// Kendals rank correlation test

mean = (4*yrs/yrs/(yrs-1)-1;
sd = 2*(2*yrs+5)/9.0/yrs/(yrs-1);
if(sd>0){
     mean = mean/sqrt(sd);
     fprintf(f2,"Z of Kendals test =
%.3f',mean);
     if(mean>2.576)
          fprintf(f2,"**");
     else if(mean>1.96)
          fprintf(f2,"*");
     fprintf(f2,"\n");
     }
else
     fprintf(f2,"Variance zero, Z not
possible Kendals test\n");

// Serial correlation coefficient

float *y;
y = new float [number+2];
fprintf(f2,"Serial correlation coefficients\n");
line(f2,50);
fprintf( f2,"Leg\tCorrelation\tCovariance\n");
line(f2,50);
int l;
for(l=0,k=0; k<yrs; k++)
```

```
		for(j=0; j<weeks; j++ ;l++)
			y[l] = x[j][k];
	for(i=l; i<=legs; i++){
		double mx,my,ssx,ssy,scp,f;
		mx=my=ssx=ssy=scp=f=0;
		for(l=0; l<number-i; l++){
			mx += y[l];
			my += y[l+i];
			ssx += y[l]*y[l];
			ssy += y[l+i]*y[l+i];
			scp += y[l]*y[l+i];
			f++
			}
		scp = scp-mx*my/f;
		ssx = (ssx-mx*mx/f)*{ssy-my*my/f);
		if (ssx>0) {
			ssx = scp/sqrt(ssx);
		fprintf(f2,"%3d\t%.3 f\t%.3 f\n ",i,ssx,s
	cp);
			}
		else{
			fprintf{f2,"%3d\tNot
	possible\t%.3f\n",i,scp);
			}
		}
	line(f2,50);
	double vars,a,b;
	for(a=b=0,i=0; i<number; i++){
			a += y[i];
			b += y[i]*y[i];
			}
	vars = (b-a*a/number)/(number-1);

// Periodic components
	fprintf(f2,"Periodic componebts\n");
	line(f2,50);
	fprintf(f2,"Order\tAk\tBk\tAmptd\tThat
	var\tCum Var\n");
	line(f2,50);
	double prevce = 100;
	float term;
	double *ak,*bk;
	ak = new double[weeks/2+2];
	bk = new double[weeks/2+2];
	printf("\nCumulative var. terminating point
	%% increase?");
	scanf("%f',&term);
```

```
          mean = x[weeks][yrs]/yrs/weeks;
          point = weeks/2;
          for(sd=l=0,i=1; i<=point; i++){
               double a,b,c,d;
               for(a=0,j=0; j<number; j++)
                    a +=
          y[j]*cos(2*22/7 .0*i*(j + l)/weeks);
               a = 2*a/number;
               fprintf(f2,"%d\t%.3f",i,a);
               for(b=0,j=0; j<number; j++)
                    b+=
          y[j]*sin(2*22/7.0*i*(j+ 1)/weeks);
               b = 2*b/number;
               fprintf(f2,"\t%.3f",b);
               c = sqrt(a*a+b*b);
               fprintf(f2,"\t%.3f",c);
               d = atan(a/b);
               fprintf(f2, "\t%.3f",d);
               c = 100*c/vars;
               fprintf(f2, "\t%.3f",c);
               sd += c;
               fprintf(f2,"\t%.3 f\n",sd);
               if(fabs(prev-sd)>term) {
                    prev=sd;
                    l++;
                    ak[l]=a;
                    bk[l]=b;
                    }
               }
          line (f2,50);
//        fprintf(f3,"%d\t%.3f\t ",l,mena);
          point = l;
// Fourier decomposition of Periodic components of
          Mean series
          fprintf(f2, "Fourier decomposition of Periodic
          components of mean series \n");
          line(f2,50);
          fprintf(t2,"Order\tA lpha\tBeta\tAm p l itude\tEx
          p var\tCum Var\n");
          line(f2,50);
          double *msdj, var;
          msdj = new double[weeks/2+2];
          for(var=0,j=0; j<weeks; j++){
               double temp;
               temp =x[j][yrs]/yrs-mean;
               if(temp !=0) .
```

```
                var += temp*temp;
          }
        var = var/(weeks-1);
        double *alpha, *beta;
        alpha = new double[weeks/2];
        beta = new double[weeks/2];
        for(sd=l=0,i=1; i<=weeks/2; i++){
             double a,b,c,d;
             for(a=0, j=0; j<weeks; j++)
                  a +=
        x[j] [yrs]/yrs* cos(2*22/7.0*i*(j+ l)/weeks);
             alpha[i] = a = 2*a/weeks;
             fprintf(f2,"%d\t%.3f',i,a);
             for(b=0,j=0; j<weeks; j++)
                  b +=
        x[j][yrs]/yrs*sin(2*22/7 .0*i*(j+ l )/weeks);
             beta[i] = b = 2*b/weeks;
             fprintf(f2,"\t%.3f',b );
             c = sqrt(a*a+b*b);
             fprintf(f2,"\t%.3f",c" );
             msdj[i] = d = (a*a+b*b)/2.0;
             d = 100*d/var.
             fprintf(f2,"\t%.3f',d);
             sd+=d;
             fprintf(f2,"\t%.3 f\n ",sd);
             }
        line(f2,50);
        var = var*(weeks-1 )/weeks;
// Cumulative periodogramme test
        fprintf(f2,"Cumulative periodogramme
        test\n");
        line(f2,50);
        fprintf(f2, "Order\tMSDU\tMSDJ\tCum
        MSDJ\tPi\n");
        line(f2,50);
        for(i=l; i<=l; i++){
             double a,b;
             for(a=0, j=1; j<=weeks/2; j++){
                  a += msdj[j];
                  b = a/var;

             fprintf(f2,"%d\t%.3 f\t%.3 f\t%.3 f\t%.
        3f\n", j, var,msdj[j],a,b);
                  }
             }
        line(f2,50);
        delete [ ] msdj;
```

```
// Residual series correlation
        float *sr,*y1,*pdc;
        sr = new float[legs+2];
        y1 = new float[number+2];
        pdc = new float[number+2];
        for(i=0; i<number; i++){
            double a;
            for(a=0,j=1; j<=point; j++)
                a+=
        ak[j]*cos(2*22/7.0*j*(i+ l )/weeks) +
        bk[j]*sin(2*22/7.0*j*(i+ l )/weeks);
            pdc[i] = mean+a;
            y1[i] = y[i]-pdc[i];
            }
        fprintf(f4,"Residual Series\n");
        line(f4,20);
        for(i=0; i<number; i++)
            fprintf(f4,"%.3f\t\n",y 1 [i]);
        line(f4,20);

        fprintf(f2,"Residual Serial correlation
        coefficients\n");
        line(f2,50);
        fprintf(f2,"Leg\tCorrelation\tCovariance n" ,
        line(f2,50);
        for(i=1; i<=legs; i++){
            double mx,my,ssx,ssy,scp,f;
            mx=my=ssx=ssy=scp=f=0;
            for(l=0; l<number-i; l++){
                mx+=y1[l];
                my += y1[l+i];
                ssx += y1[l]*yl[l];
                ssy+=y1[l+i]*y1[l-+il};
                scp += y1 [l]*y1 [l+i];
                f++;
                }
            scp = scp-mx*my/f;
            ssx = (ssx-mx*mx/f)*(ssy-my*nl\
            if(ssx>0) {
                sr[ i-1] = ssx = scp/sqrtl ~

            fprintf( f2,"%3d\t%.3f\t%.3f\n ", i
        cp);
                }
            else{
                fprintf(f2,"%3d\tNot
```

```
possible\t%.3f\n",i,scp );
        }
    }
for(i=1; i<=legs; i++){
    double ssX,scp;
    ssx=scp=0;
    for(l=0; l<number-i; 1++)
        scp += (y[l]-mean)*(y[l+i]-
mean);
    if(vars>0) {
        sr[i-l] = ssx = scp/vars;

    fprintf( f2, "%3d\t%.3 t\t6/o. 3 t\n", i,ssx,s
cp);
        }
    else{
        fprintf(f2,"%3d\tNot
possible\t%.3t\n'~,i,scp );
        }
    }
    line(f2,50);
// Selection of order of Armodeel
fprintf(f2,"Selectioin of order of AR
model\n");
line(f2,50);
fprintf(f2,"Leg\tSr\tAP Corr\tRes var\n");
line(f2,50);
int small,small 1;
double *phi;
phi = new double [legs+2];
if(vars>0)
    sd = sqrt(vars);
double c,temp= 1;
fscanf(f5,"\n%s",&title);
int xx=0;
for(i=1; i<=legs; i++){
    double a,b,aic[4];
fscanf(f5,"%1f',&phi[i-1 ]);
    temp = temp*(1-phi[i-1])*(1+phi[i-
l]);
    b = number-2*i-1;
    if(b>0)
        a = (number-i)*temp/b;
    fpri ntf(f2 ,"%3d\t%.3f\t%.3f\t%.3f\n",
i,sr[i-1 ],phi[i-1 ],a);
    if(i==1)
        aic[0]=a;
```

```
	else if (i==2)
		aic[1]=a;
	else if (i==3)
		aic[2]=a;
	else{
		aic[0]=aic[1];
		aic[1]=aic[2];
		aic[2]=a;
		}
	if(i>2 && aic[0]>=aic[1] &&
aic[ 1 ]<=aic[2] && xx<1){
		xx++;
		small = i-1;
		}
	}
fprintf{f2."Smallest residual variance at order
No = %3d\n",small);
line(f2,50);
printf("\nSmallest residual variance at at order
No = %d\n",small);
small 1 = small;
printf("\nUpto which order No <=3 ? ")
scanf("%d",&small);

fprintf(f2," AR parameter of selected
model\n");
line(f2,50);
fprintf(f2,"SN\tPhi\tV alue\n ");
line(t2,50);
double **phi1.phai[5];
phi1 = alloc_d2(small+ 1 ,small+ 1);
for (i=0;i<=small;i++)
	phi 1 [i][i] = phi[i];
if(small== 1)

	fprintf(f2,"%d\t%d, %d\t%.3f\n".l.
all,1,phi[0]);
if(small==2){
	phi 1 [1 ][0]=(sr[0]-sr[0]*sr[ 1 ])/( l.
sr[0]*sr[0]);

	fprintf(f2,"%d\t%d,%d\t%.3f\n".
all,1,phi1[1][0]);

	fprintf(f2,"%d\t%d, %d\t%.3 t\n"2
all,2,phi[1]);
	phai[0]=phi 1 [1 ][0];
```

```
		phai[1]=phi[1];
		}
	if(small==3){
		phi 1 [2][0]=½ l-sr[0]*sr[0])*(sr[0]
	sr[2])-( 1-sr[ 1 ])*(sr[0]*sr[ 1 ]-sr[2])/½ 1-
	sr[ 1 ])*( 1-2* sr[0]*sr[0]+sr[ 1 ]));
		phi 1 [2][ 1 ]=(sr[ 1 ]+sr[ 1 ]*sr[ 1]-
	sr[0]*sr[0]-sr[0]*sr[2])/(l-
	2*sr[0]*sr[0]+sr[ 1 ]);

		fprintf(f2,"%d\t%d, %d\t%.3f\n", 1
	all,1,phi1[2][0]);

		fprintf(f2,"%d\t%d, %d\t%.3f\n",2 '.
	all,2,phi 1 [2][1]);

		fprintf(f2,"%d\t%d, %d\t%.3f\n".
	all,3,phi[2]);
		phai[0]=phi 1 [2][0];
		phai[1 ]=phi l [2][1];
		phai[2]=phi[2];
		}
	float *y2,*y3,*st,*z[2];
	y2 = new float[number+2];
	y3 = new float[number+2];
	st = new float[number+2];
	z[0] = new float[number+2];
	z[1] = new float[number+5];
	for (i=0; i<2; i++)
		for (J=0; j<small;j++){
			printf("\nlnitial value of
	year %d value %d ? ",i,j);
			scanf("%f',&z[i]J]);
			}
	for (i=small-1 ;i<number;i++){
		double b;
		for(b=j=0; j<small; j++)
			b += yl[i-j]*phi1[small-
	1][small-1-j];
		st[i] = b;
		y2[i] = pdc[i] -f b;
		for(k=0; k<2; k++){
			for(b=j=0; j<small; j++)
				b += z[k][i-
	j]*phi 1 {small-1 ][small-1-j];
			z[k][i+1] = pdc[i+1] + b;
			}
		}
```

```
free_d2(phi l, small+ 1);
fprintf(t2,"Different serieses and residuals\n");
line(f2,70);
fprintf( f2," SN\tOBs\tPDC\tST\tGenerated\
tResidual\t 1 994\t 1 995\n ");
line(t2,70);
for (i=0;i<number;i++){
     if(i<small){

     fprintf(f2,"%d\t%.3f\t%.3f\t\t\t",i+ 1,y
[i],pdc[i]);
          for(k=0; k<2; k++)

     fprintf( f2,"\t%.3f',z[k] [i]);
          fprintf(f2,"\n");
          }
     else{
          y3[i] = y[i]-y2[i];

     fprintf(f2, "%d\t%.3f\t%.3 t\t%. 3f\t%.
3f\t%.3f",i+ 1 ,y[i],pdc[i],st[i],y2[i],y3[i]);
          for(k=O; k<2; k++)

     fprintf(f2,"\t%.3 f",z[k] [i]);
          fprintf(t2,"\n"};
          }
     }
delete [ ] z[0];
delete [ ] z[ 1];
line(f2,70);
double d,f;
a=b=c=d=f=0;
for(1=small; 1<number; 1++){
     a += y3[1];
     b += y3[1]*y3[1];
     c += y2[1];
     d += y2[1]*y2[1];
     f++;
     }
if(f'>1){
     b = (b-a*a/f)/(f-1);
     a = alf;
     d = (d-c*c/f)/(f-1.);
     c = c/f;
     }
fprintf(f2, "Mean \t%.3f\t\t\t%.3f\t%.3 f\n ".11
n,c,a);
```

```
        fprintf(f2,"Var\t%.3f\t\t\t%.3f\t%.3 t\n ", Var
        );
        if (d>0)
            d=sqrt(d);
        if(b>0)
            b=sqrt(b);
// Skewness and Kurtosis
        float sk;
        for(sk=j=0; j<number; j++)
            sk+=pow(y[j]-rnean,3);
            sk = number*sk/(number-
        1)/(number-2)/(sd*sd*sd);
        fprintf( f2, "Skew\t%.3f\t\t" ,sk);
        for(sk=j=small;<number;++ )
            sk+=pow(y2 [j]-c,3);
            sk = (nurnber-small)*:sk/number
        srnall)-1 )/(number-small)-2)/(d*d*d);
        fprintf(f2,"\t%.3f',sk);
        for(sk=j=small;<number;j++ )
            sk+=pow(y3[j]-a,3);
            sk = (number-small)*sk/number
        small)-1)/( (number-small)-2)/(b*: b* b);
        fprintf(f2,"\t%.3f\n",sk);
        for(sk=j=0j<number;j++ )
            if (sd>0)
                sk+=pow/y[j]-mean)sd ,l
            sk=
        number*(number+ 1 )*sk/(number-1 )/(number
        2)/(number-3)-(3*pow(number-l,2)/(number
        2)/(number-3));
        fprintf(f2,"Kurt\t%3f\t\t",sk);
        for(sk=j=smallj<number;j++)
            if (d>0)
                sk+=pow((y2[j]-c )/d,4):
            sk = (number-small)*number-
        small)+ 1 )*sk/((number-small)-1 )/((number
        small)-2)/( (number-small)- 3)-
        (3*pow((number-small)-1, 2)/((number-small
        2)/((number-small)-3));
        fprintf(f2,"\t%3f",sk);

        for(sk=j=small; j<number; j++)
            if (b>0)
                sk+=pow( (y3 [j]-a)/b,4);
        sk = (number-small)*number-
        small)+ 1 )*sk/number-small)-1 )/number-
        small)-2)/( (number-small)-3)-
```

```
(3*pow((number-small)-1,2)/number-small)-
2)/((number-small)-3));
fprintf(f2,."\t%.3f\n",sk);
line(f2,50);
delete [1] yl;
delete [1] st;
delete [1] pdc; ,
fprintf(f2,"Serial correlation coefficients\n");
line(f2,50);
fprintf(f2,"Leg\tResidual\tGenerated\n ");
line(f2,50);
double mean2, var2;
mean2=var2=0;
for(i=1; i<=legs; i++){
      double a,b,c,d,e,f,a1,b1,c1,d1,el;
      a=b=c=d=e=f=a1=b1=c1=d1=e1=0;
      for(1=small; 1<number-i; 1++){
            a += y3[1];
            b += y3[l+i];
            c += y3[1]*y3[1];
            d += y3[1+i]*y3[1+i];
            e += y3[1]*y3[1+i];
            al += y2[1];
            bl += y2[l+i];
            cl += y2[1]*y2[1];
            dl += y2[1+i]*y2[1+i];
            e l += y2[1]*y2[1+i];
            f++;
            }
      e = e-a*b/f
      a = (c-a*a/f)*(d-b*b/f);
      el = el-a1*bl/f;
      al = (c1-a1*a1/f)*(dl-bl*bl/f);
      if(a>0)
            a = e/sqrt(a);
      if(al>0)
            a1 = el/sqrt(a1);

      fprintf(f2,"%3d\t%.3f\t%.3f\n",i,a,a 1)
;
      mean2 += a;
      var2 += a*a.
      }
var2 = (var2-mean2*mean2/legs)/(legs-1);
mean2 = mean2/legs;
fprintf( f2,"Mean\t%.3f\n ",mean2);
fprintf(f2,"Var. \t%.3f\n", var2);
```

```
for(d=0,1=small; 1<number; 1++)
      d += (y[1]-mean)*(y[1]-mean);
b = (d-b*(f-1))/d;

fprintf{ f'2."SS of residual series\l%.31\n".b ):
for(a=b=0,1=small; 1<number; 1++){
      a += (y[1]-y2[1])*(y[1]-y2[1));
      b += y[1]*y[1];
      }
b = 100*a/b;
fprintf(f2."ISE\t%.3f\n",b );

line(f2,50);
delete [ ] sr;
delete [ ] phi;
delete [ ] y3;
delete [ ] y2;
fprintf(f2,"Mean weekly\n");
line(f2,50);
fprintf(f2,"SN\tMean\tPeriodic\tStocastic'll j,
nerated\n");
line(f2,50);
for(i=0; i<weeks; i++){
      double a;
      for(a=j=0; j<small1; j++)
            a+=
alpha[j+ 1 ]*cos(2*22/7 .0*j* i/weeks) +
beta[j+ 1]*sin(2*22/7 .0*j*i/weeks);
      a = mean+a;
      double b,c;
      for(b=j=0; j<small; j++)
            b += week_m[i-j-
1 ]*phai[small-j-1];
      c = a + b;
      if(i<small)
            b = c = week_m[i];

      fprintf(f2, "%d\t%/o. 3f\t%.3 t\t%/i.. 3 f.1
3f\n",i+ 1,week_m[i].a.b.c);
      }
line(f2.50);
delete [ ] alpha;
delete [ ] beta;

// Analysis of variance
      fprintf(f2."Analysis of variance of Fourier
      coefficient of weekly deficit series\n"):
```

```
line(f2.50);
fprintf(f2."SN\tHamonics\tD.F. \tSS\tM S\t I
tab\n");
line(f2,50);
double var1 [6];
int df[6];
for(var=mean=i=0; i<number; i++){
     mean += y[i];
     var += y[i]*y[i];
     }
     var = (var-mean*mean/number)
     mean = mean/number;

for(i=0; i<6; i++)
     varl[i] =0;

for(i=1; i<=weeks/2; i++){
     double a,b,c,d;
     for(a=0,j=0; j<weeks; j++)
          a+=
x[j][yrs ]/yrs*sin(2*22/7 .0* i* j/weeks);
          a = 2*a/weeks;
     for(b=0; j=0; j<weeks; j++) .
          b+=
x[j][yrs ]/yrs*'cos(2*22/7 .0* i*j/weeks);
     b = 2*b/weeks.
     a = number*(a*a+b*b)/2.0;
     if(i==1)
          var1[5] = a;
     else if(i>1)
          var1[4] += a;
     if(i==2)
          var1[3] = a;
     else if(i>2)
          var1[2] += a;
     if(i==3)
          var1[1] = a;
     else if(i>3)
          var1[0] += a;
     }
double ms = var1[0]/45.0;
fprintf(f2," 1.\t4,5,...26\t45\t%.3f\t%.3f',var 1[0],ms);
double ss = var-var1[0];
double mse = ss/(number-46);
if (mse>0)
     f = ms/mse;
else
```

```
        f= -1;
fprintf(f2,"\t%.3f\n",f);
fprintf(f2.\tResidual\t%d\t%.3f\t%.3f\n",nu
mber-16,ss,mse);
a=var1[1]/2.0;
fprintf( f2,"3. \t3\t2\t%.3 f\t%.3f", var1[1 ], a);
a = var-var1[1];
f= var1[0]/2.0/(a/(number-3));
fprintf(f2,"\t%.3 f\n ", f);
fprintf(f2, "4. \tResidual\t%d\t%.3 f\t%.3 f\n ",nu
mber-3 ,a,a/(number- 3)/
line(f2,50);
a = var1[2]/47.0;
fprintf(f2,"1.\t3,4....26\t47\t%.3f",var1[2],a);
a = var-var1[2]-var1[3];
f = var1[2]/47.0/(a/(number-48));
fprintf(f2,"\t%.3f\n",f);
fprintf(f2,"2.\tResidual\t%d\t%.3f\t%.3f\n",
mber-48,a,a/(number-48));
a = var1[3]/2.0;
fprintf(f2,"3.\t2\t%.3f\t%.3f\t%.3f",var1[3]a);
a = var-var1[3];
f = var1[3]/2.0/(a/(number-3));
fprintf(f2,"\t%.3f\n",f);
fprintf(f2,"4.\tResidual\t%d\t%.3f\t%.3f\n",
mber-3, a,a/(number-3));
line(f2,50);
a = var1[4]/49.0;
fprintf(f2,"1. \t2,3,...26\t49\t%.3f\t%.3f".var1[1],a);
a = var-var1[4];
f= var1 [2]/49.0/(a/(number-50));
fprintf(f2,"\t%.3f\n",f);
fprintf(f2,"2.\tResidual\t%d\t%.3f\t%.3f\n".
mber-50,a,a/(number-50));
a = var1[5]/2.0;
fprintf(f2,"3.\tl\t2\t%.3f\t%.3f", var1 [5].a)
a = var-var1 [5];
f = var1 [5]/2.0/(a/(number-3));
fprintf(f2,"\t%.3f\n",f);
fprintf(f2,"4.\tResidual\t%d\t%3f\t%.3f\n"
mber-3,a,a/(number-3 ));
line(f2,50);
delete [ ] y;
free_f2(x, weeks+2);
float average[10];
for (i=0; i<=9;i++)
        average[i]=week_m[i];
```

```
        delete [ ] week_m;
        if(yy=='Y')
        get3(small,mean,ak,bk,phai,average);
        delete [ ] ak;
        delete [ ] bk;
}
void get3(int small, double mean, double *ak,double
        *bk,double phai[5], float average[10])
{
        int i,j,k,number,n_yrs;
        float sx,**y,**pdc;
        double cv,sd;
        printf("\nFor How many years? ");
        scanf("%d",&n_yrs);
        printf("\nHow many data to be generated
        scanf("%d",&number);
        y = alloc_f2(n_yrs+1,number+2);
        pdc = alloc_f2(n_yrs+ 1,number+2);
        for (j=0; j<small;j++)
            y[0][j]=average[j];
        for (i=1;i<=n_yrs;i++)
        for (j=0;j<small;j++){
            printf("\nInitial data for year %d an
        week %d? ",i,j+1);
            scanf("%f',&y[i][j]);
            }
        for (k=0;k<=n_yrs;k++)
        for(i=small; i<number; i++){
            double a;
            for(a=j=0;j<small;j++)
                a+=
        iak[j+ 1 ]*cos(2*22/7.0*j*i/number) +
        bk[j+ 1 ]*sin(2*22/7.0*j*i/number);
            pdc[k][i] = a = mean+a;
            double b;
            for(b=j=0; j<small; j++)
                b += y[k][i-j-1 ]*phai[small-
        j-l];
            y[k][ij=a + b;
            }
        fprintf(f2,"Validation of AR mode1\n");
        line(f2, 10+(n_yrs*10));
        fprintf(f2,"Week\tPredicted value\t\tPDC\n");
        for (k=0;k<=n_yrs;k++)
            fprintf(f2,"\tY(%d)",k+1);
        for (k=0;k<=n_yrs;k++)
            fprintf(f2,"\tY(%d)",k+1);
```

```
        fprintf(f2,"\n");
        line(f2,10+(n_yrs* 10));
        for(i=0; i<number; i++){
            fprintf(f2,"%d",i+ 1);
            for (k=0;k<=n yrs;k++)
            fprintf( f2,"\t%.3 f",y[k][i]);
            for (k=0;k<=n_yrs;k++)
            fprintf( f2,"\t%.3 f',pdc[k][i]);
            fprintf(f2,"\n");
            }
        line(f2, 10+(n_yrs*10));
        free_f2(y, n_yrs+1);
        free_f2(pdc, n_yrs+1);
}
void line(FILE *f2, int 1)
{
        while (--1)
            fprintf(f2,"-");
        fprintf(f2,"\n");
}
```